Integrated Active Antennas and Spatial Power Combining

WILEY SERIES IN MICROWAVE AND OPTICAL ENGINEERING

KAI CHANG, Editor
Texas A&M University

Integrated Active Antennas and Spatial Power Combining

JULIO A. NAVARRO
Epsilon Lambda Electronics Corporation

KAI CHANG
Texas A&M University

A WILEY-INTERSCIENCE PUBLICATION
JOHN WILEY & SONS, INC.
NEW YORK / CHICHESTER / BRISBANE / TORONTO / SINGAPORE

This text is printed on acid-free paper.

Library of Congress Cataloging in Publication Data:
Navarro, Julio A.
    Integrated active antennas and spatial power combining / Julio A.
Navarro, Kai Chang.
        p. cm. -- (Wiley series in microwave and optical
engineering)
    "A Wiley-Interscience publication."
    Includes index.
    ISBN 0-471-04984-0 (cloth : alk. paper)
    1. Antennas (Electronics)--Design and construction.   2. Microwave
integrated circuits.   3. Solid state electronics.   I. Chang, Kai,
1948- . II. Title.   III. Series.
TK7871.6.N39   1996
621.382′4--dc20   95-46168

Printed in the United States of America

10 9 8 7 6 5 4 3 2 1

Contents

Foreword

Integrated active antenna elements and spatial power combining techniques have been developed in order to overcome the fundamental limitations on output power of semiconductor circuits at higher frequencies, particularly frequencies in the millimeter range and higher. Transmission lines suitable for integrated circuits become very lossy at higher frequencies due to radiation losses, substrate losses, and increased skin-effect ohmic losses. By integrating active elements directly in the antenna element of a single antenna or an array, these losses can be significantly reduced. Also, as the frequency of operation increases for most semiconductor oscillator or amplifier devices, the size of the devices must decrease in order to reduce capacitive and transit time effects and to prevent destructive interference of the positive and negative parts of the electromagnetic wave. The reduced size reduces the maximum output power of the device proportional to a factor between $1/f$ and $1/f^2$, with the result that for moderate or high-power applications in the millimeter wave frequency range, vacuum tube circuits are required. Techniques to spatially combine the power from many small semiconductor devices promise the advantages of semiconductor integrated circuit technology: reducing systems weight, size, and cost as well as increased reliability and manufacturability; and overcoming transmission line losses inherent in corporate power combining circuits. In addition, spatial power combining of arrays of active antenna elements results in an inherently nonlinear response from the antenna array itself and in the elimination of propagation delay from the antenna. This results in the potential for entirely new functionality in antenna array circuits.

There are many commercial and industrial applications for these techniques. For example, automotive radar systems, such as collision avoidance and intelligent highway communications, are applications with huge potential markets and very active commercial development. Active antennas also offer potential advantages for personal communications systems, sensors, RFID systems, and space systems. Cost, reliability, and size constraints are absolutely critical to these applications, and active antennas and spatial power combining are very promising approaches. The potential for lower weight, smaller, less expensive, more reliable, and more efficient circuits for millimeter wave applications is extremely important for military applications. The US Army has a particular interest, due to the dispersed nature of its weapons systems and the frequent requirement for man portability. Army systems tend to be distributed over many vehicles or operators and army targets tend to be smaller and more dispersed, resulting in the requirement for many small, inexpensive systems. A missile seeker is an example. An army missile must be significantly less expensive than the value of the vehicle it targets. Furthermore, its size and weight trade off against warhead payload in a critical manner. Other examples include man portable satellite communications terminals, communications to missiles or unpiloted air vehicles in flight, and the communications and target acquisition requirements for the Twenty First Century Land Warrior (a self-contained, high-technology fighting system for the individual ground soldier). Of critical importance to any man portable system is the weight of the prime power supply, which depends on the system efficiency. For these reasons, the US Army Research Office has supported the pioneering work in this field, such as the research of the authors of this book.

Until now, no good and comprehensive treatment of these techniques has existed in textbooks or in technical reference books. This book fills the deficiency with a very readable tutorial review of the overall technology. The book is unique in its comprehensive and tutorial approaches, and complements other more specialized references. It is also unique in its depth of focus on integrated active antenna elements. The authors are well qualified to undertake this work. Professor Chang has been a pioneer in these technologies for two decades. He was the first to use FET devices in active patch antennas and the first to do spatial power combining with patch antenna elements. Dr. Navarro and Professor Chang were the first to report broadband electronically tunable active antennas using notch antenna elements and were the first to use them for power combining. They were also the first to use inverted stripline in active antenna elements and to do spatial power combining with them. They have pioneered the recent successes in using many different types of solid-state devices in active antenna elements. The result of their expertise and clarity of written expression is a valuable and understandable text covering a vital area of technology which has not been previously addressed.

JAMES F. HARVEY
JAMES W. MINK

US Army Research Office
Research Triangle Park, NC
North Carolina State University
Raleigh, NC

Preface

For the past decade, we have witnessed a rapid development in active antennas, integrated antennas, spatial power combining, and quasi-optical techniques for microwave and millimeter-wave applications. Recent advances in solid-state devices and microwave/millimeter-wave integrated circuits have made it possible to combine the solid-state devices with the planar antennas to form integrated and active antennas. Many active antenna elements can be combined to build an active phased array or a spatial power combiner.

The integrated and active antennas have many applications as low-cost transmitters and transceivers in radar and communications, active decoys for electronic warfare systems, and transceivers for low-cost sensors. The output power from a single solid-state device is limited by fundamental thermal and impedance problems. To meet the high power requirements for many applications, it is necessary to combine many devices to achieve high power output. Spatial and quasi-optical power combining techniques have been developed to combine power from active antennas in free space or in an open resonator at high frequencies. For efficient power combining, all source elements must be coherent. Injection locking through open resonators, space, or mutual coupling can be used to achieve coherency.

The purpose of this book is to introduce the theory and practical use of integrated antennas, active antennas, and spatial power combining. The subject involves an interdisciplinary background in microwave circuits, antennas, and solid-state devices. Although many papers have appeared in the literature, no single book currently covers this subject.

The book should be useful for engineers and scientists interested in microwaves, antennas, solid-state devices, and systems. This book pieces together information from different areas which are necessary to understand integrated and integrated active antennas. It emphasizes active antennas and power-combining applications and consolidates the work from many researchers and investigators. Several early chapters lay the foundation for oscillator, antenna, array, and power-combining theory to acquire an understanding of this broad topic. Active antennas have been classified according to function to facilitate some comparison. Investigators have found many ways to solve the power deficiencies at higher frequencies that make a precise classification of all the methods difficult. Overall, this book attempts to give the reader a complete review of previous integrated and active integrated antenna work and the knowledge necessary to undertake its integration.

The book was written to include as completely as possible the published work up to December 1994. Since the subject is a very active area of research and development, it is rapidly changing. The reader is advised to consult various journals and conference digests for recent results after this date.

Chapter 1 briefly introduces microwaves, integrated circuits, devices, substrates, integrated antennas, and active antennas. It presents the basic knowledge required to understand how and why integrated antennas are designed and used.

Chapter 2 discusses general oscillator theory. The chapter includes the classic derivations of oscillator conditions, Q-factor, stability, noise, and load pulling. Synchronization is also shown for injection locking and power-combining applications.

Chapter 3 introduces general antenna parameters, such as directivity, efficiency, half-power beamwidth, and polarization. Array formulations are shown along with mutual coupling effects. This chapter sets up the active antenna power combining in Chapter 4, where methods for power combining and synchronization are analyzed. Several examples are used to demonstrate the concepts.

Chapter 5 discusses the important testing parameters and techniques for active antenna measurements. Definitions for equivalent isotropic radiated power, locking gain, and locking bandwidth are presented. Measurement of oscillator characteristics such as noise, spectral quality, stability, and load pulling are discussed.

Chapters 6 through 10 discuss in detail the work done on active antennas and power combining. Specifically, these chapters review all active integrated devices on microstrip patches, grids, notches, and other antennas. The applications of these active antennas to power combining arrays are discussed. Chapter 11 describes other integrated antenna components. In these chapters, various active and integrated antenna components are discussed, including oscillators, amplifiers, mixers, detectors, rectennas, multipliers, phase shifters, and switches.

Chapter 12 sheds light on beam steering, a more recent development in active antenna arrays. Beam steering of active antennas is demonstrated; preliminary theory and possible phased-array applications are explored.

We would like to thank the Army Research Office under the leadership of Dr. James Mink and Dr. James Harvey, who sponsored most of the developments described in this book, including the work done in our laboratory. The support of the NASA Center for Space Power, under the direction of Dr. A. D. Patton, and the Texas Higher Education Coordinating Board is also acknowledged. Julio Navarro would also like to acknowledge the support for his graduate studies by a NSF fellowship and a NASA Lewis Research Center Training Grant. Finally, we wish to thank Lu Fan, Mingyi Li, James McCleary, and Mary Drummond for critical review of the manuscript and Vivian Gonzalez and Brad Heimmer for drawing many of the figures.

JULIO A. NAVARRO
KAI CHANG

CHAPTER ONE

Introduction

1.1 INTRODUCTION

This book describes and reviews the efforts of academia and industry in the area of integrated and active integrated antennas. Since the integrated antenna concept bridges together circuits, devices, and antennas, the authors briefly discuss these topics in the first five chapters, using traditional definitions. Also, since the majority of integrated antenna applications has emphasized active spatial power combining, this topic predominates throughout the book. In the last few years, other applications for active antennas have stirred interest in the marketplace. Most major publications for these types of investigations are reviewed in the appropriate chapters of this book.

1.2 BRIEF HISTORY OF MICROWAVES

In 1864, James Clerk Maxwell elegantly pieced together the works of Lorentz, Faraday, Ampere, and Gauss into what are called *Maxwell's equations* [1]. Maxwell presented the first unified theory of electricity and magnetism and founded the science of electromagnetics. He predicted transverse propagation of waves at a finite speed (i.e., the speed of light). He postulated that light was an electromagnetic phenomenon of a particular wavelength and predicted that radiation would occur at other wavelengths as well.

Maxwell's theory was not accepted until it was validated by Heinrich Rudolf Hertz in the late 1880s [2]. Hertz demonstrated radio frequency (rf) generation, propagation, and reception of electromagnetic waves at 4 m, using an end-loaded dipole transmitter and a resonant square-loop antenna receiver. His basic radio system experiments at wavelengths of 4 m and at 30 cm, using a cylindrical parabolic reflector, proved that light and electromagnetic radiation were the same [3]. For his work, Hertz is known as the father of radio, and frequency is described in units of Hertz (Hz). This work, however, remained a laboratory curiosity for almost two decades until a young man, Guglielmo Marconi, came across his experiments. In them Marconi envisioned a method for transmitting information. His efforts would later earn him the Nobel Prize of 1909.

Marconi's wireless system commercialized the use of electromagnetic waves for communications and allowed the transfer of information from one continent to another without any physical connections. Distress signals from the *SS Titanic* and *SS Republic* gave the public a great demonstration of the usefulness of wireless communications. Marconi's wireless system meant that a ship was no longer isolated in the open seas and could have continuous contact to report emergencies or position.

In the early 1900s, most wireless transmissions occurred at very long wavelengths. Transmitters consisted of Alexanderson alternators, Poulsen arcs, and spark gaps. Receivers used coherers, Fleming valves, and DeForest audions. With the advent of DeForest's triode vacuum tube in 1907, continuous waves (CW) replaced spark gaps and more reliable frequency and power output was obtained for radio broadcasting at frequencies below 1.5 MHz.

Frequencies of several hundred megahertz were ignored for the most part and not needed until World War II for the detection of enemy ships and aircraft. The method of *radio detection and ranging* became known as *radar*. The acronym radar has since become a common term describing the use of the reflections from objects to detect and determine distance to and relative speed of a target. A radar's resolution (i.e., the minimum object size which can be detected) is inversely proportional to wavelength. Therefore, shorter wavelengths (microwave frequencies and above) are required to detect smaller objects. Output power, antenna gain, receiver sensitivity, and object radar cross section (RCS) determine the longest range for detecting a particular object.

The inability to generate enough rf power at microwave frequencies was a major hurdle for early radar systems. Many attempts at solving these problems using vacuum tubes fell short. During World War II, the British used long-wave radar to detect German bombers approaching the English shore. They could then scramble fighters to intercept the aircraft. Similar long-wave radar systems were developed in the United States. The Naval Research Laboratory in Washington developed the FD 500-MHz radar, while the Army Signal Corps developed the SCR-268 radar operating at 300 MHz. They were limited by poor resolution and large ground interference, and both could benefit greatly with an increase in operating frequency.

The British recognized the need for higher operating frequencies to detect smaller objects and more power to increase the detection range. Towards this end, Randall and Boot developed the internal cavity magnetron which demonstrated 10 kW of pulsed power at the S-band. This technology was carried over to the United States through the radiation laboratory of the Massachusetts Institute of Technology. The development of the magnetron and the klystron in the early 1940s gave the allies a definitive edge in detecting enemy attacks and shortened the duration of the war considerably.

Microwave circuits have since become an integral part of modern society. Initially implemented in military radar, microwaves have been used in space, scientific, and commercial applications. The term *microwave* refers to that part of the frequency spectrum from 300 MHz to 300 GHz. Microwaves have been implemented for point-to-point communication links, wireless communication, motion sensing for security systems, remote sensing and imaging, selective heating of tumors in medicine, spectroscopy for material identification, and police radar. The use of "microwave" ovens at 2.45 GHz has made the term a

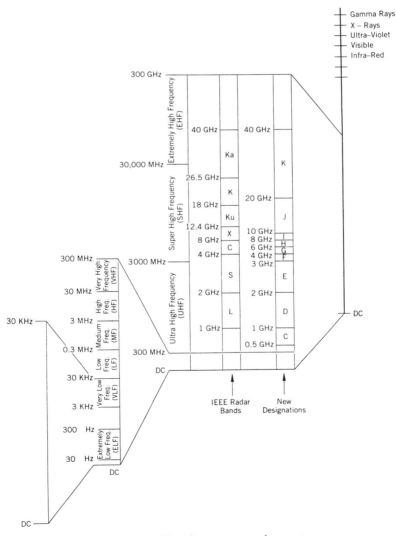

FIGURE 1.1. The electromagnetic spectrum.

household word. The frequency spectrum is divided into many different bands. These bands and some common applications are shown in Figure 1.1. The Federal Communications Commission (FCC) allocates frequency ranges and specifications for different applications in the United States, including television, radio, satellite communications, cellular phone, police radar, burglar alarms, and navigation beacons. Performance of each application is strongly affected by the atmospheric absorption curve shown in Figure 1.2. For example, a secure local area network would be ideal at 60 GHz due to the high attenuation caused by O_2 resonance.

As more applications spring up, overcrowding and interference at lower-frequency bands pushes applications toward higher operating frequencies. Higher-frequency operation has several advantages, including

1. Larger instantaneous bandwidth for greater transfer of information
2. Higher resolution for radar and more detailed imaging and sensing
3. Reduced dimensions for resonant antennas
4. Less interference from nearby applications

At higher frequencies, however, devices perform poorly and require more expensive materials with complicated fabrication methods. Furthermore, high-

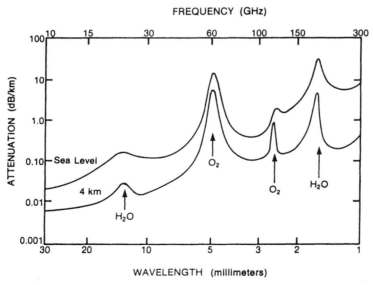

FIGURE 1.2. Average atmospheric absorption of millimeter waves for horizontal polarization.

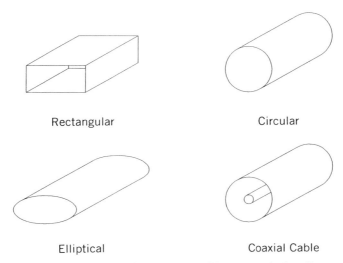

Rectangular   Circular

Elliptical   Coaxial Cable

FIGURE 1.3. Coaxial and waveguide transmission lines.

frequency circuit models tend to be oversimplified and inaccurate. As these technologies mature, operating frequencies will increase without compromising system performance.

From the beginning, most microwave components were designed using the waveguides and coaxial cables shown in Figure 1.3. Today, coaxial cables are still used in interconnects due to their flexibility. Circular and rectangular waveguides, on the other hand, are best suited for high-power and/or low-loss applications but work over a specified bandwidth with a low cutoff frequency. Waveguides are also heavy, bulky, and require costly high-tolerance machining. In spite of good performance, waveguides and coaxial configurations are not suited for improving reproducibility, reducing size and weight, and lowering overall system costs. Furthermore, modern trends have pushed microwave circuits toward miniaturization, improved reliability and greater functionality. Such improvements require a complete change in topology and methods of fabrication.

Starting from conventionally assembled coaxial and waveguide systems, hybrid microwave integrated circuits (MIC) evolved. Methods of integration include hybrid MICs and miniature hybrid MICs. These hybrid techniques have since evolved to monolithic MICs (MMICs). These methods achieve greater functionality and reproducibility by using photolithographic techniques for circuit fabrication. These approaches, however, require planar transmission lines to replace waveguides and coaxial cables. Planar transmission lines are, for the most part, open structures which tend to have lower Q-factors and higher crosstalk between lines.

The overall characteristics of the integrated circuit design depends on several major factors:

1. MIC transmission line configurations
   a. Mode of propagation (i.e., Z_0, λ_g, dispersion, etc.)
   b. Radiation and conductor losses
   c. Integration of series/shunt devices and biasing network designs
2. Substrate materials
   a. Dielectric losses and propagation effects
   b. Mechanical strength and heat expansion
   c. Methods of fabrication
3. Solid-state devices
   a. Microwave component function (i.e., switching, tuning, mixing, amplifying, oscillating, etc.)
   b. Operation frequency, power, and conversion efficiency
   c. Materials and methods of fabrication available

1.3 MICROWAVE INTEGRATED CIRCUIT TRANSMISSION LINES

MIC transmission lines consist of metallized circuit patterns supported by a dielectric substrate as shown in Figure 1.4. These *planar* lines can be fabricated with photolithographic techniques. Photolithography improves reproducibility and allows mass production of integrated circuits. The circuit pattern layout used to transport energy throughout a circuit determines many characteristics of the transmission line.

Unlike the TEM modes of coaxial lines, planar MIC lines are often non-TEM. The characteristics of some MIC lines, however, can be approximiated by assuming quasi-TEM propagation. Quasi-TEM lines include microstrip, inverted microstrip, trapped inverted microstrip, suspended stripline, coplanar waveguide, and coplanar strips.

Non-TEM MIC lines include slotline, finline, and imageline. Waveguides are non-TEM in nature, but, unlike waveguides, MIC lines suffer from dielectric and radiation losses and crosstalk between lines in a dense circuit. These transmission lines cannot handle high-power like waveguides, but they are more than suitable for low-to-medium power applications. The reduction in size, weight, and overall cost offsets the slight drop in performance. A qualitative comparison of these transmission lines is given in Table 1.1.

The *microstrip* line [4, 5] has been the most used planar transmission line in MICs and MMICs. Synthesis and analysis formulas are well documented, and many discontinuities have been characterized. Commercial programs are available in order to use these models for circuit prediction and optimization.

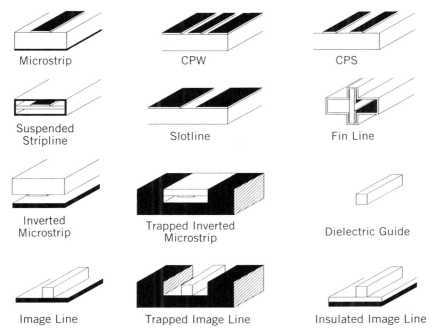

FIGURE 1.4. Microwave integrated circuit (MIC) transmission lines.

Devices can be integrated in series using planar packages, but shunt connections require drilling through the substrate in MICs and via-hole processing in MMICs. Compared to other open structures, microstrip is a proven technology which has higher power handling, lower loss, and is easily packaged. In spite of microstrip's advantages, other transmission lines offer viable alternatives to conventional microstrip integration.

A derivative of microstrip, *inverted microstrip* (IM) [6, 7] line, uses a metalized circuit pattern suspended with spacers over a ground plane support. Unlike microstrip, the ground plane has been removed from the backside of the substrate. Since most of the transmission occurs in air below the strip, the guided wavelength and characteristic impedance of inverted microstrip are larger than a similar line using conventional microstrip. Series devices are connected just as in microstrip, while shunt devices do not require drilling through the substrate. This characteristic allows nondestructive device testing and position optimization in inverted microstrip. However, inverted microstrip is very prone to surface mode excitations which would cause considerable crosstalk in a dense circuit. Thinner substrates and reduced strip-to-ground height reduce the excitation of surface modes but limit its usefulness. Smaller dimensions place higher tolerances on spacers used to separate substrate and ground support.

TABLE 1.1. Transmission Line Comparisons

Transmission Line	Useful Freq. Range (GHz)	Impedance Range (Ω)	Cross-Sectional Dimensions	Q-Factor	Power Rating	Active Device Mounting	Potential for Low-Cost Production
Rectangular waveguide	<300	100–500	Moderate to large	High	High	Easy	Poor
Coaxial line	<50	10–100	Moderate	Moderate	Moderate	Fair	Poor
Stripline	<10	10–100	Moderate	Low	Low	Fair	Good
Microstrip line	≤100	10–100	Small	Low	Low	Easy	Good
Suspended stripline	≤150	20–150	Small	Moderate	Low	Easy	Fair
Finline	≤150	20–400	Moderate	Moderate	Low	Easy	Fair
Slotline	≤60	60–200	Small	Low	Low	Fair	Good
Coplanar waveguide	≤60	40–150	Small	Low	Low	Fair	Good
Image guide	<300	30–30	Moderate	High	Low	Poor	Good
Dielectric line	<300	20–50	Moderate	High	Low	Poor	Fair

Source: K. Chang, *Microwave Solid-State Circuits and Applications*, Wiley, New York, 1994.

A further modification to inverted microstrip, the *trapped inverted microstrip* (TIM) [8] line, uses a channel to choke out surface wave modes which may otherwise propagate along the inverted microstrip configuration. This modification raises the characteristic impedance of TIM and relaxes the tolerance requirement of the spacers used in inverted microstrip. The enclosure allows the use of thicker substrates and larger ground-to-strip separations without surface wave effects. Shunt and series devices are connected just as in inverted microstrip. The channel allows the use of larger strip-to-ground spacing and provides more metal for improved heat dissipation in active applications.

Suspended stripline (SS) [9] is the last of the microstrip derivatives. Similar to IM, the backside ground plane is removed from the substrate. The strip and substrate are totally enclosed and oriented along the *H*-plane of a rectangular waveguide. Operation frequencies are limited to avoid excitation of waveguide modes. Although the operation bandwidth is limited, this variation of microstrip achieves lower losses than all other two-conductor MIC lines. Although series device integration is straightforward, shunt insertion is similar to IM. The characteristic impedance and guided wavelength depend on the waveguide dimensions, substrate dielectric constant, substrate thickness, strip width, and position in the waveguide.

Coplanar waveguide (CPW) [10] is quite different from microstrip and its derivatives. Referred to as a uniplanar line, CPW has both the conductor and ground plane on a single side of the substrate. Drilling is not required for shunt connections, and both shunt and series devices can be integrated by using planar packages. Propagation is achieved via two slots etched on a single-sided substrate. The impedance and guided wavelength characteristics depend on the dielectric constant, substrate height, slot separation, and slot width. The structure supports even- and odd-mode propagation when the ground planes on either side of the conductor are at different potentials. Even-mode propagation can be choked out by maintaining these ground planes at the same potential by using bridges or bond wires. Failure to completely choke out the even mode causes an inductive effect on standard odd-mode operation. Unlike microstrip, packaging often causes parallel plate mode problems for CPW when the package walls are too close to the circuit. This problem is relieved somewhat with a grounded version of CPW. A further modification of CPW is the channelized version which parallels what TIM does for IM.

Coplanar strips (CPS) [11] consist of two lines etched on the dielectric substrate. CPS is uniplanar and can be integrated with planar devices in either series or shunt connections. This line suffers from higher radiation losses than other planar lines. The characteristic impedance and guided wavelength depend on the strip widths, separation, substrate thickness, and dielectric constant. Similar to other uniplanar lines, enclosure proximity affects performance.

Slotline [12] is a non-TEM uniplanar MIC line using a single slot etched on a dielectric-supported layer of metal without a backside ground plane. The characteristic impedance and guided wavelength depend on the dielectric constant, substrate height, and slot width. Series and shunt devices can be incorporated for integration. Shunt mounting is straightforward, while series insertion requires some modifications. Low values of characteristic impedance ($\leqslant 60\,\Omega$) are often impractical. Low values of Z_0 require very narrow slots on thin substrates with high dielectric constants. For circuit interconnections, slotline also suffers from high radiation losses and packaging difficulties.

Finline [13, 14] is a combination of waveguide and slotline. A slot is etched on a dielectric-supported layer of metal and oriented along the *E*-plane of a rectangular waveguide. The waveguide encloses either side of the slotline. This structure combines the high performance of waveguides with the integration capabilities of uniplanar lines. However, the structure tends to be just as bulky and heavy as standard waveguide. A short is needed at the edges of the substrate, which is often disturbed for device biasing. Finlines generally have about a third of the loss of conventional microstrip and are useful up to 100 GHz.

Dielectric guide [15], *imageline* [16], and *insulated imageline* [17] are made up of a dielectric slab for wave propagation. Imagelines use a dielectric over a ground plane. This is in contrast to all previously mentioned transmission lines, which require some metallic conductor for propagation. Power transmission occurs through the dielectric. It exhibits low loss and is extremely useful at frequencies above 100 GHz. However, integration with devices and biasing must be addressed as well as high radiation at discontinuities. Crosstalk with other lines is another issue which must be overcome before it can be used for MICs. An enclosed version of this line, *trapped imageline*, attempts to correct crosstalk problems in dense circuits.

The properties of the MIC lines described are greatly dependent on the characteristics of the supporting dielectric substrate. The substrate greatly affects electromagnetic wave propagation and its associated transmission losses. The substrate material also determines the process of fabrication and several other characteristics critical for MICs and MMICs. Different dielectrics and their properties are discussed below.

1.4 DIELECTRIC SUBSTRATES

The dielectric substrate supports each of the MIC lines discussed, and its properties affect the overall performance of the ICs. Just as in each type of transmission line configuration, each substrate material possesses characteristics which may make it better suited for an application. For instance, antennas require lower dielectric constants to ensure good radiation, while higher-dielectric-constant materials would ensure smaller circuit size. The cost, frequency of operation, loss tangent, mechanical strength, and surface finish of a material are all important for hybrid MIC designs. For hybrid MICs, there is a choice between organic and inorganic materials. Organic materials include a wide range of plastics or soft substrates. Inorganic or hard substrates include ceramics, monocrystalline materials, ferromagnetic materials, and semiconductors.

In general, ε_r and μ_r should be homogeneous (independent of position) and isotropic (independent of propagation direction). Furthermore, these parameters should have very small variation with temperature to ensure circuit stability. The substrate thermal conductivity should be high enough to ensure efficient removal of heat from power transistors, attenuators, and loads in high-power applications. In high-power applications, a high breakdown voltage is also desirable. The thermal expansion coefficient of the substrate material should be similar to that of the deposited conductors and housing to withstand temperature fluctuations and improve reliability. The material must allow drilling, cutting, and machining for easy workability and lower production costs. Also important is a good surface finish (0.05 to 0.1 μm) to ensure good conductor adhesion and reduce conductor loss. These substrates should be available in sheets of various sizes to accommodate large circuit layouts and integration of many component functions.

Plastics often have low loss tangents and low ε_r. Some typical soft-substrate materials include polystyrene, polyolefin, and polytetrafluoroethylene (PTFE) substrates. To overcome mechanical instability and cold-flow problems, these substrates are reinforced with either glass fibers or ceramic particles. The former maintains the dielectric constant but adds loss, while the latter tends to increase the dielectric constant. Two common combinations for low ε_r include woven PTFE/fiberglass and microfiber PTFE/fiberglass, while a high ε_r soft substrate is ceramic PTFE. The choice depends on electrical, mechanical, chemical, and thermal conditions encountered during fabrication as well as its intended working environment. The material must also be amenable to low-cost construction methods. However, soft substrates exhibit some undesirable characteristics, including creep under stress (solid-state deformation), high coefficients of thermal expansion, and low thermal conductivity. Low thermal conductivity coefficients create thermal management problems in high-power applications, while substrate deformation presents problems for clamping or other stresses on the substrate. High thermal expansion coefficients cause fatigue at joints between the substrate and housing. Reinforcing glass fibers can reduce deformation and thermal expansion. Typically, however, this reduction occurs in the x-y plane of the substrate, while the z-direction's thermal expansion remains unchanged, causing circuit reliability and durability concerns. Table 1.2 lists the characteristics of a wide range of organic substrates.

Ceramic substrates include aluminum oxide (Al_2O_3), sapphire, quartz, titanium oxide (TiO_x), beryllium oxide (BeO), and aluminum nitride (AlN). Al_2O_3, or alumina, is an inexpensive material used for thin- and thick-film hybrid ICs. The grade of purity (0.95 to 0.995) determines the dielectric constant (9.5 to 10.5), low loss tangent (0.0001 to 0.0005), and surface finish (to 1 μm). Alumina exhibits high ε_r, low deformation, good thermal expansion coefficient (i.e., matches base metals and GaAs), and high insulation resistance. Sapphire (also called corundum or alpha-alumina) is the single-crystal form of

TABLE 1.2. Soft Microwave Substrate Properties

Laminate/Substrate	Dielectric Constant (X-Band)	Loss Tangent (X-Band)	Dimensional Stability	Chemical Resistance	Temperature Range (°C)
PTEE unreinforced	2.1	0.0004	Poor	Excellent	−27 to +260
PTFE glass woven web	2.17–2.55	0.0009–0.0022	Excellent	Excellent	−27 to +260
PTFE glass random fiber	2.17–2.35	0.0009–0.0015	Fair	Excellent	−27 to +260
PTFE quartz reinforced	2.47	0.0006	Excellent	Excellent	−27 to +260
Ceramic PTFE composite	10.2	0.002	Excellent	Good	−15 to +170
Cross-linked polystyrene	2.54	0.0005	Good	Good	−27 to +110
Cross-linked polystyrene/glass reinforced	2.62	0.001	Good	Good	−27 to +110
Cross-linked polystyrene/quartz material	2.6	0.0005	Good	Good	−27 to +110
Cross-linked polystyrene/woven quartz	2.65	0.0005	Good	Good	−27 to +110
Cross-linked polystyrene/ceramic power-filled	3–15	0.0005–0.0015	Fair to Good	Fair	−27 to +110
Teflon/unreinforced (unclad)	2.1	0.0004	Poor	Excellent	−27 to +260
Teflon/glass reinforced	2.55	0.0015	Good	Excellent	−27 to +260
Teflon/ceramic reinforced	2.3	0.001	Fair to Good	Excellent	−27 to +260
Teflon/quartz reinforced	2.47	0.0006	Good	Excellent	−27 to +260
Teflon/ceramic filled	10.3	0.002	Good	Excellent	−27 to +260
Polyphenylene oxide (PPO)	2.55	0.0016	Good	Poor	−27 to +193
Irradiated polyolefin	2.32	0.0005	Poor	Excellent	−27 to +100
Irradiated polyolefin/glass reinforced	2.42	0.001	Fair	Excellent	−27 to +100
Powder-filled Polyolefin/ceramic	3–10	0.001	Poor	Excellent	−27 to +100

Source: I. J. Bahl and K. Ely, "Modern Microwave Substrate Materials," *Microwave Journal 'State of the Art Reference*,' pp. 131–146 (1990).

Al_2O_3, and it also exhibits low loss tangent, high ε_r, good surface finish, high insulation resistance, and a thermal expansion coefficient which matches well with GaAs. However, sapphire is anisotropic with permittivity, which varies from 9.4 to 11.5 for different directions of propagation. Fused quartz offers low loss tangent, relatively low ε_r, good surface finish, dielectric strength, and repeatability. However, fused quartz is brittle, difficult to drill through, and costly. Table 1.3 lists the properties of these three popular microwave substrate materials.

Ceramics which have very high ε_r values (15 to 240) allow miniaturization of circuits (especially at low frequencies) and often provide lower loss and better temperature stability. Properties of several high-ε_r materials are listed in Table 1.4. Materials developed specifically for dielectric resonators (DR) are generally based on titanates. These titanates provide very high quality resonators with low thermal expansion and temperature coefficients.

For high thermal conductivity, beryllium oxide and aluminum nitride are available. BeO is very toxic and costly, which has limited its use but provides low ε_r, good dielectric strength, and high bulk resistivity. Aluminum nitride compares favorably with beryllia plus it maintains a more constant thermal conductivity coefficient at higher temperatures. However, large single crystals of AlN are not readily available. AlN powder must be processed in an oxidizing

TABLE 1.3. Hard Microwave Substrate Properties (at 25°C)

Property	Alumina	Sapphire	Fused Quartz
Dielectric constant at 10 GHz	9.8	11.5	3.78
Loss tangent at 10 GHz	0.0002	<0.0001	0.0001
Resistivity (Ω-cm)	10^{14}	10^{16}	10^{14}
Dielectric strength (kV/mm)	7.9	48	100
Temp. Coeff. of ε_r (ppm/°C)	+136	—	—
Coefficient of thermal expansion (ppm/°C)	6.7	5.3	0.55
Thermal conductivity (W/mK)	37	46	1
Melting point (°C)	2030	2053	—
Density (g/cm³)	3.9	3.97	2.203
Bending strength (N/mm²)	245	—	—
Modulus of elasticity (kN/mm²)	340	413	72
Tensile strength (kg/mm²)	21	260	8
Rockwell hardness (15N scale)	97	—	—
Flexural strength (kg/mm²)	34	64	7
Surface finish (μm)	<0.1	<0.1	—
Grain size (μm)	<1.5	—	—

Source: I. J. Bahl and K. Ely, "Modern Microwave Substrate Materials," *Microwave Journal 'State of the Art Reference,'* pp. 131–146 (1990).

TABLE 1.4. High ε_r Substrate Properties

Dielectric Designation Code	CF	CB	CD	CG	NR
Dielectric Constants					
at 5 GHz	21.6 ± 0.6	29 ± 0.7	37 ± 1	67.5 ± 2	152 ± 5
at 1 MHz	19.3 ± 0.6	29 ± 0.7	37 ± 1	67.5 ± 2	152 ± 5
Maximum loss tangent	0.0003	0.0004	0.0004	0.0008	0.001
Temp. coefficient (ppm/K)	$+17 \pm 5$	-10 ± 2	-30 ± 3	-23 ± 3	-1700 ± 120
Minimum resistivity (Ω-cm)	10^{14}	10^{14}	10^{14}	10^{14}	10^{14}
Thermal expansion (ppm/K)	7.8	6.3	5.8	9	10
Flexural strength (kgf/cm^2)	1492	1590	1535	1910	1548
Density (g/cm^3)	3.89	4.32	4.75	5.56	3.91
Heat conductivity (W/cm-K)	0.06	0.033	0.018	0.02	0.047
Specific heat (J/g-K)	0.8	0.66	0.58	0.48	0.77
Maximum water absorption (%)	0.01	0.01	0.01	0.01	0.01

Sources: I. J. Bahl and K. Ely, "Modern Microwave Substrate Materials," *Microwave Journal 'State of the Art Reference,'* pp. 131–146 (1990); K. Wakino, T. Nishikawa, H. Tamura, and T. Sudo, "Dielectric Resonator Materials and Their Applications," *Microwave Journal*, Vol. 30, No. 6, pp. 133–150, June 1987.

atmosphere with compounds such as yttrium oxide, yttrium fluoride, and calcium carbide. This processing lowers the thermal conductivity, but metalization is difficult and long-term reliability is questionable. Multilayer ceramics overcome some metalization problems encountered. BeO and AlN characteristics are listed in Table 1.5.

Overall, hard substrates allow precise dimensional control of circuit layout patterns and tend to have lower thermal expansion coefficients and higher coefficients of conductivity than do soft substrates. Welding or device soldering is done easily on ceramic substrates. These properties make ceramics more useful than plastics for high-power, high-temperature, and high-stress applications. However, unlike plastics, ceramics are expensive and not easily amenable to the construction of intricate circuit periphery, holes, slots, machined depressions, and so on. Ceramics are available in much smaller substrate sizes than are plastics, which makes integration of several functions more difficult. In the end, the choice between hard and soft substrates in hybrid MICs depends on cost and performance requirements discussed previously.

Monolithic MICs require substrates on which devices and circuits can be "grown" through fabrication processes. Different materials available for MMICs include gallium arsenide (GaAs), indium phosphide (InP), and silicon (Si). The properties of the materials used are listed in Table 1.6.

GaAs is the most prominent high-frequency substrate used in MMICs. Electrons in GaAs have high saturation velocities and mobilities achieved with relatively low electric fields. The small transit times for electrons flowing across

TABLE 1.5. High-Thermal-Conductivity Ceramics

Property	BeO	AlN
Dielectric constant		
at 1 MHz	6.5	8.9
at 10 GHz	6.7	—
Loss tangent		
at 1 MHz	0.0004	0.0005
at 10 GHz	0.004	—
Resistivity (Ω-cm)	10^{15}	10^{11}
Dielectric strength (kV/mm)	9.6	>5
Coefficient thermal expansion (ppm/°C)	9	4.4
Thermal conductivity (W/mK)	260	140–230
Melting point (°C)	2530	2400
Density (g/cm^3)	2.9	3.26
Modulus of elasticity (kN/mm^2)	345	300–310
Bending strength (N/mm^2)	200	315
Rockwell strength (15N scale)	91	94
Flexural strength (MPa)	241	28–320
Surface finish (μm)	<0.4	<0.5
Grain size (μm)	9–16	5–10

Source: I. J. Bahl and K. Ely, "Modern Microwave Substrate Materials," *Microwave Journal 'State of the Art Reference,'* pp. 131–146 (1990).

TABLE 1.6. Monolithic Integrated Circuit Substrate Properties

Property	Silicon	Silicon-on-Sapphire	GaAs	InP
Semi-insulating	No	Yes	Yes	Yes
Resistivity (Ω-cm)	10^3–10^5	>10^{14}	10^7–10^9	~10^7
Dielectric constant	11.7	11.6	12.9	14
Electrical mobility[a] (cm^2/V-s)	700	700	4300	3000
Saturation electrical velocity (cm/s)	9×10^6	9×10^6	1.3×10^7	1.9×10^7
Radiation hardness	Poor	Poor	Very good	Good
Density (g/cm^3)	2.3	3.9	5.3	4.8
Thermal conductivity (W/cm-°C)	1.45	0.46	0.46	0.68
Operating temperature (°C)	250	250	350	300
Handling	Very good	Excellent	Good	Poor
Cost (2-ft diameter)	$15	$50 (base sapphire)	$100	$300

[a]at 10^{17} cm^{-3} doping level.

Source: I. J. Bahl and K. Ely, "Modern Microwave Substrate Materials," *Microwave Journal 'State of the Art Reference,'* pp. 131–146 (1990).

a device allow high-frequency operation. GaAs can be prepared in a semi-insulating form with a resistivity of $\sim 10^9\,\Omega$-cm and $\varepsilon_r = 12.9$. Semi-insulating GaAs is used for microstrip line interconnections allowing the fabrication of very densely packed monolithic circuits. Although silicon has saturation velocities lower than GaAs, it is a very mature fabrication technology used exclusively at lower frequencies. Unlike semi-insulating GaAs, semi-insulating silicon is too lossy for high-frequency applications. One solution has been to use the silicon-on-sapphire (SOS) approach to overcome the semi-insulating loss problems encountered with silicon. As listed earlier, sapphire has many attractive qualities for circuits, including surface finish, low loss, and good thermal conductivity. If one accounts for its anisotropic nature, Si circuits can be aligned along the sapphire substrate C-axis to obtain an $\varepsilon_r = 11.6$ and a loss tangent of 0.002. This combination has regenerated interest in silicon bipolar transistor technology for lower microwave frequencies.

There are many different techniques for deposition of metallic and/or dielectric layers. Regardless of whether one uses organic or inorganic substrates, one may use one or several of the following for the conductor: silver, copper, gold, and aluminum. Dielectric films used include SiO, SiO_2, Si_3N_4, and Ta_2O_5. Resistive films used include NiCr, Ta, Ti, TaN, cermet, and GaAs. All of these materials are used with different fabrication processes to provide an arsenal with which to develop solid-state circuits and systems. Clearly, hybrid integration is not limited to a particular technology because it can use previously fabricated devices. However, if the materials, devices, and fabrication methods are suitable for monolithic implementation, great improvements in size and weight can be made over a hybrid circuit. Furthermore, MMICs avoid many parasitics which limit hybrids as well as costly postfabrication tuning to substantially reduce costs in mass production.

1.5 ACTIVE SOLID-STATE DEVICES

Although limited in power output, solid-state device technology has matured to provide repeatability, low cost, and good performance. There are two- and three-terminal types of solid-state devices. Typical packages for solid-state devices used in hybrid MICs are shown in Figure 1.5. Two-terminal devices available for integration include Gunn, IMPATT, varactor, *pin*, and Schottky-barrier diodes. Three-terminal transistor devices to consider include bipolar junction transistors (BJT), metal-semiconductor field-effect transistors (MESFET), heterojunction bipolar transistors (HBT), and high-electron-mobility transistors (HEMT). These devices form the basic building blocks for most circuit applications.

The diodes and transistors listed can be used to provide voltage controlled reactance, negative resistance, and current controlled reactance. Voltage or current control over reactance or negative resistance provides the mechanism

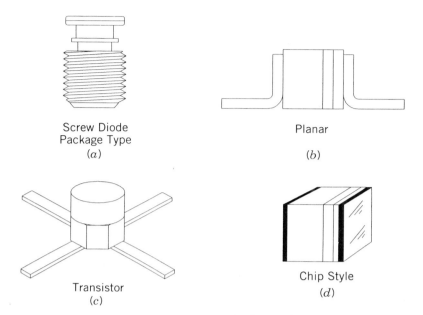

Screw Diode
Package Type
(a)

Planar

(b)

Transistor
(c)

Chip Style
(d)

FIGURE 1.5. Solid-state device packages.

for developing circuits which tune, switch, mix, amplify, or oscillate. The choice of diodes or transistors determines the cost, power output, operating frequency, dc-to-rf conversion efficiency, stability, noise, and biasing requirements. This choice is not always clear-cut or obvious.

Diodes are two-terminal devices that require less complex loading and biasing schemes than do three-terminal transistors. However, transistors can be used to provide many different functions such as switching, tuning, and amplifying, which would require several different types of diodes. Diodes reach higher operating frequencies with higher rf output power, but transistors have higher dc-to-rf conversion efficiencies and operate at lower dc input levels.

1.6   INTEGRATED CIRCUITS

With a combination of solid-state devices, dielectric material, and MIC lines, high-performance circuits for many applications can be developed. Each MIC line provides a starting point toward basic element designs for use in many components. These components include filters, couplers, transformers, and others. A basic element for the design of these components is a resonator. At lower frequencies ($\leqslant 300\,\text{MHz}$), lumped inductor/capacitor (LC) elements are suitable resonators. As the frequencies increase into the microwave range, such resonant elements are not easily constructed and tend to lose power due to radiation.

At microwave frequencies, these resonant elements are developed using distributed transmission lines, short sections of waveguides, or dielectric cavities. Shorted or opened sections of transmission lines can serve as inductors or capacitors at a frequency point in a circuit. Once the resonator configuration is established, several resonators can be cascaded to develop a filter. Similarly, oscillators, amplifiers, active antennas, and other components can be designed by imbedding the solid-state devices discussed in Section 1.5 into these resonant elements. Methods of hybrid integration can be classified into hybrid MICs and miniature hybrid MICs.

Hybrid MICs integrate prefabricated solid-state devices, resistors, inductors, and capacitors with the circuit layout pattern. Miniature hybrid MICs fabricate inductors, capacitors, and resistors within the circuit layout pattern and later integrate solid-state devices. Hybrid MICs require system assembly and tend to be labor intensive in production. Hybrid MICs are often limited at higher frequencies by the effects from device integration discontinuities and package parasitics. Miniature hybrid MICs alleviate some of the problems of hybrid MICs by reducing the circuit components which require manual integration. In MMICs, assembly and integration discontinuities are avoided by fabricating solid-state devices as well as capacitors, inductors, and resistors within the circuit layout. The MIC transmission line and device package play a major role in the success of hybrid technology circuits. Figure 1.6 shows schematic differences in two-terminal device integration between several common planar transmission lines.

MMICs overcome hybrid MIC integration problems by simultaneously fabricating solid-state devices and circuits on a single substrate. Since the device and circuit share the same substrate, there are no bonding, hybrid integration discontinuities, or package parasitics to take into account. However, the development cost of MMICs is usually very high, requiring accurate modeling software and stringent fabrication standards. Due to the small dimensions of the devices, special testing and troubleshooting techniques are needed to determine which components meet specifications. Due to small device-to-chip area, yield (i.e., components which meet specifications) depends mainly on the success of the device fabrication. Errors in modeling or fabrication tolerances affect circuit performance and increase system development times. Some clever tuning schemes have been devised to compensate for these errors. However, postfabrication tuning not only defeats the purpose of monolithic implementation (i.e., avoid postassembly tuning) but also raises costs and uses up valuable wafer space.

There are many considerations for monolithic fabrication of circuits [18]. These are discussed with microstrip line in mind, since it is the most widely used MIC transmission line. As stated previously, reducing circuit size, weight, and cost are very important. However, many problems arise as a result of

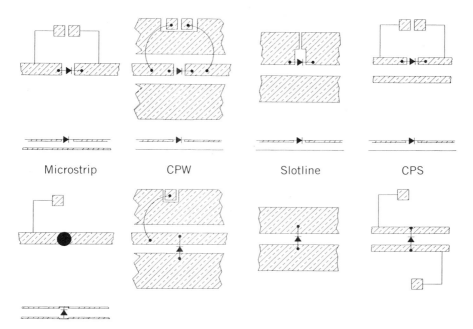

FIGURE 1.6. Solid-state device integration with common MIC lines.

miniaturization. Coupling or crosstalk between adjacent lines limits the packing density of miniaturized circuits. Wider spacings between adjacent lines increase circuit size, while enclosed lines increase costs. Other problems include packaging, circuit losses, and device reproducibility. For active applications (such as amplifiers, oscillators, etc.), energy which is not converted to rf power is lost to heat. Heat generated must be removed efficiently to maintain performance and reliability. Thermal conductivity is a measure for heat conduction, which increases for thinner substrates, but thinner substrates require narrower microstrip lines to maintain the same system impedance level. Narrower microstrip lines suffer from higher resistive losses and greater parasitic capacitance to ground. These resistive losses and parasitic capacitances result in lower quality factors. Such low Q-values prevent the realization of low-loss, narrow-band filters and low-noise oscillators. Other complications arise from the fabrication of via holes for shunt connections which lower yield. Finally, MMIC designs carry an expensive initial development cost, which can only be recovered from large production quantities. Depending on the fabrication methods, MMICs can produce high volumes and high yields without labor-intensive tweaking. This is the essence of MMICs using microstrip.

Alternative MIC configurations alleviate some of the problems mentioned but introduce others. In MMICs the transmission line, device, and component performances must be accomplished through the properties of a particular material and fabrication method. This constraint often limits the component, device, or transmission line optimal performances. For most circuit applications, MICs serve to provide very small reliable circuits with improved reproducibility. However, wafer costs, substrate characteristics, small production quantities, and the large real estate occupied by passive elements can make the use of hybrid MICs more attractive for some applications. One such application may be integrated and active integrated antennas.

1.7 INTEGRATED AND ACTIVE INTEGRATED ANTENNAS

Similar to ICs, integrated antennas are those radiators directly integrated with solid-state devices such as *pin*, varactor, or Schottky-barrier diodes to provide switching, tuning, or mixing functions. Integrated antenna applications require substrates and antenna configurations which allow efficient radiation. Packaged devices created in Si or GaAs can be conveniently integrated within the antenna on substrates which optimize the antenna radiation performance. In hybrid MICs, special low-loss substrates can be used to minimize dielectric losses and optimize antenna efficiencies. Packaging methods can be improved to reduce parasitics and integration discontinuities for higher frequencies of operation. Complete monolithic implementation of integrated antennas may not yield optimal results because MMIC substrates tend to be of higher ε_r, which reduces radiation efficiency. Also, antennas are typically much larger than the integrated solid-state device, which would leave a large amount of valuable wafer real estate unused.

Active integrated antennas are those integrated antennas which use active devices (Gunn diodes, FETs, etc.) to convert dc energy or rf power (such as oscillators, amplifiers, etc.). Integrated and active integrated antennas have the potential of reducing the size, weight, and cost of conventional transmitter, receiver, and transceiver designs by incorporating the circuit component functions at the antenna terminals.

Conventional transmitter, receiver, or transceiver systems use several different circuit components connected to an antenna via a transmission line. At the junction between circuit and transmission line, a transition attempts to provide a smooth change in mode of propagation. Similarly, the transition from the transmission line to the antenna transfers the guided wave to the radiating mode of the radiator. Although this method allows separate optimization of antenna, transmission line, and circuit components, it also limits by certain important factors. The transitions at the component and antenna to transmission line increase the size, weight, and cost of the overall system. The transitions also add circuit complexity and introduce discontinuities which cause losses and limit frequency of operation. These losses add to the losses incurred in the transmission line, which often worsen with increasing frequency. Conventional application of MICs and MMICs to miniaturize the circuit

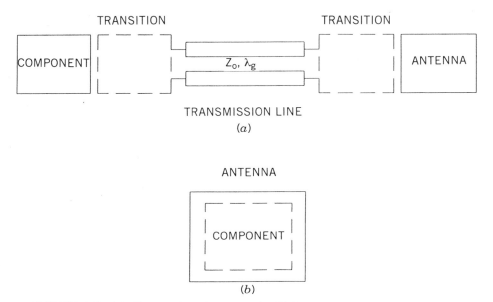

TRANSMISSION LINE
(a)

(b)

FIGURE 1.7. (a) Conventional approach; (b) integrated antenna approach.

components, transmission lines, and antenna only reduce these losses. In contrast, integrated antennas attempt to eliminate these losses altogether by integrating solid-state devices at the antenna. This integration, however, affects the performance of the antenna and the component function. One obvious example of these trade-offs is the use of active integrated antennas for rf transmitter applications.

Unlike conventional transmitters, an active antenna integrates active devices directly at the antenna terminals. By designing the antenna and oscillator on a single substrate, one avoids transition/transmission line losses from power distribution networks. Figure 1.7 shows conventional and integrated antenna approaches for transmitters. Ideal active integrated antennas would provide good component characteristics without compromising antenna performance. Although the concept is straightforward, successful implementation of this method has been difficult to achieve.

Active integrated antenna design requires knowledge in several areas of microwave engineering, including solid-state devices, circuits, and antennas. Experts in circuits and oscillator design seldom have antenna experience, and vice versa. An obvious trade-off of the active integrated antenna approach is that optimization of the oscillator and antenna must be made simultaneously on the same structure. This is sometimes self-defeating since characteristics which improve circuit oscillators often degrade antenna radiation performance. As can be seen in later chapters, active integrated antennas have been realized using two- and three-terminal devices for rf power generation, low-cost sensors, decoys, modulators, amplifiers, and sources for power combining.

1.8 SYSTEM APPLICATIONS

Active and integrated antennas can be used as an alternative to conventional approaches. Functions which would normally occur in the circuit away from the antenna (such as detection, tuning, modulation, mixing, amplifying, etc.) can be included with the antenna to reduce size, weight, and cost. If the component and antenna provide nearly the same performance as that using a conventional approach, the reduction of size and cost would greatly enhance the system's usefulness.

Pin integrated antennas can serve as modulators at the antenna terminals. Such a modulator would avoid long rf line losses. The amplitude modulator (AM) could be set up to transmit information using digital coding or analog transmission techniques. Similarly, varactor-loaded antennas can be used as frequency modulators (FM). Again, this method would avoid long rf lines to reduce carrier losses. From a receiver standpoint, *pin*s can be used as frequency switches or as a squelch setting. Varactors can be used as a front-end filter for band or channel selection as well as to increase a narrow-band antenna's operating bandwidth. In array applications, these solid-state devices can be used to increase the usable bandwidth and/or provide an amplitude- or phase-tapering mechanism. In phased arrays, the usable scanning range can be increased by tuning (or detuning) individual antenna elements.

Active antennas as oscillators make very inexpensive microwave sources. Given a dc power supply and a modulating circuit, such an inexpensive source may become a useful decoy for electronic warfare and jamming applications or for FM communication links. Decoy systems have been incorporated using monolithic techniques and conventional transceiver approaches. Further improvements could be accomplished by using an active integrated antenna approach. Communication applications normally require a transmitter with low oscillator noise and high stability. Active antennas require further development to meet these requirements.

Active antennas are inherently good Doppler sensors because radiating sources are sensitive to Doppler return from moving objects. The antenna serves as a self-oscillating mixer to mix the local oscillating frequency with the low-level frequency reflection from the moving object. Extracting this Doppler return allows relative motion to be detected. Such an application could greatly reduce the current size, weight, and cost of radar, rf identification systems, automotive rf systems, automatic door openers, perimeter monitoring systems, and burglar alarms.

Schottky-detector diodes can be integrated with the antenna for rf detection. Antenna selectivity determines the rf band or channel being monitored. This detector can be used for many different monitoring or imaging applications.

This approach also applies directly to rectifying antenna work for microwave power transmission. To illustrate, power lines would not be an effective method of distributing power in space. It would also be too costly for each satellite to have its own generating plant. However, a centralized power source could distribute power to its neighboring satellites via microwaves using large antennas. Beams would focus on the receiving satellite which would have a large rectenna array to convert the rf to dc. Positioning equipment and computer monitoring systems would be required to maintain beam taper and phasing, but such complex automated systems already keep F-16s up in the air. As device conversion efficiencies improve and fabrication costs are reduced, such a system may become a reality.

Active antennas are ideally suited for spatial power combining sources in the millimeter and submillimeter bands. At these high frequencies, solid-state devices produce very little power and tend to be very inefficient. To compensate, power-combining techniques have been developed. Power combining can be accomplished at either the chip, circuit, or spatial level. Spatial power combining techniques are specifically suited for active antennas because they incorporate many low-power radiating sources. Techniques are applied to ensure coherent operation of each source to obtain a large amount of combiner rf power. All these topics will be discussed in detail in later chapters.

REFERENCES

1. W. E. Gordon, "A Hundred Years of Radio Propagation," *IEEE Transactions on Antennas and Propagation*, Vol. 33, No. 2, pp. 126–130, February 1985.

2. J. D. Kraus, "Antennas Since Hertz and Marconi," *IEEE Transactions on Antennas and Propagation*, Vol. AP-33, No. 2, pp. 131–137, February 1985.

3. H. R. Hertz, *Electric Waves*, Macmillan, New York, 1893; reprinted Dover, New York, 1962.

4. R. M. Barrett, "Microwave Printed Circuits — A Historical Survey," *IRE Transactions on Microwave Theory and Techniques*, Vol. 3, No. 2, pp. 1–9, March 1955.

5. H. A. Wheeler, "Transmission-Line Properties of Parallel Strips Separated by a Dielectric Sheet," *IEEE Transactions on Microwave Theory and Techniques*, Vol. 13, No. 2, pp. 172–185, March 1965.

6. M. V. Schneider, "Microstrip Lines for Microwave Integrated Circuits," *Bell System Technical Journal*, Vol. 48, No. 5, pp. 1421–1444, May–June 1969.

7. M. V. Schneider, B. Glance, and W. F. Bodtmann, "Microwave and Millimeter Wave Hybrid Integrated Circuits for Radio Systems," *Bell System Technical Journal*, Vol. 48, No. 6 pp. 1703–1726, July–August 1969.

8. A. K. Ganguly and B. E. Spielman, "Dispersion Characteristics for Arbitrarily Configured Transmission Media," *IEEE Transactions on Microwave Theory and Techniques*, Vol. 25, No. 12, pp. 1138–1141, December 1977.

9. J. P. Villotte, M. Aubourg, and Y. Garault, "Modified Suspended Striplines for Microwave Integrated Circuits," *Electronics Letters*, Vol. 14, No. 18, pp. 602–603, August 1978.

10. C. P. Wen, "Coplanar Waveguide: A Surface Strip Transmission Line Suitable for Non-Reciprocal Gyromagnetic Device Applications," *IEEE Transactions on Microwave Theory and Techniques*, Vol. 17, No. 12, pp. 1087–1090, December 1969.

11. V. F. Hanna, "Finite Boundary Corrections to Coplanar Strip Line Analysis," *Electronics Letters*, Vol. 16, No. 15, pp. 604–606, July 1980.

12. S. B. Cohn, "Slot Line on a Dielectric Substrate," *IEEE Transactions on Microwave Theory and Techniques*, Vol. 17, No. 10, pp. 768–778, October 1969.

13. P. J. Meier, "Integrated Fin-line Millimeter Components," *IEEE Transactions n Microwave Theory and Techniques*, Vol. 22, No. 12, pp. 1209–1216, December 1974.

14. P. J. Meier, "Millimeter Integrated Circuits Suspended in the E-plane of Rectangular Waveguide," *IEEE Transactions on Microwave Theory and Techniques*, Vol. 26, No. 10, pp. 726–733, October 1978.

15. E. A. J. Marcatili, "Dielectric Rectangular Waveguide and Directional Coupler for Integrated Optics," *Bell System Technical Journal*, Vol. 48, No. 7, pp. 2071–2102, September 1969.

16. R. M. Knox, "Dielectric Waveguide Microwave Integrated Circuits — An Overview," *IEEE Transactions on Microwave Theory and Techniques*, Vol. 24, No. 11, pp. 806–814, November 1976.

17. R. M. Knox and P. P. Toulios, "Integrated Circuits for the Millimeter through Optical Frequency Range," *Proceedings of the Symposium Submillimeter Waves*, Polytechnic Press of the Polytechnic Institute of Brooklyn, New York, pp. 497–516, 1970.

18. D. K. Ferry, *Gallium Arsenide Technology*, Howard W. Sams & Co. Inc., Indianapolis, 1985.

CHAPTER TWO

Oscillators and Synchronization

2.1 INTRODUCTION

For the most part, integrated antennas have emphasized active diode and transistor integrations for transmitters and power combiners. The combiners involve typically oscillators for spatial and quasi-optical power combining. The following sections review important oscillator theory and characteristics. These characteristics include derivation of the oscillation conditions and load pulling. Also included are descriptions of spectral properties such as quality factor, stability, and noise. Synchronization techniques are also discussed.

2.2 OSCILLATION CONDITIONS

Solid-state sinusoidal oscillators are resonant circuits integrated with active solid-state devices which convert dc energy to rf power. The resonant circuit or resonator can take many shapes and forms and, because of its importance, usually determines the name of the resulting oscillator (i.e., waveguide and microstrip oscillators). Resonators include short lengths of transmission lines, rings, dielectric disks, rectangular and circular waveguide cavities, and yttrium iron garnet (YIG) spheres. In active integrated antennas, the antenna serves as the resonant circuit as well as the radiator. A dc bias line delivers the proper voltage and current for the active device to generate electromagnetic energy. The resonator circuit provides the necessary reactive storage which compensates for the active device reactance. Oscillations occur at the frequency where the overall circuit reactances cancel out.

Oscillation start-up occurs due to noise or dc bias transients. The power level builds up until the active device is saturated. The device achieves a steady-state equilibrium by continually restoring the power delivered to and dissipated in the circuit. Since the device impedance is a function of the rf current, the oscillation frequency may slightly change while the device reaches its steady state. At equilibrium, the oscillation frequency and amplitude remain unchanged. For a free-running, steady-state oscillator, the sum of the circuit (Z_c) and device (Z_d) impedances must be zero at the device's operating point, as shown by

$$Z_d(V_{dc}, I_{dc}, \omega_n, I_{rf}, T, \ldots) + Z_c(\omega_n) = 0 \qquad (2.1)$$

where the circuit impedance varies only with frequency and the device impedance is a function of its dc operating point (V_{dc}, I_{dc}), operating frequency components (ω_n), rf current amplitude (I_{rf}), and temperature (T), etc. The subscript n refers to the nth harmonic frequency, which can take values $1, 2, \ldots$.

Types of sinusoidal oscillators include subharmonic, fundamental, and harmonic. Oscillators provide power at each of its harmonics. If the embedding circuit is designed to operate at the first harmonic and suppress others, it is called a fundamental oscillator. Similarly, subharmonic oscillators operate at fractional multiples below the fundamental, and harmonic oscillators operate at integer multiples above the fundamental. Since the impedance characteristics of the devices change with dc operating point, the bias voltage can be used to tune the oscillation frequency. This characteristic is called bias tuning (or pushing), which is used for frequency modulation (FM) and has created a large commercial market for the voltage controlled oscillator (VCO).

2.3   VOLTAGE CONTROLLED OSCILLATORS

The VCO uses voltage to control the output frequency of the oscillator. This characteristic is used for frequency modulation in communication links. A change in voltage can change the device or circuit reactance. One can change the dc operating point (V_{dc}, I_{dc}) of the active device, which changes the device impedance Z_d and pushes the circuit to the new operating frequency of Equation (2.1). One drawback of bias tuning is that the modulating circuit must be able to drive the dc power drained in the active device, which may be several watts. Also, since the operating point is altered, the corresponding change in the negative resistance may lead to wide rf power deviations.

Another method relies on a second solid-state device called a varactor (variable capacitor) to alter the reactance of the circuit. The varactor is strategically placed to alter the circuit reactance $X_c(\omega_n)$. Unlike bias tuning, the active device operating point remains the same and the output power remains more constant over a wider frequency tuning range. Furthermore, the varactor consumes little power, which puts few constraints on the modulating circuit. However, a second device and another control voltage increases circuit complexity and system cost. Depending on the cost allowed, performance requirements, space available, materials provided, and power allocated to the rf section in a system, one method may be better suited than the other. Microwave designers evaluate all of these trade-offs and attempt to achieve the best possible design for a set of specifications.

Voltage controlled oscillators (VCOs) form an integral part of many military and commercial microwave systems for local oscillator and transmitter applications. Oscillator specifications give system designers the ability to categorize oscillators for different applications. Several characteristics used to describe oscillators are

1. *Stability:* ability of an oscillator to return to the original operating point after experiencing a slight electrical or mechanical disturbance

2. *Noise*
   - Amplitude modulation (AM) noise: amplitude variations of the output signal
   - FM noise: unwanted frequency variations
   - Phase noise: phase variations

3. *Quality (Q) Factor*
   - Unloaded: accounts for resonator losses R_{loss} only
   - External: accounts for the load resistor R_{load} only and assumes $R_{loss} = 0$
   - Loaded: accounts for both resonator losses and external loading

4. *Frequency*
   - *Jumping:* discontinuous change in oscillator frequency due to non-linearities in the device impedance.
   - *Pulling:* change in oscillator frequency versus a specified load mismatch over 360° of phase variation
   - *Pushing:* change in oscillator frequency versus dc bias point variation.

5. *Spurious Signals:* output signals at frequencies other than the desired oscillation carrier

TABLE 2.1. Typical Commercial VCO Specifications

Frequency (f_0)	35 GHz	94 GHz
Power (P_0)	250 mW	50 mW
Bias pushing range (typical)	50 MHz/V	200 MHz/V
Varactor tuning range	± 250 MHz	± 250 MHz
Frequency drift over temperature	-2 MHz/°C	-4 MHz/°C
Power drop over temperature	-0.03 dB/°C	-0.03 dB/°C
Q_{ext}	800–1000	800–1000
Harmonics level	-20 dB	-20 dB
Modulation bandwidth	dc -50 MHz	dc -50 MHz
Modulation sensitivity (MHz/V)	25–50	50–500
FM noise @ 100-kHz offset	-90 dBc/kHz	-80 dBc/kHz
AM noise @ 100-kHz offset	-155 dBc/kHz	-150 dBc/kHz

6. *Post-tuning Drift:* frequency and power drift of a steady-state oscillator due to heating of the solid-state device

7. *Thermal Stability:* change in output power and frequency versus temperature

Table 2.1 gives data for typical commercial Gunn VCOs operating at 35 and 94 GHz.

2.4   DERIVATION OF OSCILLATION CONDITIONS

There are several ways to study oscillators, including using scattering parameters, stability regions, and Nyquist theory applied to feedback amplifiers and a negative-resistance network. Kurokawa used a negative resistance in a two-terminal network to describe oscillator characteristics [1]. The following derivation follows his work closely to lead to oscillation conditions, stability, noise, and Q-factor formulations. We begin with the one-port equivalent circuit of Figure 2.1. The forcing function $e(t)$ is used to model noise for oscillation start-up or an external injection-locking signal used for synchronization.

In the circuit, an rf current flows through the device. The rf current has magnitude $A_n(t)$, frequency ω_n, and phase $\phi_n(t)$:

$$I(t) = \sum_{n=1}^{\infty} A_n(t)e^{j(\omega_n t + \phi_n(t))}, \qquad n = 1, 2, 3, \ldots \tag{2.2}$$

The magnitude $A_n(t)$ and phase $\phi_n(t)$ are functions of time because they can be subjected to noise and dc bias transients. These transients are often present at very low frequencies so that $A_n(t)$ and $\phi_n(t)$ are slowly varying with respect to the carrier frequency ω_n. Each frequency harmonic contributes to the overall current $I(t)$.

Each of the harmonics where oscillations are occurring must satisfy Equation (2.1). Using Kirchhoff's voltage law for the loop, we can write equations for each harmonic as follows:

$$
\begin{aligned}
e_1(t) &= A_1(t)e^{j(\omega_1 t + \phi_1(t))}[Z_d(A_1, \omega_1) + Z_c(\omega_1)] \\
e_2(t) &= A_2(t)e^{j(\omega_2 t + \phi_2(t))}[Z_d(A_2, \omega_2) + Z_c(\omega_2)] \\
&\vdots \\
e_n(t) &= A_n(t)e^{j(\omega_n t + \phi_n(t))}[Z_d(A_n, \omega_n) + Z_c(\omega_n)] \\
&\vdots
\end{aligned}
\tag{2.3}
$$

where the device and circuit impedances are

$$
\begin{aligned}
Z_d(A_n, \omega_n) &= R_d(A_n, \omega_n) + jX_d(A_n, \omega_n) \\
Z_c(\omega_n) &= X_d(\omega_n) + jX_c(\omega_n)
\end{aligned}
\tag{2.4}
$$

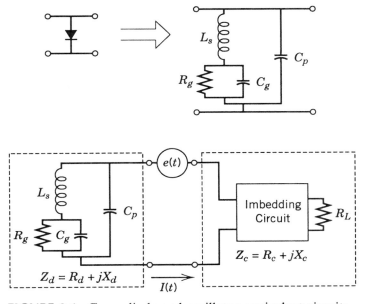

FIGURE 2.1. Gunn diode and oscillator equivalent circuits.

The device impedance $Z_d(A_n, \omega_n)$ is a function of both the magnitude of the rf current and oscillating frequency, while the circuit impedance $Z_c(\omega_n)$ only depends on the frequency. Equations (2.3) imply that the sum of the circuit and device voltages must equal the forcing function $e_n(t)$ at each harmonic and can be rewritten in the more compact form

$$e_n(t) = A_n(t)e^{j(\omega_n t + \phi_n(t))}(Z_d(A_n, \omega_n) + Z_c(\omega_n)), \qquad n = 1, 2, 3, \ldots \quad (2.5)$$

If the change in frequency $\delta\omega_n$ is small with respect to the carrier ω_n, the circuit impedance $Z_c(\omega_n)$ can be expanded in a Taylor series about $\omega_n + \delta\omega_n$. We also expand the device impedance $Z_d(A_n, \omega_n)$ about $A_n + \delta A_n, \omega_n + \delta\omega_n$ with a two-variable expansion:

$$Z_c(\omega_n + \delta\omega_n) = Z_c(\omega_n) + \delta\omega_n \left.\frac{d}{d\omega} Z_c\right|_{\omega_n} + \cdots + \frac{(\delta\omega_n)^m}{m!} \left.\frac{d^m}{d\omega^m} Z_c\right|_{\omega_n} + \cdots \quad (2.6)$$

$$Z_d(A_n + \delta A_n, \omega_n + \delta\omega_n) = Z_d(A_n, \omega_n) + \delta\omega_n \left.\frac{\partial}{\partial\omega} Z_d\right|_{A_n, \omega_n} + \delta A_n \left.\frac{\partial}{\partial A} Z_d\right|_{A_n, \omega_n}$$

$$+ \cdots + \frac{(\delta A_n)^m}{m!} \left.\frac{\partial^m}{\partial A^m} Z_d\right|_{A_n, \omega_n} + \frac{(\delta\omega_n)^m}{m!} \left.\frac{\partial^m}{\partial\omega^m} Z_d\right|_{A_n, \omega_n} + \cdots$$

$$(2.7)$$

For small perturbations, we can neglect higher powers of the Taylor series and products of small quantities. The result is a compact approximation for the perturbed circuit and device impedances.

$$Z_c(\omega_n + \delta\omega_n) \approx Z_c(\omega_n) + \delta\omega_n \left.\frac{d}{d\omega} Z_c\right|_{\omega_n} \tag{2.8}$$

$$Z_d(A_n + \delta A_n, \omega_n + \delta\omega_n) \approx Z_d(A_n, \omega_n) + \delta\omega_n \left.\frac{\partial}{\partial\omega} Z_d\right|_{A_n,\omega_n} + \delta A_n \left.\frac{\partial}{\partial A} Z_d\right|_{A_n,\omega_n}$$

$$\tag{2.9}$$

We replace the exponential (i.e., $e^{j(\omega_n t + \phi_n(t))} = \cos(\omega_n t + \phi_n(t)) + j\sin(\omega_n t + \phi_n(t)))$ and include the real and imaginary parts of the impedances. Keeping only the real part of the resulting voltage drops around the current loop of Eq. (2.5), results in the equation

$$e_n(t) = A_n(t)\cos(\omega_n t + \phi_n(t))(R_d(A_n, \omega_n) + R_c(\omega_n))$$
$$- A_n(t)\sin(\omega_n t + \phi_n(t))(X_d(A_n, \omega_n) + X_c(\omega_n)), \quad n = 1, 2, 3, \ldots \tag{2.10}$$

We then replace the real part of the circuit and device impedance with the appropriate approximations of Equations (2.8) and (2.9). The time dependence of $e_n(t)$, $A_n(t)$, and $\phi_n(t)$ is implied. The notation Z_{c_n} stands for the localized circuit impedance at the nth harmonic frequency ω_n. Similarly, the notation Z_{d_n} denotes the localized device impedance at the frequency ω_n and a current magnitude A_n. The forcing function $e_n(t)$ can be used to model noise fluctuation during oscillator start-up or an external injection-locking signal. Equation (2.10) becomes

$$e_n(t) = A_n\cos(\omega_n t + \phi_n)\left(R_{d_n} + R_{c_n} + \delta\omega_n \frac{\partial}{\partial\omega}(R_{c_n} + R_{d_n}) + \delta A_n \frac{\partial}{\partial A}(R_{d_n})\right)$$

$$- A_n\sin(\omega_n t + \phi_n)\left(X_{d_n} + X_{c_n} + \delta\omega_n \frac{\partial}{\partial\omega}(X_{c_n} + X_{d_n}) + \delta A_n \frac{\partial}{\partial A}(X_{d_n})\right)$$

We multiply the result by $\cos(\omega_n t + \phi_n)$ and $\sin(\omega_n t + \phi_n)$ and integrate over a single period of oscillation. Because of the orthogonal properties of sine and cosine functions, the following two equations are formed:

$$\frac{A_n}{2}\left(R_{d_n} + R_{c_n} + \delta\omega_n \frac{\partial}{\partial\omega}(R_{c_n} + R_{d_n}) + \delta A_n \frac{\partial}{\partial A}(R_{d_n})\right) = \int_{T_n} e_n(t)\cos(\omega_n t + \phi_n)\, dt$$

$$= e_{c_n}$$

$$\frac{-A_n}{2}\left(X_{d_n} + X_{c_n} + \delta\omega_n \frac{\partial}{\partial\omega}(X_{c_n} + X_{d_n}) + \delta A_n \frac{\partial}{\partial A}(X_{d_n})\right) \tag{2.11}$$

$$= \int_{T_n} e_n(t)\sin(\omega_n t + \phi_n) = e_{s_n} \tag{2.12}$$

where the right-hand sides are the cosine and sine transformations of the forcing function for each of the N harmonics. To conserve space, the cosine and sine transformations will be denoted by e_{c_n} and e_{s_n}, respectively.

Equations (2.11) and (2.12) represent general oscillation conditions accounting for the device impedance's frequency and current magnitude dependence as well as the circuit impedance's dependence on frequency. These equations also assume that harmonics are present in the circuit which can be superimposed in the frequency domain.

For a free-running, steady-state oscillator the operating frequency is stable and the output power is constant (i.e., $e_{c_n} = e_{s_n} = 0$, $\delta\omega_n = 0$, and $\delta A_n = 0$). The equations then reduce to the following conditions:

$$(R_{d_n} + R_{c_n}) = 0 \tag{2.13}$$

$$-(X_{d_n} + X_{c_n}) = 0 \tag{2.14}$$

These equations show that each of the harmonics must satisfy Equation (2.1). The resonator characteristics determine the higher-order harmonics which satisfy the oscillation conditions.

For a free-running, steady-state oscillator to operate, the magnitude of the active device's negative resistance must be equal to the circuit resistance

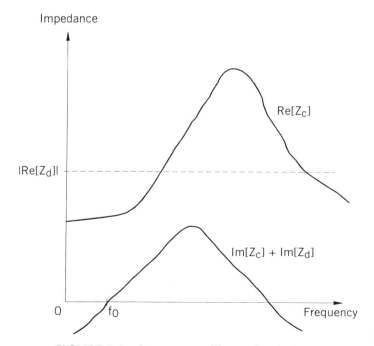

FIGURE 2.2. One-port oscillator simulation.

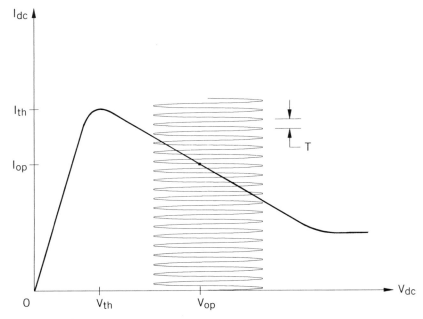

FIGURE 2.3. Gunn diode I versus V characteristics.

presented to its terminals. Oscillation frequency occurs where the reactances cancel. These conditions for oscillation are summarized here and shown in Figure 2.2 at the oscillation frequency f_o as an example:

$$R_c \leqslant |R_d|, \qquad X_d = -X_c \qquad (2.15)$$

Furthermore, the start-up is more likely to occur if the magnitude of the negative resistance is at least 20% greater than the circuit resistance (i.e., $|R_d| \geqslant 1.2R_c$).

Figure 2.3 shows the Gunn diode I-V characteristics, including the threshold voltage, operating current, and operating voltage. Impressed on the graph is an rf variation which simulates oscillation conditions.

2.5   *S*-PARAMETER FORMULATION OF THE OSCILLATION CONDITIONS

At microwave frequencies and above, it becomes impractical to measure currents and voltages, but relative power and scattering (S) parameters can be measured accurately on network analyzers, spectrum analyzers, and power meters. S-parameters measurements are very useful for MIC designs. The equations are therefore modified to use S-parameter definitions. First, the circuit and device impedances are used to obtain the corresponding reflection coefficients (Γ):

$$\Gamma_c(\omega_n) = \frac{Z_c(\omega_n) - Z_0(\omega_n)}{Z_c(\omega_n) + Z_0(\omega_n)}$$

$$\Gamma_d(A_n, \omega_n) = \frac{Z_d(A_n, \omega_n) - Z_0(\omega_n)}{Z_d(A_n, \omega_n) + Z_0(\omega_n)}$$

(2.16)

where Z_0 represents the characteristic impedance of the transmission line used, which may be a function of frequency. The reflection coefficients in Equation (2.16) are equivalent to one-port S-parameter, denoted by S_{11}. The equivalent condition to Equation (2.15) for the one-port negative-resistance oscillator becomes

$$\Gamma_d(A_n, \omega_n)\Gamma_c(\omega_n) = 1$$

(2.17)

Equation (2.11) can be derived by using definitions for the one-port S-parameter in Equations (2.16); $\Gamma_c = S_c = b_c/a_c$ is the ratio of the reflected signal b_c to the transmitted signal a_c (see Fig. 2.4). Similarly, $\Gamma_d = S_d = b_d/a_d$ for the device. Solving for b_c and b_d leads to

$$b_c = S_c a_c \quad \text{and} \quad b_d = S_d a_d$$

(2.18)

When the device and the circuit ports are connected such that $a_c = b_d$ and $a_d = b_c$, we can solve for the transmitted signal a_c:

$$\left. \begin{array}{l} a_c = b_d = S_d a_d \\ a_d = b_c = S_c a_c \end{array} \right\} \quad a_c = S_d S_c a_c$$

(2.19)

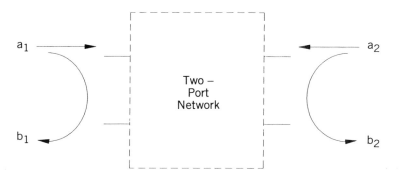

FIGURE 2.4. Scattering parameter signal definitions.

After rearranging Equation (2.19), we obtain the S-parameter condition for oscillations [2–5]:

$$(S_d S_c - 1)a_c = 0 \qquad (2.20a)$$

Since $a_c \neq 0$, the oscillation frequency will exist only if $S_d S_c - 1 = 0$, or

$$S_d S_c = 1 \qquad (2.20b)$$

Note that Equation (2.20b) is equivalent to Equation (2.17) and that the basic form of the oscillation condition in Equations (2.20a) and (2.20b) remains unchanged. This is true even for an N-port network where the quantity in parentheses becomes an $N \times N$ matrix equation. The oscillation condition is given by the determinant of the matrix.

For transistors, proper dc biasing may not be enough to guarantee a negative-resistance region. A negative-resistance region can be induced by proper feedback and termination elements. A FET circuit in a common-gate configuration forms the two-port network shown in Figure 2.5a. Similarly, common-source and common-drain configurations are possible. The two-port S-parameter oscillation condition is no longer a single multiplier as in the one-port case. The device and load network parameters are now defined by 2×2 matrices. An element coefficient S_{ij} of the matrix is the ratio of the resulting voltage at port i due to the induced voltage at port j. The oscillation condition is the determinant of the 2×2 S-matrix.

$$([S_d] \cdot [S_c] - [1]) \cdot [a] = \left[\begin{bmatrix} S_{11}^d & S_{12}^d \\ S_{21}^d & S_{22}^d \end{bmatrix} \cdot \begin{bmatrix} S_{11}^c & S_{12}^c \\ S_{21}^c & S_{22}^c \end{bmatrix} - \begin{bmatrix} 1 & 0 \\ 0 & 1 \end{bmatrix} \right] \cdot \begin{bmatrix} a_1 \\ a_2 \end{bmatrix} = 0$$

$$(2.21)$$

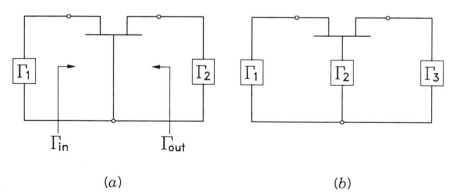

$$(a) \qquad\qquad\qquad\qquad (b)$$

FIGURE 2.5. (a) FET in a two-port configuration; (b) FET in a three-port configuration.

Equation (2.21) is solved if

$$\begin{vmatrix} S_{11}^d \cdot S_{11}^c + S_{12}^d \cdot S_{21}^c - 1 & S_{11}^d \cdot S_{12}^c + S_{12}^d \cdot S_{22}^c \\ S_{21}^d \cdot S_{11}^c + S_{22}^d \cdot S_{21}^c & S_{21}^d \cdot S_{12}^c + S_{22}^d \cdot S_{22}^c - 1 \end{vmatrix} = 0$$

$$(S_{11}^d \cdot S_{11}^c + S_{12}^d \cdot S_{21}^c - 1) \cdot (S_{21}^d \cdot S_{12}^c + S_{22}^d \cdot S_{22}^c - 1)$$
$$- (S_{21}^d \cdot S_{11}^c + S_{22}^d \cdot S_{21}^c) \cdot (S_{11}^d \cdot S_{12}^c + S_{12}^d \cdot S_{22}^c) = 0$$

For simple loads, there is no coupling or feedback between circuit ports (i.e., $S_{12}^c = S_{21}^c = 0$). The equation then reduces to

$$(S_{11}^d \cdot S_{11}^c - 1) \cdot (S_{22}^d \cdot S_{22}^c - 1) - (S_{11}^d \cdot S_{22}^c \cdot S_{12}^d \cdot S_{21}^d) = 0 \qquad (2.22)$$

Using conventional notation, setting $S_{11}^c = \Gamma_1$ and $S_{22}^c = \Gamma_2$, and remove superscripts from device parameters to obtain two well-known oscillation conditions for the two-port network:

$$S_{11} + \frac{S_{12}S_{21}\Gamma_2}{1 - S_{22}\Gamma_2} = \frac{1}{\Gamma_1} \qquad (2.23a)$$

$$S_{22} + \frac{S_{12}S_{21}\Gamma_1}{1 - S_{11}\Gamma_1} = \frac{1}{\Gamma_2} \qquad (2.23b)$$

The following equations* are given for Γ_{in} and Γ_{out}:

$$\Gamma_{in} = S_{11} + \frac{S_{12}S_{21}\Gamma_2}{1 - S_{22}\Gamma_2} \qquad (2.23c)$$

$$\Gamma_{out} = S_{22} + \frac{S_{12}S_{21}\Gamma_1}{1 - S_{11}\Gamma_1} \qquad (2.23d)$$

From the definitions of Γ_{in} and Γ_{out}, Equations (2.23) can be rewritten as

$$\Gamma_{in} = 1/\Gamma_1 \qquad (2.24a)$$

$$\Gamma_{out} = 1/\Gamma_2 \qquad (2.24b)$$

For a FET that is not in a common-gate, common-source, or common-drain configuration, the circuit becomes the three-port network shown in Figure 2.5b. If the circuit is composed of just three simple loads, then there is no coupling or feedback to contend with (i.e., $S_{12}^c = S_{13}^c = S_{21}^c = S_{31}^c = S_{23}^c = S_{32}^c = 0$) and

---

*Those equations can be found in many books (see, for example, K. Chang, *Microwave Solid-State Circuits and Applications*, Wiley, 1994).

the following condition holds:

$$
\begin{vmatrix}
S_{11}^{d}S_{11}^{c} - 1 & S_{12}^{d}S_{22}^{c} & S_{13}^{d}S_{33}^{c} \\
S_{21}^{d}S_{11}^{c} & S_{22}^{d}S_{22}^{c} - 1 & S_{23}^{d}S_{33}^{c} \\
S_{31}^{d}S_{11}^{c} & S_{12}^{d}S_{22}^{c} & S_{33}^{d}S_{33}^{c} - 1
\end{vmatrix} = 0 \qquad (2.25)
$$

The determinant of Equation (2.25) is calculated just as before, but we use the more compact notation $S_{11}^{c} = \Gamma_1$, $S_{22}^{c} = \Gamma_2$, $S_{33}^{c} = \Gamma_3$ to distinguish the load from the device. We then remove superscripts from the device S-parameter notation to adhere to conventional notation. The complexity of the three-port network can be characterized in a more intuitive fashion with S-parameters, even though the same calculations are being carried out. The determinant can then be expanded to give

$$
\frac{S_{12}S_{21}\Gamma_1\Gamma_2}{(1 - S_{11}\Gamma_1)(1 - S_{22}\Gamma_2)} + \frac{S_{13}S_{31}\Gamma_1\Gamma_3}{(1 - S_{22}\Gamma_2)(1 - S_{33}\Gamma_3)}
$$

$$
+ \frac{S_{23}S_{32}\Gamma_2\Gamma_3}{(1 - S_{11}\Gamma_1)(1 - S_{22}\Gamma_2)}
$$

$$
+ \frac{\Gamma_1\Gamma_2\Gamma_3(S_{12}S_{23}S_{31} + S_{21}S_{32}S_{13})}{(1 - S_{11}\Gamma_1)(1 - S_{22}\Gamma_2)(1 - S_{33}\Gamma_3)} = 1 \qquad (2.26)
$$

The pattern progression shown for one-, two-, and three-port S-parameter formulation of the oscillation condition continues for N-port networks. The resulting matrix determinant is formed from an $N \times N$ matrix set of linear equations. The S-parameter notation is useful, compact, and very important for microwave circuit designers. It blends well with current test measurement equipment and computer-aided design (CAD) techniques.

The two- and three-port formulations of the oscillation condition are useful for FET and BJT integrated active antennas. The one-port negative resistance is sufficient to determine other oscillator characteristics of interest, such as oscillator stability, Q-factor, AM and FM noise, and load pulling.

2.6 QUALITY FACTOR

The Q-factor defines different types of losses within the oscillating circuit. The Q-factor of a linear network is proportional to the ratio of stored energy per cycle to energy dissipated per cycle. High Q-factors are associated with low losses and pure, stable signals. In general, a circuit's Q-factor depends on the operating frequency, the load network, and the slope of the overall reactance versus frequency as shown:

$$Q = 2\pi f_0 \frac{\{\text{Time avg. energy}_{\text{stored}}\}}{\{\text{Power}_{\text{dissipated}}\}} = 2\pi f_0 \frac{\{\frac{1}{4}(\partial X_0/\partial\omega)I_0^2\}}{\{\frac{1}{2}R_{\text{loss}}I_0^2\}} = \frac{\omega_0}{2R_{\text{loss}}}\frac{\partial X_0}{\partial\omega} \quad (2.27)$$

where X_0 is the overall circuit reactance and R_{loss} is the resistance which accounts for power dissipated in the circuit.

There are three useful definitions for quality factors, namely unloaded, external, and loaded Q-factors. The loaded Q-factor depends on the external and unloaded Q-factors and is defined by

$$\frac{1}{Q_l} = \frac{1}{Q_u} + \frac{1}{Q_{\text{ext}}} \quad (2.28)$$

where the unloaded and external Q-factors are given by

$$Q_u = \frac{\omega_0}{2R_{\text{loss}}}\frac{\partial(X_d(A,\omega) + X_c(\omega))}{\partial\omega} \quad (2.29)$$

$$Q_{\text{ext}} = \frac{\omega_0}{2R_{\text{load}}}\frac{\partial(X_d(A,\omega) + X_c(\omega))}{\partial\omega} \quad (2.30)$$

The Q-factor will change with dc operating point because the device reactance X_d is a function of oscillator current. The load resistance R_{load} is usually fixed at $50\,\Omega$ as a standard for rf test equipment, and R_{loss} is the loss resistance which is related to dissipating energy in the circuit. For high quality factors it is important to have low values of R_{loss} and high values of $\partial(X_d(A,\omega) + X_c(\omega))/\partial\omega$ around the operating frequency point.

2.7 STABILITY

Slight electrical, thermal, and mechanical disturbances can cause an oscillator to cease oscillating. The disturbance may cause the device to fail or it could change the device impedance such that the conditions of Equations (2.13) and (2.14) are not satisfied. When disturbed by small changes in current magnitude and operation frequency, stable oscillators tend to return to their steady-state oscillation point. Once again, we assume that the device is a function of both current amplitude and frequency, while the circuit depends only on the operating frequency.

For a small increase of δA_n to the current amplitude A_n of the steady-state oscillator, saturation of the negative resistance reduces the output current which then damps out the initial increase. Similarly, a decrease of δA_n on the current A_n increases the negative resistance which then produces more current to restore the equilibrium. Since the device reactance $X_d(A_n, \omega_n)$ is also a function of oscillator current, current fluctuations of $\pm\delta A_n$ alter the device reactance, which changes the output frequency $\pm\delta\omega_n$. This characteristic is useful for modulation when using the device bias for tuning. However, when small fluctuations occur due to device impurities or an unwanted external source, it is called FM noise. When the fundamental frequency is disturbed, all harmonics are affected. The response of higher order harmonics is then tied to the characteristics of the fundamental.

Stability is a measure which describes an oscillator ability to return to its steady-state operating point. A stability criterion can be derived by perturbing the rf current amplitude with δA_n and the frequency with $\delta\omega_n$. The Taylor expansions are expressed by Equations (2.6) and (2.7) and approximated by Equations (2.8) and (2.9).

For a circuit without an external signal near equilibrium, the voltage perturbation or forcing function $e_n(t)$ and steady-state quantities in Equations (2.11) and (2.12) can be neglected. Separating the real and imaginary parts leaves us with a system of linear equations for the variables δA_n and $\delta\omega_n$. This system of equations is a valid approximation of the stability for each harmonic. Substituting Equations (2.13) and (2.14) into Equations (2.11) and (2.12) and setting the perturbation to zero (i.e., $e_{cn} = e_{sn} = 0$) gives [6]

$$
\begin{bmatrix}
\dfrac{\partial}{\partial A} R_{dn} & \dfrac{\partial}{\partial\omega}(R_{dn} + R_{cn}) \\
\dfrac{\partial}{\partial A} X_{dn} & \dfrac{\partial}{\partial\omega}(X_{dn} + X_{cn})
\end{bmatrix}
\begin{bmatrix}
\partial A_n \\
\delta\omega_n
\end{bmatrix} = 0
\tag{2.31}
$$

The system has the trivial solution when the oscillating circuit is stable (i.e., $\delta A_n = \delta\omega_n = 0$). For a slightly disturbed circuit, the determinant is called the stability factor S. We can ignore the subscript n, since a disturbance on the fundamental will also affect higher-order harmonics:

$$
S = \frac{\partial R_d}{\partial A} \frac{\partial(R_d + R_c)}{\partial\omega} \left(\frac{\partial(X_d + X_c)/\partial\omega}{\partial(R_d + R_c)/\partial\omega} + \frac{\partial X_d/\partial A}{\partial R_d/\partial A} \right)
\tag{2.32}
$$

At each harmonic the change in oscillating frequency due to small perturbations is assumed to be very small. This allows us to ignore the frequency dependence of the device impedance (i.e., $\partial R_d/\partial\omega = \partial X_d/\partial\omega \approx 0$). Equation (2.32) reduces to the more standard form

$$
S = \frac{\partial R_d}{\partial A} \frac{\partial R_c}{\partial\omega} \left(\frac{\partial X_c/\partial\omega}{\partial R_c/\partial\omega} + \frac{\partial X_d/\partial A}{\partial R_d/\partial A} \right)
\tag{2.33}
$$

Equation (2.33) is regarded as the stability factor for the oscillator at the operating frequency ω. It also assumes that the stability of the fundamental or higher-order frequency components can be calculated separately by using the same equation. A more general stability formulation which accounts for the effects of higher-order components on the fundamental has been derived by Bates and Khan [7].

2.8   NOISE

Random phase fluctuations and amplitude variations are defined as noise in oscillators. The spectral purity of a signal as displayed in a spectrum analyzer reveals the signal's noise content. By deriving the power versus frequency or power spectral density of a signal, we can obtain equations for oscillator noise characteristics. Kurokawa [8] developed a noise model for the oscillator, assuming a pulse train of delta functions with magnitude B_n occurring at time t_n:

$$e(t) = \sum_{n=0}^{N} B_n \delta(t - t_n) \tag{2.34}$$

The amplitude and time of the pulses are random, and further analysis deals with the average value of these quantities. The spectral density of the noise voltage requires the calculation of the autocorrelation of $e(t)$:

$$R_e(\tau) = \lim_{T \to \infty} \frac{1}{2T} \int_{-\infty}^{\infty} e(t)e(t + \tau) \, dt = m\langle \varepsilon^2 \rangle \delta(\tau) \tag{2.35}$$

where m is the average number of pulses per unit of time, T is one period, and $\langle \varepsilon^2 \rangle$ is the average of the square of the pulse strength for all the noise pulses. The power spectral density of $e(t)$ is the Fourier transform of $R_e(\tau)$, given by

$$|e(f)|^2 = \int_{-\infty}^{\infty} m\langle \varepsilon^2 \rangle \delta(\tau) e^{-j2\pi f\tau} \, d\tau = m\langle \varepsilon^2 \rangle = |e|^2 \tag{2.36}$$

where $|e(f)|^2$ is constant amplitude over frequency and ideally designated as white noise. Its constant power characteristic over frequency is ideal for testing circuit noise behavior.

We can calculate the noise perturbation $n(t)$ for the train of delta functions given in Equation (2.34). We then find the corresponding autocorrelation of both $n_c(t)$ and $n_s(t)$. Since superposition holds for the linear operations performed, we simply add the contribution of each:

$$n_c(t) = \frac{1}{T_0} \int_{t-T_0}^{t} \sum_{n=0}^{N} B_n \delta(t - t_n) \cos(\omega t + \phi)\, dt$$

$$n_s(t) = \frac{1}{T_0} \int_{t-T_0}^{t} \sum_{n=0}^{N} B_n \delta(t - t_n) \sin(\omega t + \phi)\, dt \tag{2.37}$$

Noise fluctuates rapidly with time, while the current amplitude varies slowly. Since T_0 is a fairly short period of time, $e(t)$ can be regarded as a δ-function with magnitude $(2B_n/T_0)\cos(\omega t_n + \phi)$ occurring at $t = t_n + T_0/2$. The autocorrelation in Equations (2.35) then yields $n_c(\tau) = n_s(\tau) = 2m\langle \varepsilon^2 \rangle \delta(\tau)$. From Equation (2.36), this leads to

$$|n_c(f)|^2 = |n_s(f)|^2 = 2|e|^2 \tag{2.38}$$

If we allow the amplitude A to deviate slightly (δA) from the operating point A_0, we can rewrite the equations for the real and imaginary parts of the impedance as

$$R_c(\omega_0) + R_d(A) = \frac{\delta A}{A_0} A_0 \frac{\partial R_d}{\partial A} \tag{2.39}$$

$$X_c(\omega_0) + X_d(A) = \frac{\delta A}{A_0} A_0 \frac{\partial X_d}{\partial A} \tag{2.40}$$

These equations can be used in the general oscillation conditions to obtain

$$A_0 \left(\delta A \frac{\partial R_d}{\partial A} \right) X_c' - A_0 \left(\delta A \frac{\partial X_d}{\partial A} \right) R_c' + |Z_c'|^2 \frac{d(\delta A)}{dt} = X_c' n_c(t) + R_c' n_s(t)$$

$$A_0 \left(\delta A \frac{\partial R_d}{\partial A} \right) R_c' + A_0 \left(\delta A \frac{\partial X_d}{\partial A} \right) X_c' + A_0 |Z_c'|^2 \frac{d\phi}{dt} = R_c' n_c(t) - X_c' n_s(t) \tag{2.41}$$

The autocorrelation and power spectral density of each term of Equation (2.41) leads to the solution of the noise amplitude and phase fluctuation:

$$|\delta A(f)|^2 = \frac{2|Z_c'|^2 |e|^2}{\omega^2 |Z_c'|^4 + A_0 S^2} \tag{2.42}$$

$$|\phi(f)|^2 = \frac{2|e|^2}{\omega^2 A_0^2} \frac{\omega^2 |Z_c'|^2 + A_0^2 [(\partial R_d/\partial A)^2 + (\partial X_d/\partial A)^2]}{\omega^2 |Z_c'|^4 + A_0 S^2} \tag{2.43}$$

2.9 PULLING FIGURE

As defined earlier, pulling figure determines the tendency of the oscillator to drift with load impedance. This effect can be avoided by using isolators or buffers to isolate the oscillator from the varying load and to maintain the intended frequency and power output.

Obregon and Khanna [9] have derived a nonlinear negative-resistance oscillation pulling figure by using a change-in-load admittance approach. For the admittance oscillation condition

$$Y_d + Y_c = 0 \tag{2.44}$$

where Y_d is the oscillator nonlinear device admittance and Y_c is the circuit admittance, and the load perturbation ΔY_c, the oscillation condition using admittances instead of impedances can be written as

$$Y_d + Y_c + \Delta Y_c + \frac{d}{d\omega}(Y_d + Y_c)\Delta\omega + \frac{d}{dV}(Y_d + Y_c)\Delta V = 0 \tag{2.45}$$

Since $Y_d + Y_c = 0$ at the oscillation frequency this simplifies to

$$\Delta Y_c + \frac{d}{d\omega}(Y_d + Y_c)\Delta\omega + \frac{d}{dV}(Y_d + Y_c)\Delta V = 0 \tag{2.46}$$

Equation (2.46) can be separated into its real (conductance, G) and imaginary (susceptance, B) parts to obtain

$$\Delta G_c + \frac{d}{d\omega}(G_d + G_c)\Delta\omega + \frac{d}{dV}(G_d + G_c)\Delta V = 0 \tag{2.47}$$

$$\Delta B_c + \frac{d}{d\omega}(B_d + B_c)\Delta\omega + \frac{d}{dV}(B_d + B_c)\Delta V = 0 \tag{2.48}$$

Solving for the frequency deviation $\Delta\omega$ and voltage change ΔV, we obtain

$$\Delta\omega = \frac{\Delta B_c \frac{d}{dV}(G_d + G_c) - \Delta G_c \frac{d}{dV}(B_d + B_c)}{\frac{d}{d\omega}(G_d + G_c)\frac{d}{dV}(B_d + B_c) - \frac{d}{dV}(G_d + G_c)\frac{d}{d\omega}(B_d + B_c)} \tag{2.49}$$

$$\Delta V = \frac{\Delta B_c \frac{d}{d\omega}(G_d + G_c) - \Delta G_c \frac{d}{d\omega}(B_d + B_c)}{\frac{d}{dV}(G_d + G_c)\frac{d}{d\omega}(B_d + B_c) - \frac{d}{d\omega}(G_d + G_c)\frac{d}{dV}(B_d + B_c)} \tag{2.50}$$

By writing the transferred load admittance at the output plane and normalizing the perturbating admittance using $S = 1 + \Delta G/Y_0$, we obtain

$$Y_0 = Y_0 \frac{Y_0 + \Delta G + jY_0 \tan\theta}{Y_0 + j(Y_0 + \Delta G)\tan\theta} = Y_0 \frac{S(1 + \tan^2\theta)}{1 + S^2\tan^2\theta} + jY_0 \frac{(1 - S^2)\tan\theta}{1 + S^2\tan^2\theta} \quad (2.51)$$

The change in load admittance at the output plane is then represented for the conductance and susceptance as

$$\Delta G_c = Y_0 \frac{(S - 1)(1 - S\tan^2\theta)}{1 + S^2\tan^2\theta} \quad (2.52)$$

$$\Delta B_c = Y_0 \frac{(1 - S^2)\tan\theta}{1 + S^2\tan^2\theta} \quad (2.53)$$

These equations are substituted into Equation (2.49) in order to rewrite the deviation in frequency:

$$\Delta\omega =$$

$$Y_0 \frac{(1 - S^2)\tan\theta \dfrac{d}{dV}(G_d + G_c) - (S - 1)(1 - S\tan^2\theta)\dfrac{d}{dV}(B_d + B_c)}{(1 + S^2\tan^2\theta)\left(\dfrac{d}{d\omega}(G_d + G_c)\dfrac{d}{dV}(B_d + B_c) - \dfrac{d}{dV}(G_d + G_c)\dfrac{d}{d\omega}(B_d + B_c)\right)}$$

$$(2.54)$$

The maximum frequency deviation in frequency will occur when $d(\Delta\omega)/d\theta = 0$, or

$$\frac{d(\Delta\omega)}{d\theta} = Y_0 \left[\frac{(1 - S^2)(1 + \tan^2\theta)}{(1 + S^2\tan^2\theta)^2}\right]\left[S^2\tan^2\theta + 2S\frac{d(B_d + B_c)/dV}{d(G_d + G_c)/dV}\tan\theta - 1\right] = 0$$

$$(2.55)$$

The zeros of Equation (2.55) for a nonzero disturbance are determined by the factor

$$[S^2\tan^2\theta + 2S\alpha\tan\theta - 1] \quad (2.56)$$

where $\alpha = (d(B_d + B_c)/dV)/(d(G_d + G_c)/dV)$ is a nonlinear constant for a given oscillator. The two solutions for θ are designated by θ_1 and θ_2:

$$\theta_1 = \tan^{-1}\left(\frac{\sqrt{1 + \alpha^2} - \alpha}{S}\right) \quad (2.57)$$

$$\theta_2 = \tan^{-1}\left(\frac{\sqrt{1 + \alpha^2} + \alpha}{-S}\right) \tag{2.58}$$

These solutions are substituted into Equation (2.54) to obtain the limits in the deviation frequency:

$$\Delta\omega_1 = Y_0 K \frac{S - 1}{2S} \frac{(2\alpha^2 + S + 1)\sqrt{\alpha^2 + 1} - 2\alpha(\alpha^2 + 1)}{\alpha^2 - \alpha\sqrt{\alpha^2 + 1} + 1} \tag{2.59}$$

$$\Delta\omega_2 = Y_0 K \frac{1 - S}{2S} \frac{(2\alpha^2 + S + 1)\sqrt{\alpha^2 + 1} + 2\alpha(\alpha^2 + 1)}{\alpha^2 + \alpha\sqrt{\alpha^2 + 1} + 1} \tag{2.60}$$

where

$$K = \frac{\dfrac{d}{dV}(G_d + G_c)}{\dfrac{d}{dV}(G_d + G_c)\dfrac{d}{d\omega}(B_d + B_c) - \dfrac{d}{d\omega}(G_d + G_c)\dfrac{d}{dV}(B_d + B_c)}$$

This gives rise to the maximum total deviation for the oscillator as

$$\Delta\omega_M = \Delta\omega_1 - \Delta\omega_2 = Y_0 K\sqrt{\alpha^2 + 1}\left(S - \frac{1}{S}\right) \tag{2.61}$$

2.10   OSCILLATOR SYNCHRONIZATION

Adler [10] derived the oscillator phase with respect to time formulation, giving relations between the oscillator and the external injected signal. The formulation was instrumental in describing the injection-locking process as well as the production of a distorted beat. Figure 2.6 shows the notation used for the synchronized oscillator described herein. Given the oscillator voltage E and injected voltage E_i, the resultant voltage E_g is the vector sum of E and E_i. The corresponding angular frequencies for E, E_i, and E_g are ω_0, ω_i, and ω. The phase difference between the induced voltage and the injected voltage is

$$\phi = \frac{-E_i}{E}\sin(\alpha) \tag{2.62}$$

For the typical single tuned oscillator in Figure 2.7, we can linearly approximate the change in phase versus frequency by the slope, $A = d\phi/d\omega$, as long as it is close to ω_0. Therefore, the frequency difference between E_g and E

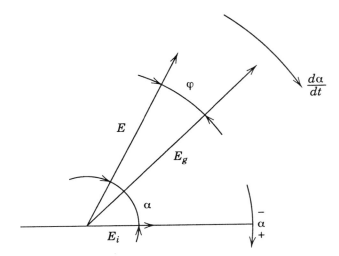

FIGURE 2.6. Vector diagram of instantaneous voltages.

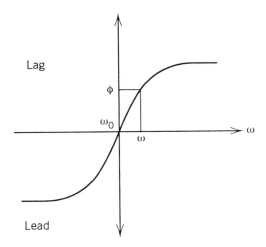

FIGURE 2.7. Phase versus frequency for a simple tuned circuit.

is used to calculate the phase by

$$\phi = A(\omega - \omega_0) \tag{2.63}$$

Since the injected signal is at a frequency different from the oscillator signal, there is an instantaneous beat frequency generated, $\Delta\omega$ which is described by the change in the angle between E and E_i versus time, $d\alpha/dt$. From Equations

(2.62) and (2.63), we have

$$\Delta\omega = \frac{d\alpha}{dt} = \frac{-E_i}{EA}\sin(\alpha) + \Delta\omega_0 \qquad (2.64)$$

where $\Delta\omega = \omega - \omega_i$ and $\Delta\omega_0 = \omega_0 - \omega_i$.

Equation (2.64) can be rewritten to show that the instantaneous frequency, ω, is shifted from the free-running frequency, ω_0, by an amount proportional to the sine of the phase angle, α, and the relative amplitude ratio of the injected to oscillator signal, E_i/E, existing at the locking instant:

$$\omega = \frac{-E_i}{EA}\sin(\alpha) + \omega_0 \qquad (2.65)$$

For a single tuned circuit operated near resonance,

$$\tan(\phi) = 2Q\frac{\omega - \omega_0}{\omega_0} \qquad (2.66)$$

and for small angles, $\phi \approx 2Q((\omega - \omega_0)/\omega_0)$. This equation is similar in form to Equation (2.63) with

$$A = \frac{2Q}{\omega_0} \qquad (2.67)$$

Substituting for A in Equation (2.64) and using the approximation for small angles in Equation (2.67) result in an expression relating the relative amplitude of the voltages, Q-factor, and differences between the free-running and injected frequencies (i.e., the undisturbed beat frequency):

$$\Delta\omega = \frac{d\alpha}{dt} = \frac{-E_i}{E}\frac{\omega_0}{2Q}\sin(\alpha) + \Delta\omega_0 \qquad (2.68)$$

Equation (2.68) must equal zero for a system which has reached steady state. We can then solve for the stationary phase angle between the oscillator and injected signal:

$$\sin(\alpha) = 2Q\frac{E}{E_i}\left(\frac{\omega_0 - \omega_i}{\omega_0}\right) \qquad (2.69)$$

Equation (2.69) leads directly to a condition for synchronization since the right side is bounded by ±1. Therefore, the condition for synchronization is

$$\frac{E_i}{E} > 2Q\left|\frac{\omega_0 - \omega_i}{\omega_0}\right| \qquad (2.70)$$

Of special interest is the locking bandwidth (BW_L), which requires finding the maximum $\Delta\omega$ for which synchronization is maintained. As long as the change in phase versus frequency is given by the slope near ω_0, the two-sided locking bandwidth is given by

$$BW_L = 2\Delta\omega_{max} = \frac{\omega_0}{Q}\frac{E_i}{E} \tag{2.71a}$$

Writing in terms of power, we have

$$\frac{\Delta f_{max}}{f_0} = \frac{1}{Q}\sqrt{\frac{P_i}{P}} \tag{2.71b}$$

Transient analysis of Equation (2.64) gives insight into the process of how an oscillator is pulled into the resulting frequency. Two cases are possible: when the injected signal is the same as the oscillator frequency and when it is not. Assuming that the injected signal is precisely at the oscillator's frequency sets $\Delta\omega_0 = 0$ in Equation (2.64). In such a case, Equation (2.64) will be satisfied for any injected signal and Q-factor of the oscillator:

$$\frac{d\alpha}{dt} = \frac{-E_i}{E}\frac{\omega_0}{2Q}\sin(\alpha) \tag{2.72}$$

Solving the differential equation gives

$$\int \frac{d\alpha}{\sin(\alpha)} = \int \frac{-E_i}{E}\frac{\omega_0}{2Q}\,dt$$

$$\ln\left(\frac{1-\cos(\alpha)}{\sin(\alpha)}\right) = \frac{-E_i}{E}\frac{\omega_0}{2Q}t \tag{2.73}$$

$$\frac{1-\cos(\alpha)}{\sin(\alpha)} = \exp\left(\frac{-E_i}{E}\frac{\omega_0}{2Q}t\right)$$

Any valid solutions are in the range $-\pi < \alpha < \pi$. For cases where α is very small, $\sin(\alpha) \approx \alpha$ and the solution is simply the exponential on the right-hand side of Equation (2.73). This states that the oscillator phase approaches the phase of the injected signal exponentially. The speed of the approach depends on the relative strength of the injected signal to oscillator signal and the Q-factor of the oscillator.

In the second case, when $\Delta\omega_0 \neq 0$, Equation (2.64) becomes

$$\frac{d\alpha}{dt} = \frac{-E_i}{E}\frac{\omega_0}{2Q}\sin(\alpha) + \Delta\omega_0 \tag{2.74}$$

We first substitute

$$K = 2Q \frac{E}{E_i} \frac{\Delta\omega_0}{\omega_0}$$

and rewrite the synchronization condition as

$$K = 2Q \frac{E}{E_i} \frac{\Delta\omega_0}{\omega_0} < 1 \qquad (2.75)$$

Integrating Equation (2.74), following the same procedure as in Equations (2.73), results in

$$\int \frac{d\alpha}{\sin(\alpha) - K} = \int \frac{-E_i}{E} \frac{\omega_0}{2Q} dt \qquad (2.76)$$

where the solution, α, is

$$\alpha = \frac{-2}{\sqrt{K^2 - 1}} \arctan\left(\frac{K \tan(\alpha/2) - 1}{\sqrt{K^2 - 1}}\right) \qquad (2.77)$$

When the synchronization condition is not met, $|K| > 1$, there will be an average angular beat frequency present, which is described by

$$\overline{\Delta\omega} = \Delta\omega_0 \frac{\sqrt{K^2 - 1}}{K} \qquad (2.78)$$

where $\Delta\omega_0$ is the beat frequency present if the oscillator maintains its free-running frequency. The term on the right side becomes zero for $K = 1$ and approaches unity for large values of K. Substituting for K gives

$$\overline{\Delta\omega} = \frac{-E_i}{E} \frac{\omega_0}{\Delta\omega_0} \frac{1}{2Q} \sqrt{\left(2Q \frac{E}{E_i} \frac{\Delta\omega_0}{\omega_0}\right)^2 - 1} \qquad (2.79)$$

Figure 2.8 shows $\overline{\Delta\omega}$ plotted versus $\Delta\omega_0$. The figure shows the reduction in beat frequency due to injection locking or synchronization of the free-running oscillator. Thus, the instantaneous beat frequency varies periodically between the free-running and maximum locking frequencies $\Delta\omega_0 \pm \Delta\omega_{max}$, for $\Delta\omega_0 > \Delta\omega$.

A study on nonlinear synchronized LC oscillators has been carried out by Odyniec [11]. The study uses harmonic balance techniques and averaging to

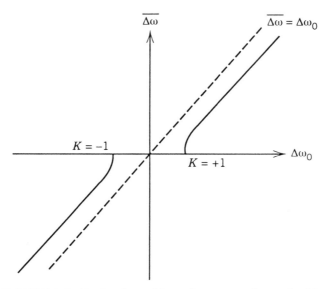

FIGURE 2.8. Reduction of beat frequency due to locking.

determine characteristics of these synchronized oscillators. A useful algorithm for finding stability zones and oscillations for typical oscillators was presented.

Further studies describing the noise properties of synchronized oscillators have been presented by Kurokawa [12] and Schlosser [13]. An extension of these studies was shown by Schunemann and Behm in 1979 [14].

2.11 ANALYSIS OF A WEAK COUPLING TWO-DIMENSIONAL ACTIVE ANTENNA OSCILLATION ARRAY

Making an active array work properly requires injection locking. Injection locking refers to synchronizing a free-running oscillator to a weak injection signal at a frequency close to the operating frequency of the oscillator. The basic phenomenon of injection locking has strong historical roots [15, 16]. Recently, analyses of injection-locking active antenna arrays have been reported [17, 18, 19]. A modified form of Adler's equation that introduces a coupling coefficient for multiple devices was presented [17]. A simple chain of four similar oscillators was analyzed based on the theory. The results indicated that the elements should be placed in multiples of one wavelength for a stable, in-phase mode.

This section describes the analysis for a two-dimensional array [20, 21]. For two-dimensional mutually synchronized arrays, let the coupling coefficient between elements mn and ij be written as $\mathbf{C}_{mn,ij} = C_{mn,ij} \exp(-j\Phi_{mn,ij})$, where

$C_{mn,ij} \ll 1$. A general form of Adler's equation for the ijth oscillator can be obtained in a complex form as

$$S = S_{ij} \left[1 - \sum_{\substack{m=1 \\ m \neq i}}^{M} \sum_{\substack{n=1 \\ n \neq j}}^{N} \frac{1}{j4Q_{ij}A_{ij}^2} (\mathbf{A}_{mn} \mathbf{C}_{ij,mn} \mathbf{A}_{ij}^* - \mathbf{A}_{mn}^* \mathbf{C}_{ij,mn}^* \mathbf{A}_{ij}) \right],$$

$$i = 1,2,\ldots,M \quad j = 1,2,\ldots,N \qquad (2.80)$$

where

$\mathbf{A}_{ij} = A_{ij} \exp(-j\varphi_{ij})$
A_{ij} = free-running amplitude of the ijth oscillator
φ_{ij} = free-running phase of the ijth oscillator
Q_{ij} = external Q of the ijth oscillator circuit
$S_{ij} = j\omega_{ij}$, free-running complex frequency of the ijth oscillator
$S = j\omega$, complex frequency of a injection-locked array
* denotes complex conjugate

Some simple results can be derived from Equation (2.80). A modified form suitable for a one-dimensional array is

$$S = S_i \left[1 - \sum_{\substack{m=1 \\ m \neq i}}^{M} \frac{1}{j4Q_iA_i^2} (\mathbf{A}_m \mathbf{C}_{i,m} \mathbf{A}_i^* - \mathbf{A}_m^* \mathbf{C}_{i,m}^* \mathbf{A}_i) \right], \quad i = 1,2,\ldots,M \qquad (2.81)$$

For a two-element linear array $M = 2$, we get the coupled equations

$$S = S_1 \left[1 - \frac{1}{j4Q_1A_1^2} (\mathbf{A}_2 \mathbf{C}_{1,2} \mathbf{A}_1^* - \mathbf{A}_2^* \mathbf{C}_{1,2}^* \mathbf{A}_1) \right] \qquad (2.82)$$

$$S = S_2 \left[1 - \frac{1}{j4Q_2A_2^2} (\mathbf{A}_1 \mathbf{C}_{2,1} \mathbf{A}_2^* - \mathbf{A}_1^* \mathbf{C}_{2,1}^* \mathbf{A}_2) \right] \qquad (2.83)$$

Assuming that the amplitudes of the oscillators are approximately the same, we have $\mathbf{C}_{1,2} = \mathbf{C}_{2,1}$ by reciprocity.

Let $\mathbf{C} = \mathbf{C}_{1,2} = \mathbf{C}_{2,1} = C\exp(-j\Phi)$, $\mathbf{A}_1 = A\exp(-j\varphi_1)$, $\mathbf{A}_2 = A\exp(-j\varphi_2)$, and $Q = Q_1 = Q_2$. Then Equations (2.82) and (2.83) can be written as

$$S = S_1 \left[1 - \frac{C}{2Q} \sin(\varphi_1 - \varphi_2 - \Phi) \right] \qquad (2.84)$$

$$S = S_2 \left[1 - \frac{C}{2Q} \sin(\varphi_2 - \varphi_1 - \Phi) \right] \qquad (2.85)$$

For two radiating elements operating at the same oscillation frequency $(S_1 = S_2)$ in a two-element linear array (either E-plane or H-plane), from Equations (2.84) and (2.85), it is required that

$$\sin(\varphi_1 - \varphi_2 - \Phi) = \sin(\varphi_2 - \varphi_1 - \Phi) \tag{2.86}$$

Therefore, $\varphi_1 - \varphi_2 - \Phi = \varphi_2 - \varphi_1 - \Phi + 2n\pi$, and thus $\varphi_1 - \varphi_2 = n\pi$. This result states that for coherent combining the phase relation of the elements must satisfy the equation

$$\varphi_1 - \varphi_2 = n\pi \tag{2.87}$$

where $n = 0, \pm 1, \pm 2, \ldots$. This result agrees with Ref. 18.

Consider a more complex case of a two-by-two element array, $M = 2$ and $N = 2$. From Equation (2.80) the following coupled equations are derived:

$$S = S_{11}\left[1 - \frac{1}{j4Q_{11}A_{11}^2}(A_{12}C_{11,12}A_{11}^* + A_{21}C_{11,21}A_{11}^* + A_{22}C_{11,22}A_{11}^*\right.$$
$$\left. - A_{12}^*C_{11,12}^*A_{11} - A_{21}^*C_{11,21}^*A_{11} - A_{22}^*C_{11,22}^*A_{11})\right]$$
$$\tag{2.88}$$

$$S = S_{12}\left[1 - \frac{1}{j4Q_{12}A_{12}^2}(A_{11}C_{12,11}A_{12}^* + A_{21}C_{12,21}A_{12}^* + A_{22}C_{12,22}A_{12}^*\right.$$
$$\left. - A_{11}^*C_{12,11}^*A_{12} - A_{21}^*C_{12,21}^*A_{12} - A_{22}^*C_{12,22}^*A_{12})\right]$$
$$\tag{2.89}$$

$$S = S_{21}\left[1 - \frac{1}{j4Q_{21}A_{21}^2}(A_{11}C_{21,11}A_{21}^* + A_{12}C_{21,12}A_{21}^* + A_{22}C_{21,22}A_{21}^*\right.$$
$$\left. - A_{11}^*C_{21,11}^*A_{21} - A_{12}^*C_{21,12}^*A_{21} - A_{22}^*C_{21,22}^*A_{21})\right]$$
$$\tag{2.90}$$

$$S = S_{22}\left[1 - \frac{1}{j4Q_{22}A_{22}^2}(A_{11}C_{22,11}A_{22}^* + A_{12}C_{22,12}A_{22}^* + A_{21}C_{22,21}A_{22}^*\right.$$
$$\left. - A_{11}^*C_{22,11}^*A_{22} - A_{12}^*C_{22,12}^*A_{22} - A_{21}^*C_{22,21}^*A_{22})\right]$$
$$\tag{2.91}$$

For the same reason as before and for phase-coherent power combining, allowing the spaces between the elements to be the same, we have $C_{mn,ij} = C_{ij,mn}$ by reciprocity. Let

$$\mathbf{A}_{11} = A \exp(-j\varphi_{11}), \quad \mathbf{A}_{12} = A \exp(-j\varphi_{12}), \quad \mathbf{A}_{21} = A \exp(-j\varphi_{21})$$

$$\mathbf{A}_{22} = A \exp(-j\varphi_{22}), \quad Q = Q_{11} = Q_{12} = Q_{21} = Q_{22}$$

and

$$C_a = C_{11,12} = C_{11,21} = C_a \exp(-j\Phi_a), \quad C_b = C_{11,22} = C_{22,11} = C_b \exp(-j\Phi_b).$$

One can obtain

$$S = S_{11}\left[1 - \frac{C_a}{2Q}\left(\sin(\Delta_1 - \Phi_a) + \sin(\Delta_2 - \Phi_a) + \frac{C_b}{C_a}\sin(\Delta_3 - \Phi_b)\right)\right] \quad (2.92)$$

$$S = S_{12}\left[1 - \frac{C_a}{2Q}\left(-\sin(\Delta_1 + \Phi_a) + \frac{C_b}{C_a}\sin(\Delta_2 - \Delta_1 - \Phi_b) + \sin(\Delta_3 - \Delta_1 - \Phi_a)\right)\right]$$

$$(2.93)$$

$$S = S_{21}\left[1 - \frac{C_a}{2Q}\left(-\sin(\Delta_2 + \Phi_a) + \frac{C_b}{C_a}\sin(\Delta_1 - \Delta_2 - \Phi_b) + \sin(\Delta_3 - \Delta_2 - \Phi_a)\right)\right]$$

$$(2.94)$$

$$S = S_{22}\left[1 - \frac{C_a}{2Q}\left(-\frac{C_b}{C_a}\sin(\Delta_3 + \Phi_b) + \sin(\Delta_1 - \Delta_3 - \Phi_a) + \sin(\Delta_2 - \Delta_3 - \Phi_a)\right)\right]$$

$$(2.95)$$

where $\Delta_1 = \varphi_{11} - \varphi_{12}$, $\Delta_2 = \varphi_{11} - \varphi_{21}$, $\Delta_3 = \varphi_{11} - \varphi_{22}$.

For these elements operating at the same frequency for coherent combining ($S_{11} = S_{12} = S_{21} = S_{22}$), from Equations (2.92) to (2.95), it is required that $\Delta_1 = n\pi$, $\Delta_2 = m\pi$, $\Delta_3 = k\pi$, $n, m, k = 0, \pm 1, \pm 2, \ldots$. In other words, for a symmetrical 2×2 array, the phase difference of a phase-coherent mode must be multiples of π for any spacing between the elements. The theory is suitable for a weakly coupled ($C_{mn,ij} \ll 1$) array, and the spacing between elements has an effect on the radiation pattern of the array.

We can conclude that

1. For a weakly coupled coherent power combining of a 2×2 symmetrical array, the phase relation must be $n\pi$ for any spacing between the elements.

2. Any coherent power combining array has an optimal element spacing that produces the lowest side lobes for that array.

REFERENCES

1. K. Kurokawa, "Some Basic Characteristics of Broadband Negative Resistance Oscillator Circuits," *Bell System Technical Journal*, Vol. 48, No. 6, pp. 1937–1955, July–August 1969.

2. K. M. Johnson, "Large Signal GaAs MESFET Oscilator Design," *IEEE Transactions on Microwave Theory and Techniques*, Vol. 27, No. 3, pp. 217–227, March 1979.

3. G. R. Basawapatna and R. B. Stancliff, "A Unified Approach to the Design of Wide-Band Microwave Solid-State Oscillators," *IEEE Transactions on Microwave Theory and Techniques*, Vol. 27, No. 5, pp. 379–385, May 1979.

4. A. P. S. Khanna and J. Obregon, "Microwave Oscillator Analysis," *IEEE Transactions on Microwave Theory and Techniques*, Vol. 29, No. 6, pp. 606–607, June 1981.

5. D. J. Esdale and M. J. Howes, "A Reflection Coefficient Approach to the Design of One-Port Negative Impedance Oscillators," *IEEE Transactions on Microwave Theory and Techniques*, Vol. 29, No. 8, pp. 770–776, August 1981.

6. S. Yngvesson, *Microwave Semiconductor Devices*, Kluwer, Boston, 1991.

7. B. D. Bates and P. J. Khan, "Stability of Multifrequency Negative-Resistance Oscillators," *IEEE Transactions on Microwave Theory and Techniques*, Vol. 32, No. 10, pp. 1310–1318, October 1984.

8. E. L. Holzman, *Solid-state Microwave Power Oscillator Design*, Artech House, Norwood, MA, 1992.

9. J. Obregon and A. P. S. Khanna, "Exact Derivation of the Nonlinear Negative-Resistance Oscillator Pulling Figure," *IEEE Transactions on Microwave Theory and Techniques*, Vol. 30, No. 7, pp. 1109–1111, July 1982.

10. R. Adler, "A Study of Locking Phenomena in Oscillators," *Proceedings of the I. R. E. and Waves and Electrons*, pp. 351–357, June 1946.

11. M. Odyniec, "Nonlinear Synchronized LC Oscillators: Theory and Simulation," *IEEE Transactions on Microwave Theory and Techniques*, Vol. 41, No. 5, pp. 774–780, May 1993.

12. K. Kurokawa, "Nose in Synchronized Oscillators, *IEEE Transactions on Microwave Theory and Techniques*, Vol. 16, No. 4, pp. 234–240, April 1968.

13. W. O. Schlosser, "Noise in Mutually Synchronized Oscillators," *IEEE Transactions on Microwave Theory and Techniques*, Vol. 16, No. 9, pp. 732–737, September 1968.

14. K. F. Schunemann and K. Behm, "Nonlinear Noise Theory for Synchronized Oscillators," *IEEE Transactions on Microwave Theory and Techniques*, Vol. 27, No. 5, pp. 452–458, May 1979.

15. R. Adler, "A Study of Locking Phenomena in Oscillators," *Proceedings of the IEEE*, Vol. 61, pp. 1380–1385, October 1973.

16. K. Kurokawa, "Injection-Locking of Solid-State Microwave Oscillators," *Proceedings of the IEEE*, Vol. 61, pp. 1386–1410, October 1973.

17. R. A. York and R. C. Compton, "Quasi-Optical Power Combining Using Mutually Synchronized Oscillator Arrays," *IEEE Transactions on Microwave Theory and Techniques*, Vol. 39, No. 6, pp. 1000–1009, June 1991.

18. K. D. Stephen and S. L. Young, "Mode Stability of Radiation-Coupled Inter-injection-locked Oscillators for Integrated Phased Arrays," *IEEE Transactions on Microwave Theory and Techniques*, Vol. 36, No. 5, pp. 921–924, May 1988.

19. K. D. Stephan, "Inter-injection-locked Oscillators for Power Combining and Phased Arrays," *IEEE Transactions on Microwave Theory and Techniques*, Vol. 34, No. 10, pp. 1017–1025, October 1986.

20. X. Wu and K. Chang, "Novel Active FET Circular Patch Antenna Arrays for Quasi-Optical Power Combining," *IEEE Transactions on Microwave Theory and Techniques*, Vol. 42, No. 5, pp. 766–771, May 1994.

21. X. Wu and K. Chang, "Dual FET Active Patch Elements for Spatial Power Combiners," *IEEE Transactions on Microwave Theory and Techniques*, Vol. 43, No. 1, pp. 26–30, January 1995.

CHAPTER THREE

Antennas and Arrays

3.1 INTRODUCTION

This chapter discusses the radiating properties of integrated and active integrated antennas. Since arrays of active antennas are similar to their passive counterparts under certain conditions, array theory is very useful in obtaining the overall active antenna patterns. It begins with basic antenna definitions and characteristics and ends with array theory and implementation.

3.2 ANTENNAS

For integrated and active integrated antenna systems, the antennas perform the circuit, component, and radiating functions. A full understanding of what makes antennas useful is critical in using them for these integrations. This chapter defines antenna terminology and views antennas from the circuit and radiator points of view.

Antennas are structures which provide transitions between guided and free-space waves. Guided waves are confined to the contours of a transmission line to transport signals from one point to another, while free-space waves radiate unbounded. A transmission line guides waves with little radiation leakage, but an antenna optimizes radiation. The contrast in propagation modes between guided and free-space waves makes an efficient MIC design difficult to realize. Typically, antennas and transmission lines are optimized separately and later connected together.

Guided waves in circuits go through antennas to generate free-space waves. Antennas can be as simple as wires, rods, and loops. They are used for radios, televisions, satellites, radar, and telephones as well as many military applications. Figure 3.1 shows some basic antenna types.

Since antennas interface circuits to free space, they share both circuit and radiation qualities. From a circuit point of view, an antenna is merely a

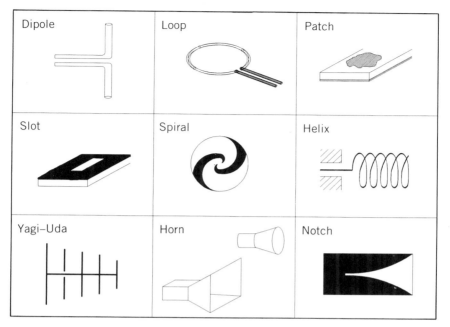

FIGURE 3.1. Several types of wire, MIC, and waveguide antennas.

one-port device with an associated impedance over frequency. As a one-port network, the antenna receives guided-wave power through some form of transmission line. Once received, the antenna then converts the power to a radiating wave. The characteristics and efficiency of this conversion can be obtained from the radiation patterns of the antenna. Several key properties of antennas are input VSWR, bandwidth (BW), directivity (D_0), gain (G_0), efficiency half-power beamwidth (HPBW), radiation patterns, side lobes, cross-polarization levels, and field polarization. The definitions of these parameters can be found in many antenna books. A designer needs these parameters to develop a particular antenna.

3.3 CIRCUIT CHARACTERISTICS

As a one-port device or circuit, an antenna is described by a single scattering parameter, S_{11}, which describes the reflected signal from the one-port network and quantifies the impedance mismatch between the source and the antenna: S_{11} is defined as

$$S_{11} = \frac{b}{a} = \Gamma_1(\omega) \qquad (3.1)$$

where b is the reflected wave and a is the incident wave. This ratio is computed over a range of frequencies and is a function of the source impedance and the antenna input impedance. This turns out to be very useful for integrating devices. Not only must the device package be compatible with the radiating structure, but also the circuit must have impedance and field matching to maintain good component and antenna performance.

The return loss is commonly referred to as the reflection coefficient in Equation (2.16). It is a function of both the source and device-under-test (DUT) impedances:

$$\Gamma_1(\omega) = \frac{Z_0 - Z_{in}(\omega)}{Z_0 + Z_{in}(\omega)} \equiv \rho \angle \phi \qquad (3.2a)$$

$$\text{Return loss} \equiv -20 \log(\rho) \qquad (3.2b)$$

where Z_0 is the characteristic impedance of the transmission line or source impedance, Z_{in} is the antenna input impedance, ρ is the magnitude of Γ_1, and ϕ is the phase of Γ_1. The reflection coefficient is commonly converted to a voltage standing-wave ratio (VSWR) of the voltage or current wave. The

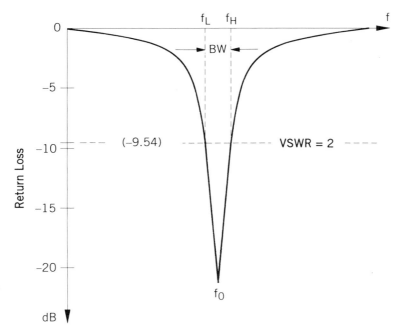

FIGURE 3.2. VSWR $= 2$ bandwidth.

VSWR is the peak-to-peak ratio of these quantities and is given by

$$\text{VSWR} = \frac{1 + \rho}{1 - \rho} \tag{3.3}$$

Optimal VSWR occurs when $\rho = 0$ or VSWR = 1. Typically, VSWR = 2 is acceptable for most applications. The power coupled to the antenna is $1 - \rho^2$ times the power available from the source. The power reflected back from the antenna is ρ^2 times the power available from the source.

Network analyzers and other coaxial test sets are commonly 50-Ω systems which measure the reflection coefficient or return loss over frequency. The network analyzer sends a frequency sweep to the antenna input port and measures the amplitude and phase of the reflected signal. The return loss plotted versus frequency shows the ratio of the reflected power and the incident power, as shown in Figure 3.2. The power entering the device depends on the input impedance locus of the antenna. Therefore, the impedance BW is the range of frequencies over which the input impedance conforms to a specified standard. This standard is commonly taken to be where VSWR $\leqslant 2$ (i.e., $\rho \leqslant 1/3$) and translates to a reflection of about 11%. Alternatively, approximately 89% of the incident energy reaches the antenna. Some applications may require a more stringent specification, such as a VSWR of 1.5 or less.

Methods of widening the impedance BW include impedance-matching networks, tuning devices, and nonresonant antennas. These methods attempt

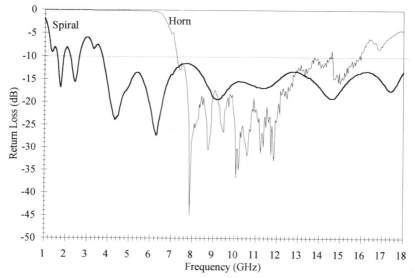

FIGURE 3.3. Measured return loss for a horn antenna and a spiral antenna.

to maintain a constant input impedance over a wide frequency range. Although the impedance BW may be large, the usable bandwidth may be limited by the radiation characteristics of the antenna. The antenna efficiency, gain, polarization purity, and HPBW are also functions of frequency and may deteriorate over the impedance BW. The return loss and impedance of an X-band horn antenna and of a spiral antenna are shown over a wide frequency range in Figure 3.3. It can be seen that the spiral antenna has a wider bandwidth than the horn antenna.

Unlike passive circuits which are matched to a real 50-Ω line, solid-state device impedances are generally complex functions of frequency and dc operating point. This characteristic as well as the package configuration must be accounted for during integration.

3.4   RADIATION CHARACTERISTICS

A low VSWR ($\leqslant 2$) means that more than 90% of the guided wave energy reaches the antenna structure for possible radiation. How the antenna transforms this guided wave depends on its material and configuration. Antennas are constructed with metal and dielectrics, using various shapes to obtain a particular performance. Many configurations have been investigated, each having its own merits and deficiencies. A good antenna transforms the guided wave efficiently into free space for radiation. An antenna is rated according to its bandwidth (BW), HPBW, directivity, gain, and cross-polarization level. This information is contained in the radiation pattern, which plots either power or field strength as a function of space coordinates. With a calibrated antenna range, the gain can also be determined. Other information from the radiation patterns includes major, minor, side-, and back-lobe positions, side-lobe level (SLL), and first-null beamwidth (FNBW). Figure 3.4 shows several definitions on a radiation pattern.

Antenna radiation is categorized by using a hypothetical radiator called an isotropic point source, which has a complete spherical coverage. It radiates equally well in all directions with all polarizations. A physical antenna that radiates or receives better in some directions than others is defined as a directional antenna. Physical antennas are also identified with some polarization which describes the orientation of the radiated field as a function of time.

A dipole or monopole is a linearly polarized radiator which has directional properties in one plane but not in another, and is called an omnidirectional antenna. The radiation patterns of dipole, horn, and helix antennas are shown in Figure 3.5 for comparison. The application determines which antenna is best suited for use.

The complete radiation properties of the antenna require that the electric or magnetic fields be plotted over a sphere surrounding the antenna. However, it is often enough to take principal pattern cuts. Antenna pattern cuts are shown in Figure 3.6. As shown, the antenna has E-plane and H-plane patterns with

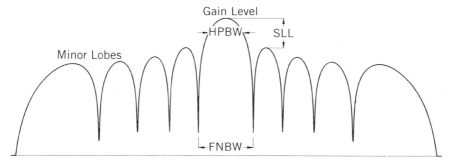

FIGURE 3.4. Antenna pattern characteristics.

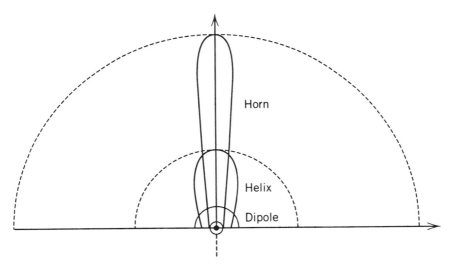

FIGURE 3.5. Directivity comparison.

co- and cross-polarization components in each. The E-plane pattern refers to the plane containing the electric field vector (E_θ) and the direction of maximum radiation. (E_ϕ is the cross-polarization component.) Similarly, the H-plane pattern contains the magnetic field vector and the direction of maximum radiation.

The gain of an antenna is the directivity multiplied by the efficiency of the antenna to radiate the energy presented to its terminals. The directivity of an antenna is referenced to an isotropic point source. An antenna loses gain to ohmic and dielectric losses and polarization quality. The half-power beamwidth is the angle coverage over which the radiated power is at least half of the maximum power. The HPBW may be different for the E- and H-plane

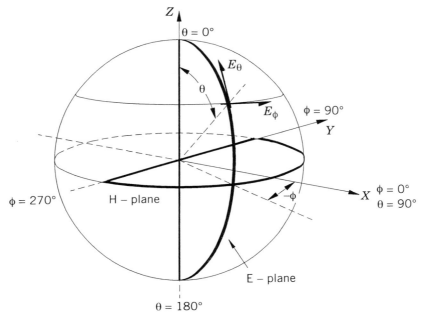

FIGURE 3.6. Antenna pattern coordinate convention.

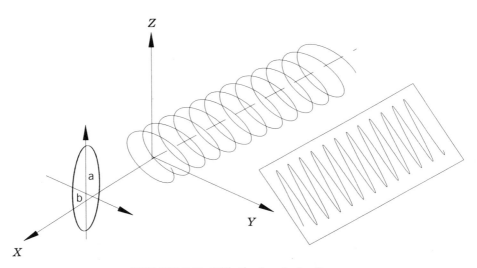

FIGURE 3.7. Elliptical polarization.

pattern cuts of the antenna. HPBW measures the efficiency of the antenna to focus at least half of the radiated power in a particular direction.

Polarization refers to the orientation of the electric field as observed from the source versus time. In general, polarization is elliptical as shown in Figure 3.7 with two special cases worth mentioning. An ellipse with equal dimensions is circular, while an ellipse with one axis equal to zero is linear. For elliptical and circular polarizations a ray trace of the electric field as observed from the source can be rotating clockwise (CW) or counterclockwise (CCW). This fact gives rise to CW and CCW circular polarization components. Ideally, linear polarization means that the electric field is in only one direction, but this is seldom the case. For linear polarization, the cross-polarization level (CPL) determines the amount of polarization impurity.

For many integrations the component (i.e., microstrip oscillator) and antenna offer very good performance separately. However, during integration with the antenna several factors contribute to the deterioration of the antenna performance. Usually, the device package and size disturb the typical currents which flow along the antenna, causing distorted radiation patterns.

3.5 ONE-DIMENSIONAL ARRAYS

Single antennas are often limited for many applications because of a large HPBW and, consequently, a lower gain. For many applications, a high-gain, narrow pencil beam is required. Since most antennas have dimensions of about one wavelength and since beamwidth is inversely proportional to antenna size, more than one antenna is required to sharpen the radiation beam. An array of antennas working simultaneously can focus the reception or transmission of energy in a direction, which increases the useful range of a system.

For active antennas, arrays have been used for beam sharpening as well as power combining. In power combining, distributed sources of relatively low power are made to work coherently to appear as a high-power source. The method has many advantages over a conventional high-power source whose power is distributed to individual antenna elements. To use these techniques, one- and two-dimensional array concepts must be reviewed. In a linear array shown in Figure 3.8, the radiated field from a set of sources can be described by

$$E_{\text{total}} = I_1 f_1(\theta, \phi) \rho_1 \frac{e^{-j(k_0 r_1 + \beta_1)}}{4\pi r_1} + I_2 f_2(\theta, \phi) \rho_2 \frac{e^{-j(k_0 r_2 + \beta_2)}}{4\pi r_2}$$

$$+ \cdots + I_i f_i(\theta, \phi) \rho_i \frac{e^{-j(k_0 r_i + \beta_i)}}{4\pi r_i} + \cdots \qquad (3.4)$$

where I_i, ρ_i, and β_i are the ith element's magnitude, polarization, and phase, respectively, $f_i(\theta, \phi)$ is the radiation pattern of the ith element, and r_i is the distance from the ith element to an arbitrary point in space.

Far Field Amplitude Variations

$$r_1 \approx r_2 \approx r_3 \approx \cdots \approx r_N \approx r$$

Far Field Phase Variations

$$r_1 = r$$
$$r_2 = r + d_z\cos\theta$$
$$r_3 = r + 2d_z\cos\theta$$
$$\vdots$$
$$r_N = r + (N-1)d_z\cos\theta$$

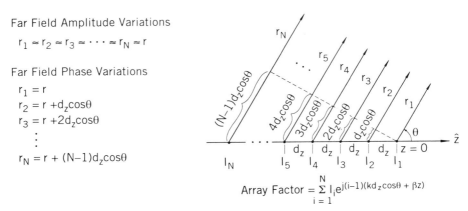

$$\text{Array Factor} = \sum_{i=1}^{N} I_i e^{j(i-1)(kd_z\cos\theta + \beta z)}$$

FIGURE 3.8. An N-element linear array along z-axis.

Typically, the polarization of every element is aligned for copolarization (i.e., $\rho_i \approx \rho \equiv 1$). The array has N elements with uniform spacing (d_z). It is oriented along the z-axis with a phase progression β_z. The first element is placed at the origin, and the distance r_i in the phase term is approximated with the following equations:

$$
\begin{aligned}
r_1 &\cong r \\
r_2 &\cong r + d_z \cos(\theta) \\
&\vdots \\
r_N &\cong r + (N-1)d_z \cos(\theta)
\end{aligned}
\tag{3.5}
$$

These approximations allow the total field to be given by

$$E_{\text{total}} = f(\theta, \phi)\, \frac{e^{-jk_0 r}}{4\pi r} \sum_{i=1}^{N} I_i e^{j(i-1)(k_0 d_z \cos(\theta) + \beta_z)} \tag{3.6}$$

$$= \text{Element pattern} \times \text{Array factor}$$

The total field from the array described by these equations consists of an element pattern $(f(\theta, \phi))\, e^{-jk_0 r}/4\pi r$ and the array factor (AF). The concept of isotropic sources at each antenna location allows separate computation of the array factor from the element pattern.

If an M-element linear array is oriented along the x-axis, the array factor can be described by [1]

$$AF_x = \sum_{m=1}^{M} I_m \exp[j(m-1)(k_0 d_x \sin(\theta)\cos(\phi) + \beta_x)] \tag{3.7}$$

where I_m are the magnitude coefficients, d_x is the uniform spacing between elements, β_x is progressive phase shift, and $k_0 = 2\pi/\lambda_0$. Similarly, the array factor for an N-element linear array oriented along the y-axis is given by

$$AF_y = \sum_{n=1}^{N} I_n \exp[j(n-1)(k_0 d_y \sin(\theta)\sin(\phi) + \beta_y)] \tag{3.8}$$

where I_n, d_y, and β_y are the magnitude coefficients, uniform spacing and progressive phase shift for an N-element linear array along the y-axis, respectively.

3.6   TWO-DIMENSIONAL ARRAYS

Several linear arrays can be arranged side by side to create an $M \times N$ planar array. The radiation pattern of a planar array which lies on the $x - y$ plane has uniform spacing d_x and progressive phase shift β_x in the x-direction, and d_y and β_y in the y-direction [1] can be written as

$$AF = \sum_{n=1}^{N} I_n \left[\sum_{m=1}^{M} I_m e^{j(m-1)(k_0 d_x \sin(\theta)\cos(\phi) + \beta_x)} \right] e^{j(n-1)(k_0 d_y \sin(\theta)\sin(\phi) + \beta_y)} \tag{3.9}$$

Equation (3.9) can be separated into a product of x and y linear arrays as

$$AF = \sum_{m=1}^{M} I_m e^{j(m-1)(k_0 d_x \sin(\theta)\cos(\phi) + \beta_x)} \sum_{n=1}^{N} I_n e^{j(n-1)(k_0 d_y \sin(\theta)\sin(\phi) + \beta_y)} \tag{3.10}$$

Figure 3.9 shows the planar array coordinates and configuration. Although Equation (3.10) simplifies array analysis, it does not account for differences in the performance of individual antennas or interactions between elements due to mutual coupling. Mutual coupling can become a serious problem for densely packed arrays. A simple description and example are given in Section 3.7. For many spatial power combiners, Equation (3.10) is useful in predicting the overall radiating performance. The current coefficients in Equation (3.10) can be used to modify or alter the performance of the array aperture.

The current coefficients of the x- and y-directed linear arrays can be tapered to control the characteristics of the radiation patterns. These amplitude tapers can be achieved via many different functions, which include uniform, cosine, Gaussian, Dolph-Chebyshev, and binomial [2]. A uniform distribution is simply equal amplitude coefficients at each antenna location ($I_i = I_0 \equiv 1$).

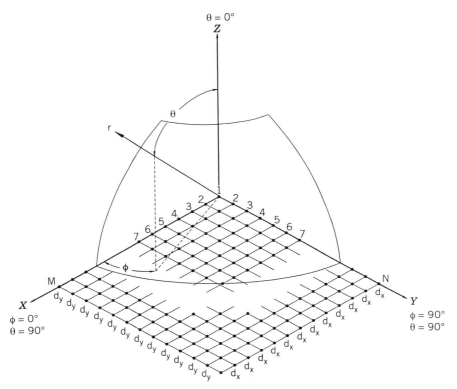

FIGURE 3.9. Antenna pattern coordinate convention.

For an N-element linear array the magnitude coefficients for the cosine distribution are

$$I_i = I_0 \cos^x\left(\pi \, \frac{2i - N - 1}{2N}\right), \qquad i = 1, 2, \ldots, N \tag{3.11}$$

where x determines the smoothness of aperture distribution taper and, consequently, the beamwidth of the main beam and side-lobe level. Similarly, a Gaussian distribution is defined by the equation

$$I_i = I_0 \exp\left(-\frac{1}{2}\left(\sigma \, \frac{2i - N - 1}{N}\right)^2\right), \qquad i = 1, 2, \ldots, N \tag{3.12}$$

where σ is the standard deviation. As σ increases, side-lobe levels decrease and HPBW increases.

An alternative distribution for the array element amplitudes provides the narrowest beamwidth for a given SLL. In a Dolph-Chebyshev distribution, all of the side lobes are at a specified level, which is chosen in dB (α) to obtain

$$x_0 = \frac{1}{N-1} \cosh^{-1}(10^\alpha) \tag{3.13}$$

The actual Chebyshev polynomial ($T_n(x)$) is defined by the following set of equations:

$$T_n(x) = \begin{cases} (-1)^n \cosh[n \cosh^{-1}|x|], & x < -1 \\ \cos[n \cos^{-1}(x)], & |x| \leqslant 1 \\ \cosh[n \cosh^{-1}(x)], & x > 1 \end{cases} \tag{3.14}$$

The individual coefficients are given for an even number of elements ($N = 2K$) by

$$|I_i| = \frac{2}{N}\left\{10^\alpha + 2\sum_{m=1}^{K} T_{N-1}\left[x_0 \cos\left(m\frac{\pi}{N}\right)\right]\cos\left((2i-1)m\frac{\pi}{N}\right)\right\}, \quad 1 \leqslant i \leqslant K \tag{3.15}$$

and for an odd number of elements ($N = 2K + 1$), the coefficients are

$$|I_i| = \frac{2}{N}\left\{10^\alpha + 2\sum_{m=1}^{K} T_{N-1}\left[x_0 \cos\left(m\frac{\pi}{N}\right)\right]\cos\left(2im\frac{\pi}{N}\right)\right\}, \quad 0 \leqslant i \leqslant K \tag{3.16}$$

A Dolph-Chebyshev design with SLL $= -\infty$ is called a binomial distribution. This distribution lowers the side lobes to a minimum level at the expense of some efficiency and a fairly wide HPBW. The excitation coefficients can be determined from the binomial series expansion of

$$I_m = (1 + x)^{m-1} = 1 + (m-1)x + \frac{(m-1)(m-2)}{2!}x^2$$

$$+ \frac{(m-1)(m-2)(m-3)}{3!}x^3 + \cdots \tag{3.17}$$

$$
\begin{array}{ll}
m = 1 & 1 \\
m = 2 & 1\ \ 1 \\
m = 3 & 1\ \ 2\ \ 1 \\
m = 4 & 1\ \ 3\ \ 3\ \ 1 \\
m = 5 & 1\ \ 4\ \ 6\ \ 4\ \ 1 \\
m = 6 & 1\ \ 5\ \ 10\ \ 10\ \ 5\ \ 1 \\
m = 7 & 1\ \ 6\ \ 15\ \ 20\ \ 15\ \ 6\ \ 1 \\
m = 8 & 1\ \ 7\ \ 21\ \ 35\ \ 35\ \ 21\ \ 7\ \ 1 \\
m = 9 & 1\ \ 8\ \ 28\ \ 56\ \ 70\ \ 56\ \ 28\ \ 8\ \ 1 \\
m = 10 & 1\ \ 9\ \ 36\ \ 84\ \ 126\ \ 126\ \ 84\ \ 36\ \ 9\ \ 1
\end{array}
\tag{3.18}
$$

where m is the number of antenna elements. Equation (3.18) is a list of binomial distributions for up to 10 antenna elements. As listed, it is called Pascal's triangle, which shows how the side-lobe level is lowered by reducing the contribution from edge elements. However, this distribution also widens the main beam and lowers the array efficiency.

3.7   MUTUAL COUPLING ARRAYS

For several free-running active antennas, mutual coupling can be used to synchronize the sources. When working in phase, broadside power combining is maximized. However, mutual coupling or crosstalk interferes with the antenna's impedance, causing a deviation in the expected pattern. The array factor computed earlier assumes that there is no interaction between elements. This is often not the case.

To understand the effect of mutual coupling, we can look at the input impedance of a dipole radiating in an unbounded medium. This input impedance is seen to be a function of the dipole length, wire diameter, and current distribution. Using a very thin wire or filament approximation and assuming a sinusoidal current distribution, the input impedance is [1]

$$
\begin{aligned}
Z_m = {} & j\,\frac{\eta}{4\pi}\int_{-l/2}^{0}\sin\!\left(k\!\left(\frac{l}{2}+z'\right)\right)\!\left(\frac{e^{-jkR_1}}{R_1}+\frac{e^{-jkR_2}}{R_2}-2\cos\!\left(\frac{kl}{2}\right)\frac{e^{-jkr}}{r}\right)dz' \\
& + j\,\frac{\eta}{4\pi}\int_{0}^{l/2}\sin\!\left(k\!\left(\frac{l}{2}-z'\right)\right)\!\left(\frac{e^{-jkR_1}}{R_1}+\frac{e^{-jkR_2}}{R_2}-2\cos\!\left(\frac{kl}{2}\right)\frac{e^{-jkr}}{r}\right)dz'
\end{aligned}
\tag{3.19}
$$

Where l is the length of the dipole centered at $z = 0$.

Equation (3.19) gives a typical input impedance of $73 + j42.5\,\Omega$. However, when a dipole is side by side with another dipole, the currents from one antenna affect the other and alter this input impedance. The mutual coupling is a function of the orientation and separation between the elements. Mutual coupling can be calculated by integrating the sinusoidal current distribution of one antenna over the length of the other antenna:

$$Z_{21m} = j\,\frac{\eta}{4\pi} \int_{-l_2/2}^{l_2/2} \sin\left(k\left(\frac{l_2}{2} - |z|\right)\right)\left[\frac{e^{-jkR_1}}{R_1} + \frac{e^{-jkR_2}}{R_2} - 2\cos\left(\frac{kl_1}{2}\right)\frac{e^{-jkr}}{r}\right]dz$$

$$(3.20)$$

Although the difference in a two-element array may not be obvious, the effect on an array with more elements can be quite noticeable. For example, consider a five-element $\lambda/2$ dipole array spaced 0.125λ apart with equal excitation phases. If the voltages are set at each element as shown in Equation (3.21), mutual coupling causes a change in input impedance of each element, which changes the dipole currents and ultimately the radiation pattern:

$$V_i = \begin{bmatrix} 1 \\ -3.262 \\ 4.613 \\ -3.262 \\ 1 \end{bmatrix} \Rightarrow I = Z^{-1}V \Rightarrow I_i = \begin{bmatrix} 0.005 - j0.057 \\ 0.000375 + j0.146 \\ -0.000404 - j0.189 \\ 0.000375 + j0.146 \\ 0.005 - j0.057 \end{bmatrix} \qquad (3.21)$$

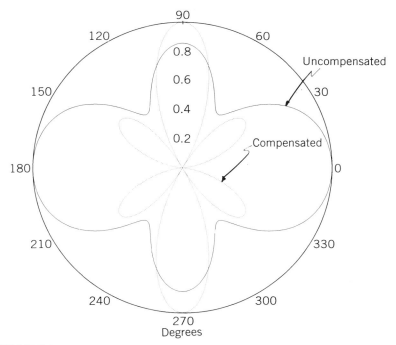

FIGURE 3.10. Mutual coupling effects on a five-element array design.

If the currents are specified instead of the voltages, however, mutual coupling effects are compensated for and one obtains the desired radiation pattern from the array as shown in Figure 3.10.

$$I_i = \begin{bmatrix} 1 \\ -3.262 \\ 4.613 \\ -3.262 \\ 1 \end{bmatrix} \Rightarrow V = ZI \Rightarrow V_i = \begin{bmatrix} 0.957 - j\,1.562 \\ 0.617 - j\,72.261 \\ -0.080 + j\,122.737 \\ 0.617 - j\,72.261 \\ 0.957 + j\,1.562 \end{bmatrix} \tag{3.22}$$

REFERENCES

1. C. A. Balanis, *Antenna Theory: Analysis and Design*, Wiley, New York, pp. 260–262, 1984.
2. M. Mikavica and A. Nesic, *CAD for Linear and Planar Antenna Arrays of Various Radiating Elements*, Artech House, Norwood, MA, 1992.

CHAPTER FOUR

Power Combining

4.1 POWER-COMBINING CONSIDERATIONS

There are many different power sources for rf power generation. These sources include klystrons, traveling-wave tubes (TWT), crossed-field amplifiers (CFA), magnetrons, and solid-state devices. Typically, tubes are very efficient and produce sufficient power to meet many system requirements. However, tubes are bulky, costly, less reliable, and require a high operating voltage. Solid-state devices, on the other hand, are fabricated photolithographically, allowing tighter dimension control. They normally operate at much lower voltages and are small in size. Large numbers of nearly identical solid-state circuits can be batch-processed to lower the cost per circuit. Solid-state devices provide several advantages over tube-type devices, such as [1]

1. No hot cathodes
   ⇒ No heater power required
   ⇒ Little warm-up delays
   ⇒ Long operating life
2. Lower operating voltages
   ⇒ Reduces restrictions on power supplies
   ⇒ Does not need oil filling, encapsulation, or a large space
   ⇒ Reduces overall system size and weight
   ⇒ Improves overall system reliability
3. Wider operating bandwidths ($\sim 50\%$) over tubes which are typically $< 20\%$
4. Can be integrated monolithically within circuits
5. Are small and compact, integrating many different types of functions within small areas
6. Greater mean time between failures (MTBF) over tubes

Solid-state devices, however, are fundamentally limited to relatively low power levels and individual devices cannot meet many system needs. They are, however, inexpensive, small, light, reproducible, and provide good performance. In order to overcome the low power output, many power-combining schemes have been devised [2, 3].

Power-combining methods include chip, circuit, and spatial power combiners. Each method can be used individually or in conjunction with the others. Chip-level combining occurs at the device level and attempts to increase the device surface area to effectively produce higher power output. Circuit power combiners use multiple devices and circuitry to increase the output power at a single port. Spatial power combining differs from the previous methods in that combining occurs in free space. Therefore, each device must be intimately attached to an antenna. Its use provides several advantages not previously available:

1. Transceiver modules at each antenna location:
   ⇒ Eliminates rf distribution network losses and complexity.
   ⇒ Functions are distributed (i.e., amplification, phase shifting, etc.).
   ⇒ Peak rf power levels occur only in space.
   ⇒ Phase shifter losses occur at lower power levels, which improves overall efficiency.
2. Graceful degradation:
   ⇒ Overall power output degrades only as $20 \log(P_{on}/P_{total})$, where P_{on} is the power from working sources and P_{total} is the combined power output from all sources.

Power combining uses the powers of a large number of distributed free-running sources and synchronizes them to obtain a single, coherent higher output power source. Many factors determine the success or failure of a power combiner. The most important characteristic of a combiner is its power-combining efficiency (η). This factor describes the extent to which the total output power approaches the arithmetic sum of the powers from the individual sources. For an N-way power combiner with individual sources of power, P_n,

$$\eta = \frac{P_{total}}{\sum_n P_n} \times 100\% \tag{4.1}$$

Power-combining efficiency is largely determined by the power-combining materials used, combiner configuration, and method of synchronization. Differences in materials and fabrication cause changes in an individual signal's frequency, phase, and amplitude. Maximum efficiency for a perfectly symmetric N-way combiner is obtained when all incoming signals are identical in frequency, amplitude, and phase. Differences in the individual power levels and phase values from the distributed sources cause considerable combining losses.

Optimal combining requires the fabrication of identical devices and circuits. This may someday be achieved with monolithic techniques, but it becomes increasingly difficult at higher operating frequencies. In the meantime, individual sources must be accessible to some form of mechanical or electronic tuning. Tuning schemes allow postassembly optimization to compensate for differences in the individual sources.

The three basic classifications of power combiners are chip, circuit, and spatial power combining. Spatial power combining can be further classified into spatial arrays and open resonators. Each will be discussed. The various schemes can be compared as to combining efficiency, packing density, degradation due to device failure, stability, noise, multimoding, and so on.

4.2 GENERAL POWER-COMBINING METHODS

Power-combining methods have been reviewed for microwave [2] and millimeter-wave [3] systems. These references deal with several circuit configurations used in power combining. Figure 4.1 shows different types of power combiners.

Chip-level power combining is accomplished during device fabrication. Several devices are built and connected to create a larger area and higher thermal dissipation capability for a more powerful source. The method can be controlled photolithographically, which provides tight control over the device and interconnection layout.

At the circuit level, several devices can be combined by using hybrid or monolithic techniques. Circuit-level techniques include resonant and nonresonant power combiners. The use of either of these methods depends on bandwidth and output power requirements.

Resonant circuit-level power combiners have been demonstrated in rectangular/circular waveguide cavities and in microstrip lines. The resonator is used to synchronize the operating frequencies of many individual devices. These power combiners attempt to combine as many devices as possible, giving resonator dimensions of the order of one to several wavelengths. Small dimensions allow sufficient mode separation and avoid multimoding effects. The small volume and device package perturbations, however, may still cause multimoding. An advantage of microstrip is that, when it is monolithically fabricated, circuit dimensions are very tightly controlled and devices are not limited by their package parasitics. This combining method attempts to operate each device at nearly identical operating points. Given nearly identical devices, materials, and circuit dimensions allow the output of each source to have nearly identical power, frequency, and phase. In such a scenario, efficient power combining can occur.

Nonresonant combiners can be classified into N-way combiners and corporate combiners. The nonresonant combiner is used for wideband systems. Dimensions may be several wavelengths for the whole structure, but individual

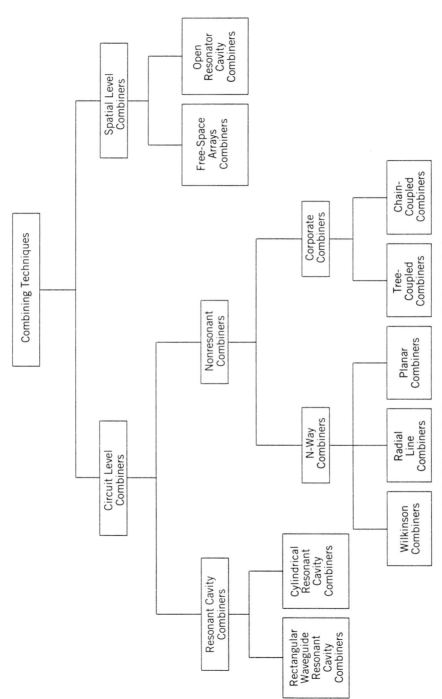

FIGURE 4.1. Different power combining techniques.

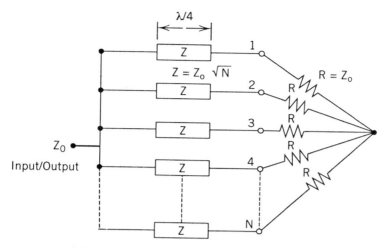

FIGURE 4.2. N-way Wilkinson combiner [5].

arms are still electrically small with respect to wavelength. When each device operates at the same power, frequency, and phase, good stability and power-combining efficiency results.

The three major types of N-way power combiners (or dividers) are Wilkinson [4,5], radial [5], and planar [6–8]. Figure 4.2 shows a Wilkinson N-way

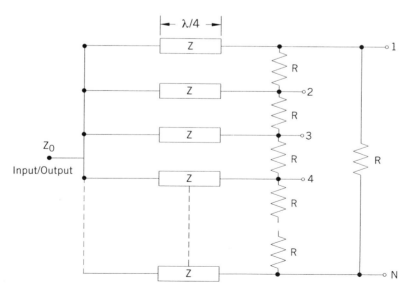

FIGURE 4.3. N-way radial combiner [5].

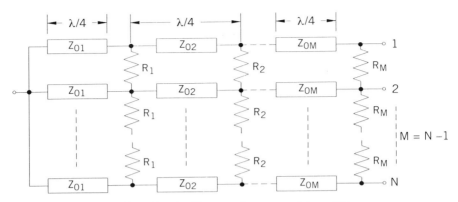

FIGURE 4.4. N-way planar combiner [5].

power combiner [5] which provides low loss, moderate bandwidth and good amplitude and phase balance. Radial line power combiners [5], as shown in Figure 4.3, exhibit low loss, inherent phase symmetry, and good isolation. The planar power combiner [5] in Figure 4.4 requires $N(N-1)$ quarter-wave sections for maximum isolation, which often makes the circuit large and inefficient. However, planar combiners provide large bandwidth and good isolation with moderate loss. A modified power-combining scheme using active devices with input and output combiners is shown in Figure 4.5 [5].

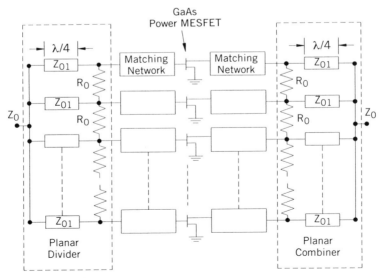

FIGURE 4.5. Modified N-way planar divider combiner [5].

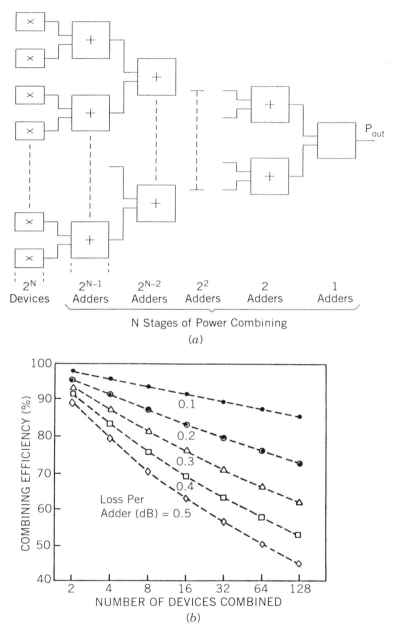

FIGURE 4.6. (*a*) Corporate combining structure, (*b*) combining efficiency for a corporate combining structure. (From Ref. 2 with permission from IEEE.)

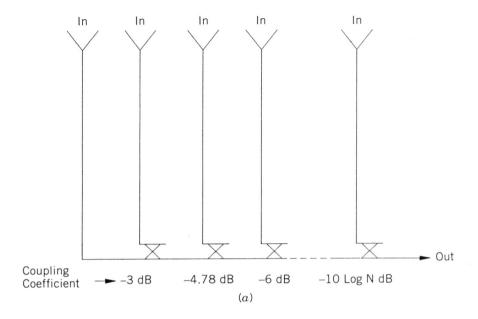

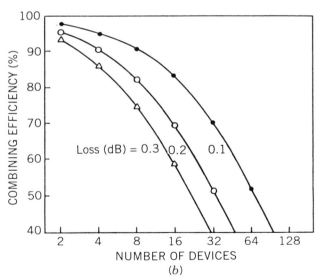

FIGURE 4.7. (*a*) Serial combining structure; (*b*) combining efficiency for the chain-combining structure. Loss in decibels refers to the loss in each power path in each stage's coupler. (From Ref. 2 with permission from IEEE.)

Corporate combiners are classified by either a chain structure [2], as shown in Figure 4.6, or the tree structure of two-way adders [2], shown in Figure 4.7. Two-way adders include directional couplers [9], hybrids [10], and two-way Wilkinson combiners [11]. The corporate structure is useful for large-bandwidth combiners with good isolation, but it is limited to a small number of branches due to losses. The combining efficiency depends on the loss per adder network. The chain or serial-combining scheme provides large bandwidth, good efficiency, and high isolation. Serial combining, however, requires very tight tolerances on circuit dimensions and is usually complex. Advantages and disadvantages are summarized in Table 4.1.

Circuit-level power combiners often suffer from large ohmic or dielectric losses. The use of rectangular and circular waveguides avoids dielectric losses, but they are often heavy, bulky, and expensive to manufacture. At higher operating frequencies, waveguide tolerances become increasingly more difficult and costly. Microstrip and other MIC line combiners lower fabrication costs and improve reproducibility but suffer from higher ohmic, dielectric, and radiation losses. In an attempt to obtain higher output power levels, resonant circuit-level combiners densely pack a large number of devices in electrically small resonator dimensions. This often causes multimoding problems and has led to an alternative approach called spatial power combining.

TABLE 4.1. Comparison of Circuit-Level Power-Combining Techniques

Combining Technique	Advantages	Disadvantages
N-way waveguide cavity	Low loss High efficiency	Nonplanar Complex assembly Narrow band
N-way Wilkinson	Low loss Moderate bandwidth Good isolation & high efficiency	Nonplanar Low power
N-way radial line	Low loss Good isolation	Nonplanar Complex assembly
N-way planar	Large bandwidth Good isolation & moderate loss	Large size Low efficiency
Corporate structure	Good isolation Large bandwidth	Impractical beyond four-way combiner due to loss
Chain structure	More flexible Octave or greater bandwidth Good efficiency & good isolation	High-resolution fabrication required Complex

Source: I. J. Bahl and P. Bhartia, *Microwave Solid State Circuit Design*, Wiley, New York, 1988.

Unlike circuit-level combiners, spatial power combiners use arrays of distributed sources which radiate freely into space. Synchronization of these sources can be accomplished with the aid of mutual coupling, external circuit networks, an external source, or an open resonator. When the array is synchronized through the modes of an open resonator, it is called quasi-optical power combining due to its similarities to Fabry-Perot laser applications in the optical region. Free-space power combining avoids ohmic and dielectric losses associated with MIC transmission lines and dielectrics. Radiation losses are reduced for sufficiently large resonator reflectors. This technique is not as limited to multimoding problems and allows the combination of a greater number of devices. Since spatial and quasi-optical combining occurs in space or in an open cavity, the individual sources are basically active integrated antennas. Consequently, active antenna research was originally spurred on for the purpose of spatial power combining. Power combining at microwave, millimeter-wave, and submillimeter-wave regions is feasible with active antennas.

4.3 SYNCHRONIZATION METHODS FOR SPATIAL POWER COMBINERS

An oscillator is described by several characteristics, such as frequency, power, noise, stability, and so forth. In large arrays of distributed sources, each oscillator will be free running with a random phase unless some mechanism is used to synchronize them. In such an array, each oscillator must be synchronized to obtain efficient power combining. For synchronization, locking bandwidth and locking gain are most important. Locking gain describes the amount of power required to externally lock an oscillator. Locking bandwidth is defined as the frequency locking range for a given relative oscillator power. There are several methods to injection-lock an array source:

1. Open-cavity resonator
2. External source
3. Mutual coupling (free-space waves, surface waves, external circuitry)

The different injection-locking schemes are shown for typical microstrip patch power-combining arrays in Figures 4.8a–c. Figure 4.8a shows an open-cavity resonator which couples an active array to beam modes set up between two electrically large reflectors [12]. The reflectors form a cavity and provide feedback to synchronize the frequencies of the distributed active antennas. The reflector separation and curvature determine the resulting power-combining frequency.

An external source can be used to synchronize the distributed sources as shown in Figure 4.8b. The external source should have enough power to

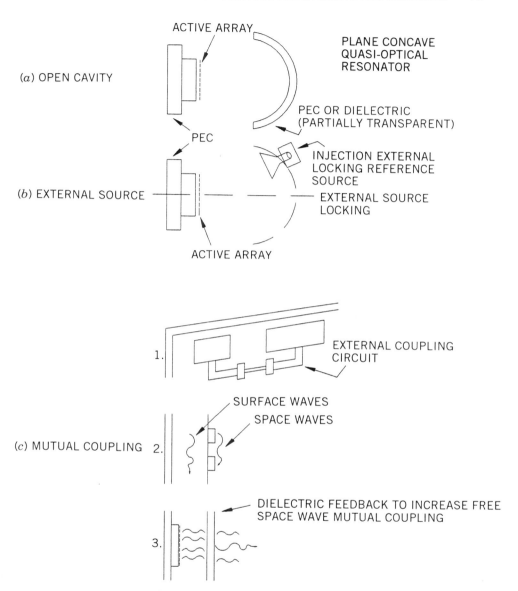

FIGURE 4.8. Synchronization methods.

injection-lock all sources in a spatial array [13]. Frequency modulation can be applied to the external source. Given sufficient power within a certain band-width, the distributed sources in the array follow the modulation, essentially amplifying the output power. The position of the external source determines phase distribution across the array aperture and the resulting combiner beam position in space. Varying the position of the external source can be used for mechanical beam steering.

Alternatively, Figure 4.8c shows mutual coupling methods which can synchronize the array of distributed sources. This method is the most compact and, in many respects, simpler to control. Mutual coupling naturally occurs due to surface or space waves. Surface wave coupling is determined by the substrate thickness and dielectric parameters as well as array separation. Space waves can be enhanced with a dielectric layer which reflects a portion of the incident energy. The dielectric layer material properties, thickness, orientation, and distance from the array determine the amount of coupling and the resulting operating frequency.

Another method of providing coupling between sources uses an external interconnecting (coupling) circuit. This approach provides the most control over the magnitude and phase of the coupling, which can be used to increase the combiner's bandwidth of operation.

In summary, all element sources can be injection-locked to one another via mutual coupling or an external source. Mutual coupling occurs in an array through free-space waves or surface waves. Coupling can be enhanced in active arrays to maintain injection locking with external circuitry or dielectric feedback. Dielectric feedback differs from the open-resonator method in that there is no open-cavity mode setup.

4.4   OPEN-RESONATOR POWER-COMBINING THEORY

Mink [12] first proposed quasi-optical power combiners that combine power from many distributed sources by locking onto quasi-optical resonant modes of the open cavity. Open resonators are called quasi-optical combiners because they are similar to laser cavities at optical frequencies. They relax fabrication tolerances of cavity combiners and avoid dielectric losses of circuit-level combiners. Open resonators replace transmission lines with free space and use radiation from the distributed sources to lock into an open-cavity mode. These open-cavity modes [14] have very high quality factors, which enhance the stability and phase spectra of the combined source signal. Since open-resonator dimensions are several wavelengths long, they can combine many more sources without the multimode problems of closed resonators.

Theoretical analyses on Fabry-Perot resonators assume that the dimensions of the reflectors and separation are large with respect to wavelength and the fields within the resonator are primarily TEM at resonance. Given those assumptions, application of the Fresnel-Kirchhoff formulation of Huygens' principle to the Fabry-Perot resonator results in integral equations which describe the field distribution on the reflectors. Assuming azimuthal field variations, Kogelnik and Li [15] gave integral equations for circular reflector geometries as

$$
p_{l1}S_{l1}(\rho_1)\sqrt{\rho_1} = \int_0^{a_2} \frac{j^{l+1}}{d} J_l\left(\frac{k\rho_1\rho_2}{d}\right)\sqrt{\rho_1\rho_2}
$$

$$
\times \exp\left(-\frac{jk}{2d}(g_1\rho_1^2 + g_2\rho_2^2)\right) S_{l2}(\rho_2)\sqrt{\rho_2}\,d\rho_2 \qquad (4.2)
$$

$$
p_{l2}S_{l2}(\rho_2)\sqrt{\rho_2} = \int_0^{a_1} \frac{j^{l+1}}{d} J_l\left(\frac{k\rho_1\rho_2}{d}\right)\sqrt{\rho_1\rho_2}
$$

$$
\times \exp\left(-\frac{jk}{2d}(g_1\rho_1^2 + g_2\rho_2^2)\right) S_{l1}(\rho_1)\sqrt{\rho_1}\,d\rho_1 \qquad (4.3)
$$

where J_l is a Bessel function of the first kind of lth order, S_{l1} and S_{l2} are eigenfunctions with eigenvalues p_{l1} and p_{l2}, $g_1 = 1 - d/R_1$ and $g_2 = 1 - d/R_2$, the resonator is stable for $0 < g_1g_2 < 1$, R_1 and R_2 are radii of curvature of the mirrors, a_1 and a_2 are the radii of the mirror apertures, and d is the mirror spacing.

The field solutions are of the form $f(\rho,\phi) = S_l(\rho)e^{-jl\phi}$. The integers l describe the angular modes of the resonator, and the field distribution on reflector 1 is calculated by integrating the field distribution on reflector 2 and vice versa. A beam propagates back and forth with slight diffraction losses out of the sides of the resonator. Figure 4.9 shows a typical configuration of an open resonator. Different modes of operation are shown in Figure 4.10 for square and circular mirrors.

The field distributions of the open-resonator modes are a first step in determining the proper configuration for a power-combining application. Exact solutions for the fields are known only for the confocal case

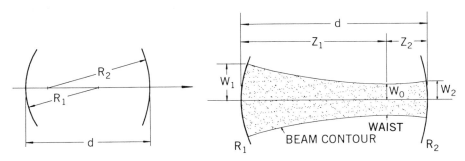

FIGURE 4.9. Quasi-optical open resonator. (From Ref. 15 with permission from IEEE.)

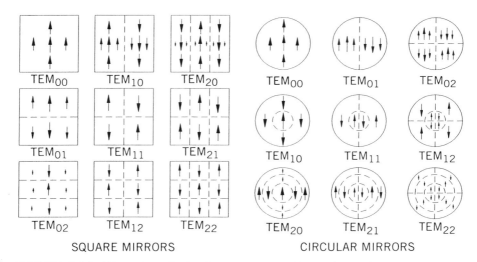

SQUARE MIRRORS   CIRCULAR MIRRORS

FIGURE 4.10. Linearly polarized resonator mode configurations. (From Ref. 15 with permission from IEEE.)

($R_1 = R_2 = d$). Products of Gaussian and Hermite functions, approximations of the exact solution for confocal rectangular geometries, can be generalized to describe the properties of nonconfocal resonators. In circular geometries the approximations are products of Gaussian and Laguerre functions. Kogelnik and Li [15] show that these approximate solutions satisfy an approximation of the scalar wave equation. Their analysis is outlined here for circular geometries.

We desire to find a field solution u to the scalar wave equation:

$$\nabla^2 u + k^2 u = 0 \tag{4.4}$$

The wave beam propagates along the z-direction and has the form $u = \psi(x, y, z)e^{-jkz}$. Substituting this into Equation (4.4) give

$$\frac{\partial^2 \psi}{\partial^2 x^2} + \frac{\partial^2 \psi}{\partial^2 y^2} - j2k\frac{\partial \psi}{\partial z} = 0 \tag{4.5}$$

where we have assumed that ψ varies so slowly with z that we can neglect its second derivative with respect to z.

A solution of Equation (4.5) is the fundamental Gaussian beam mode

$$u = \frac{w_0}{w}\exp\left[-j(kz - \Phi) - r^2\left(\frac{1}{w^2} + j\frac{k}{2R}\right)\right] \tag{4.6a}$$

where the radius away from the z-axis is

$$r = \sqrt{x^2 + y^2} \tag{4.6b}$$

The wave phase shift of the beam is

$$\Phi = \arctan\left(\frac{\lambda z}{\pi w_0^2}\right) \tag{4.6c}$$

the radius of curvature of the phase front of the beam is defined by

$$R(z) = z\left[1 + \left(\frac{\pi w_0^2}{\lambda z}\right)^2\right] \tag{4.6d}$$

The parameter w is often called the beam radius or "spot size" and $2W$ is the beam diameter. The Gaussian beam contrasts to a minimum diameter $2W_0$ at the beam waist where the phase front is plane. The beam radius is a function of z given as

$$w^2(z) = w_0^2\left[1 + \left(\frac{\lambda z}{\pi w_0^2}\right)^2\right] \tag{4.6e}$$

The beam radius and waist are illustrated in Figure 4.10.

The resonant frequencies of the resonator occur when the phase shift of the beam from one reflector to the other is an integer multiple of π. In cylindrical coordinates, the resonant frequencies are given by

$$f_{plq} = \frac{c}{2d}\left[(q + 1) + \frac{1}{\pi}(2p + l + 1)\arccos\left(\sqrt{\left(1 - \frac{d}{R_1}\right)\left(1 - \frac{d}{R_2}\right)}\right)\right] \tag{4.7}$$

The quantity $q + 1$ gives the number of half-wavelengths of the field along the resonator axis. The integers p and l describe variations in the transverse field distributions of higher-order modes.

The fundamental mode described by q occurs when $p = l = 0$. The beam waist radius as a function of the radii of curvature is

$$w_0^4 = \left(\frac{\lambda}{\pi}\right)^2 \frac{d(R_1 - d)(R_2 - d)(R_1 + R_2 - d)}{(R_1 + R_2 - 2d)^2} \tag{4.8}$$

This beam waist describes the mode footprint on the reflector for the fundamental mode. It determines the area which can be used to place active devices for efficient power combining. The low quality factors of typical MIC resonators make the high Q-factors of the open-resonator modes attractive for stabilization of solid-state power combiners in the millimeter-wave region.

The use of open resonators for power combining at millimeter wavelengths was demonstrated by Wandinger and Nalbandian [16], using two spherical metal reflectors. Two waveguide-cavity oscillators using InP Gunn diodes coupled to the open resonator through dielectric wedge launchers. A 54% rf combining efficiency was demonstrated at 60 GHz. Figure 4.11 shows the combiner configuration and test instrumentation. As shown, the oscillators used in the power combining were cavity oscillators coupled to the open resonators via the dielectric tapered rod antennas [17].

Mink analyzed the use of the Fabry-Perot resonator for millimeter-wave power combining [12]. Mink's configuration (Fig. 4.12*a*) has one metal reflector and one dielectric partially reflecting reflector. The metal reflector is flat and ideal for providing structure support and heat sinking in active arrays. The concave dielectric reflector has a slight radius of curvature to match to the Gaussian field mode within the resonator. An array of short, thin filaments of current placed near the metal reflector ($z = 0$) are used to generate rf energy. The individual elements of the active array couple to and lock onto the modes of the Fabry-Perot resonator. The concave dielectric reflector maintains the majority of the energy within the open resonator with a certain amount of leakage for radiation.

The open resonator uses reiterative wave beams or modes which exist between the reflectors. The modes described by Goubau and Schwering [18] satisfy orthogonality relationships as in conventional waveguides but have much larger dimensions. Reflector dimensions may range from 20 to 100 wavelengths or more. Cartesian coordinates are used due to the rectangular array of current sources used in the analysis. The wave beams for this case are satisfied by Hermite-Gaussian functions. The following definition is used for the Hermite polynomial and the recurrence relation:

$$H_{en}(X) = (-1)^n \left(\frac{X^2}{2}\right) \frac{d^n}{dX^n}(e^{-X^2/2}) \tag{4.9}$$

$$H_{e(n+1)}(X) = XH_{e(n)}(X) - nH_{e(n-1)}(X) \tag{4.10}$$

These polynomials form a complete system of orthogonal functions for all reals with the exponential weighting function. These functions are used to obtain the orthonormal spectrum of wave beam modes for each linearly polarized component of the wave beam. The field distributions on the reflectors are plotted in Figure 4.10 by using Equation (4.9) for several values of n. Resonator cross-sectional dimensions and field notation over the different regions are shown in Figure 4.12*b*. The fields for $z > D$ are given by

$$E_{mn}^{\pm} = \frac{\sqrt[4]{\mu/\varepsilon}}{\sqrt{\pi XY m! n!}} \left[\frac{1}{\sqrt[4]{1+u^2}\sqrt[4]{1+v^2}}\right] H_{em}\left(\frac{\sqrt{2}x}{x_z}\right) H_{en}\left(\frac{\sqrt{2}y}{y_z}\right)$$

$$\cdot e^{-[(x/x_z)^2 + (y/y_z)^2]/2}$$

$$\cdot e^{\mp j[kz + 1/2(u(x/x_z)^2 + v(y/y_z)^2) - (m+1/2)\arctan(u) - (n+1/2)\arctan(v)]} \tag{4.11a}$$

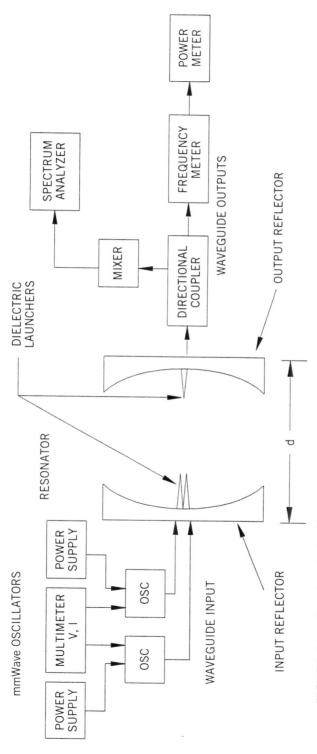

FIGURE 4.11. Quasi-optical millimeter-wave power combiner schematic. (From Ref. 16 with permission from IEEE.)

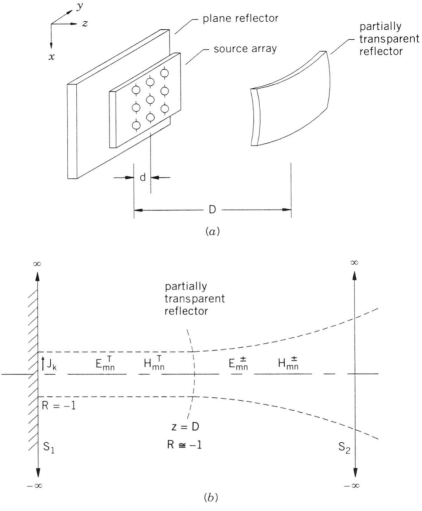

FIGURE 4.12. Mink's quasi-optical resonator: (a) Resonator-source array configuration; (b) Resonator cross section showing spatial regions. (From Ref. 12 with permission from IEEE.)

where u and v are defined using the mode parameters in the x- and y-directions:

$$u = \frac{z}{k\bar{X}^2} \tag{4.11b}$$

$$v = \frac{z}{k\bar{Y}^2} \tag{4.11c}$$

and the coordinates within the resonator is defined by

$$x_z^2 = \bar{X}^2 \left(1 + \frac{z^2}{k^2 \bar{X}^4} \right) \qquad (4.11\text{d})$$

$$y_z^2 = \bar{Y}^2 \left(1 + \frac{z^2}{k^2 \bar{Y}^4} \right) \qquad (4.11\text{e})$$

The relationships between the fields are $E_{xmn}^{\pm} = \pm\sqrt{u/\varepsilon}H_{ymn}^{\pm}$ and $E_{ymn}^{\pm} = \mp\sqrt{\mu/\varepsilon}H_{xmn}^{\pm}$. The mode parameters $\bar{X}$ and $\bar{Y}$ ensure that the wave beam mode satisfies the imposed condition that the waves repeat in amplitude and phase for each round trip along the resonator. The mode parameters depend on the wavelength of operation, the distance between reflectors (D), and the focal length in the x- and y-planes (F_x, F_y). The mode parameters are

$$k\bar{X}^2 = \sqrt{(2 - D/F_x)F_x D} \qquad (4.12$$

$$k\bar{Y}^2 = \sqrt{(2 - D/F_y)F_y D} \qquad (4.13)$$

For the region within the resonator, $0 < z < D$, forward and backward traveling waves for each mode at resonance are computed by using an incident field incoming from $z \approx \infty$. A fractional power-coupling coefficient determines the amount of power in the desired fundamental mode with respect to all excited modes within the resonator for a given current distribution. The partially transparent reflector is characterized by a two-port reflection coefficient magnitude (R) and phase (θ), $S_{11} = S_{22} = R \angle \theta$. This allows the computation of the transmission coefficient:

$$S_{12} = S_{21} = \sqrt{1 - R^2} \angle \left(\theta + \frac{\pi}{2} \right) \qquad (4.14)$$

Given the incident field arriving from $+\infty$ and the presence of the perfect conductor at $z = 0$ implies that for $z > 0$ there is no net power flow through any transverse plane. The fields within the resonator become

$$E_{mn}^T = \frac{2R \sin(\psi) + \sqrt{1 - R^2 \cos^2(\psi)}}{\sqrt{1 - R^2}} e^{j(\psi + \pi/2)} \operatorname{Re}(E_{mn}^+) \sin(kz) \qquad (4.15)$$

Resonance occurs when the fields within the resonator are real, which is satisfied for $\psi = \pi/2$. Given an arbitrary array of current sources, coupling to different resonator modes may be computed from the Lorentz reciprocity theorem. Without losing generality, we consider an array with only x-directed impressed currents, which gives

$$\oiint_{S} (\mathbf{E}_{mn}^{\pm} \times \mathbf{H}_1 - \mathbf{E}_1 \times \mathbf{H}_{mn}^{\pm}) \cdot \hat{\mathbf{n}} \, da = \iiint_{V} \mathbf{J} \cdot \mathbf{E}_{mn}^{T} \, dv \qquad (4.16)$$

where the fields due to an array of x-directed current sources are given by

$$\mathbf{E}_1 = \hat{\mathbf{x}} \sum_{kq} a_{kq} E_{kq}^+ \qquad \text{for } z > D \qquad (4.17)$$

$$\mathbf{H}_1 = \hat{\mathbf{y}} \sqrt{\frac{\varepsilon}{\mu}} \sum_{kq} a_{kq} E_{kq}^+ \qquad \text{for } z > D \qquad (4.18)$$

Here $E_{mn}^{\pm}$ describes the modal fields in space, E_{mn}^{T} describes the modal fields within the resonator, and V is the open-resonator volume. This volume used with the Lorentz reciprocity relation is bounded by two surfaces S_1 and S_2 which extend to infinity in the transverse direction: S_1 is a perfectly conducting infinitely large flat surface at $z = 0$, and S_2 is an infinitely large curved plane located at $z = D$. It is a partially transparent surface made of dielectric material which maintains a large part of the electromagnetic energy within the resonator and allows the rest of the energy to radiate. For the calculations, the closed surface integration only has a nonzero contribution on the surface of S_2 since $n \times E = 0$ on S_1. This gives the following relation:

$$\oiint_{S_2} \left[E_{mn}^{\pm} \hat{\mathbf{x}} \times \sqrt{\frac{\varepsilon}{\mu}} \sum_{kq} a_{kq} E_{kq}^+ \hat{\mathbf{y}} - \sum_{kq} a_{kq} E_{kq}^+ \hat{\mathbf{x}} \times \sqrt{\frac{\varepsilon}{\mu}} E_{mn}^{\pm} \hat{\mathbf{y}} \right] \cdot \hat{\mathbf{n}} \, da = \iiint_{V} \mathbf{J} \cdot \mathbf{E}_{mn}^{T} \, dv \qquad (4.19)$$

The equality $E_{mn}^{-} = E_{mn}^{+}{}^{*}$ is used with the orthogonality relation of wave beam modes to evaluate each term of the integrals:

$$\sqrt{\frac{\varepsilon}{\mu}} \int\int_{-\infty}^{\infty} \mathbf{E}_{mn} \cdot \mathbf{E}_{m'n'}^{*} \, dx \, dy = \delta_{mm'} \delta_{nn'} \qquad (4.20$$

To evaluate each term of the integrals as

$$\oiint_{S_2} \left[\mathbf{E}_{mn}^{\pm} \times \mathbf{H}_1 - \mathbf{E}_1 \times \mathbf{H}_{mn}^{\pm} \right] \cdot \hat{\mathbf{n}} \, da = \iiint_{V} \mathbf{J} \cdot \mathbf{E}_{mn}^{T} \, dv = 2a_{mn} \qquad (4.21)$$

Assuming the array is made up of filamentary currents which are very small with respect to the mode parameter and are aligned with the electric field, the expansion coefficients become

$$a_{mn} \approx \frac{1}{2} \Sigma I_p \Delta X_p E_{mn}^T(x_p, y_p, z_p) \tag{4.22}$$

where I_p is the current at the terminals of the pth element of effective length $\Delta X_p = (1/I_p) \int I_p(l) \cdot dl_p$, and $E_{mn}^T(x_p, y_p, z_p)$ is the electric field strength of the m, n mode at the location of the pth current element. With the aid of Equation (4.15) and the internal fields of the resonator given in Equation (4.22), one can calculate the electromagnetic fields E_1 and H_1 due to an array of current sources.

To combine the powers of many sources efficiently, we must be able to efficiently couple energy from each source to the mode of operation. This requires knowledge of the driving point impedance at the pth current element. This impedance will vary away from the center of the resonator. At resonance the reactive component will be negligible and the dipole driving point impedance for the pth current element at each mode is given by

$$Z_{pmn}^T = \frac{1}{I_p^2} \iiint_V \mathbf{J}_p \cdot \mathbf{E}_{mn}^T \, dv \tag{4.23}$$

Considering only small dipoles of equal lengths, we obtain

$$Z_{pmn} = 2A(\Delta X)^2 \sin^2(kz_p) \mathrm{Re}[E_{mn}^+(x_p, y_p)]$$

$$\times \sum_q \frac{I_q}{I_p} \mathrm{Re}[E_{mn}^+(x_q, y_q)] \tag{4.24}$$

Equation (4.24) can be written for the driving-point resistance of each source element as

$$Z_{pmn} = \frac{(\Delta X)^2}{XY} \sin^2(kz_p) \sqrt{\frac{1+R}{1-R}} \, \mathrm{Re}[\sqrt{XY} \, E_{mn}^+(x_p, y_p)] \bar{R} \tag{4.25}$$

where $\bar{R}$, the normalized resistance factor which depends only on the normalized source spacing,

$$\bar{R} = 2 \sum_q \frac{I_q}{I_p} \mathrm{Re}[\sqrt{XY} \, E_{mn}^+(x_q, y_q)] \tag{4.26}$$

The results of this derivation are used to optimize the number of current elements and spacing to efficiently couple to a wave beam mode. The fundamental mode or Gaussian wave beam mode is of special interest, and one must adjust the individual current sources so that there is little coupling to other modes. One can compute the power associated with the fundamental mode and compare it to the total power radiated:

$$P_{qk} = a_{qk} a_{qk}^* \sqrt{\frac{\varepsilon}{\mu}} \int\limits_{-\infty}^{\infty}\!\!\int E_{qk} E_{qk}^* \, dx \, dy = a_{qk} a_{qk}^* \qquad (4.27)$$

Due to the orthogonality of the modes, the fundamental-mode fractional power to the total power is

$$FP_{00} = \frac{a_{00} a_{00}^*}{\displaystyle\sum_{qk} a_{qk} a_{qk}^*} \qquad (4.28)$$

The excitation coefficient for any mode is determined only by the current distribution and the modal spectrum as if it were freely propagating in free space. Thus, FP_{00} represents the worst case since it assumes that all modes are resonating and allowed to remove energy from the fundamental mode. This assumption allows the optimization of the current distribution to couple to a given mode. Mink has analyzed the array size, spacing, and amplitude taper to optimize coupling to the fundamental mode of the resonator. Square arrays of 9, 25, 49, and 81 elements were analyzed. Graphical results show that there is an optimal source element spacing for each array size and that a Gaussian amplitude taper on the source elements can achieve nearly 100% coupling efficiency into the fundamental mode.

The graphs shown are very useful in general because the ordinate is given in normalized source spacings which are with respect to the mode parameter. The graphs are curves for arrays of 3×3, 5×5, 7×7, and 9×9 elements. Figure 4.13 shows that for a given array size FP_{00} has an optimal element separation. This optimal element separation decreases with an increase in the number of array sources. Figure 4.14 shows the fundamental-mode power as a function of element separation. Figure 4.15 shows the normalized resistance versus source spacing for the various array sizes. Figures 4.13 to 4.15 assume that the power distribution over the array aperture is constant. Similar graphs are shown in Figures 4.16, 4.17, and 4.18 for arrays of Gaussian weighted sources.

It seems that efficient coupling to the fundamental mode is possible for 25-element source arrays and larger given the proper spacing. Given a certain spacing, there appears to be a diminishing rate of return in power transfer as the array becomes larger. For power-combining applications, increasing source spacing significantly reduces output power. Some of these results are intuitive given the field distribution of the fundamental mode. Since the majority of the fields lie within the beam waist, larger arrays of devices would require denser packing to more efficiently couple to the dominant mode.

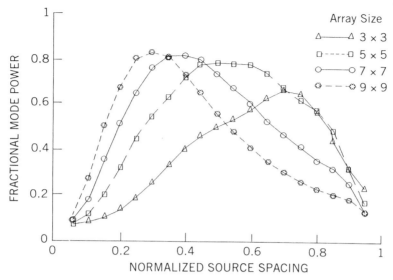

FIGURE 4.13. Fractional power into fundamental mode by equal-weight sources. (From Ref. 12 with permission from IEEE.)

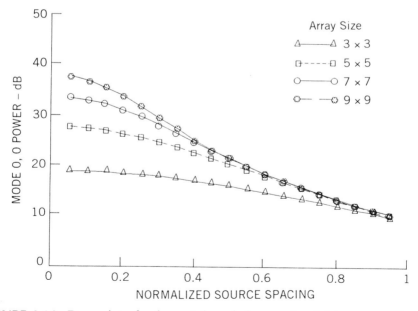

FIGURE 4.14. Power into fundamental mode by equal-weight sources. (From Ref. 12 with permission from IEEE.)

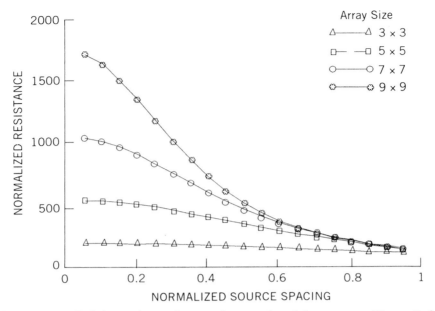

FIGURE 4.15. Driving-point resistance for equal-weight sources. (From Ref. 12 with permission from IEEE.)

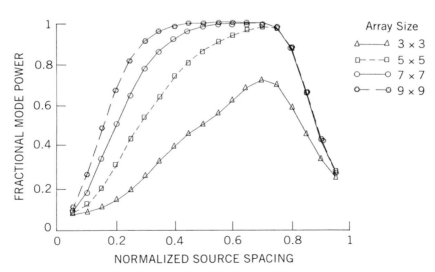

FIGURE 4.16. Fractional power into fundamental mode by Gaussian weight sources. (From Ref. 12 with permission from IEEE.)

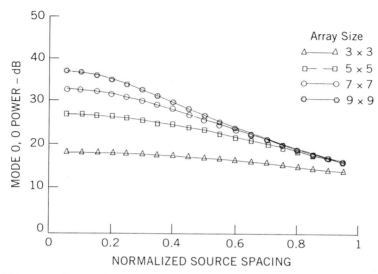

FIGURE 4.17. Power into fundamental mode by Gaussian weight sources. (From Ref. 12 with permission from IEEE.)

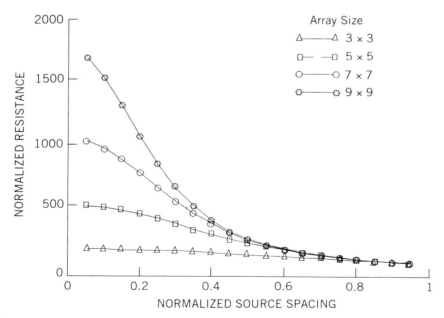

FIGURE 4.18. Driving-point resistances for Gaussian weight sources. (From Ref. 12 with permission from IEEE.)

Young and Stephen used the open-resonator technique to stabilize and improve power combining in 1987 [19]. Experiments were carried out primarily at X-band with Gunn integrated patch antennas similar to those later shown in Figure 7.5. Unlike earlier experiments, this study used microstrip active antennas, which inherently have low Q-values. The high Q-values of the open resonator drastically improve and stabilize the combiner's output spectrum.

A diode integration primarily intended for quasi-optical power combining was investigated by Cogan et al. [20]. The IMPATT diodes use a wire and a loop antenna for radiation. Biasing is accomplished through a low-pass filter network behind the ground plane. Several of these IMPATTs were used on a reflector to couple to an open quasi-optical resonator at 35 GHz. Figure 4.19 shows the adjustable quasi-optical cavity used in the experiments. Frayne and Riddaway also demonstrated high performance and efficiency in quasi-optical combiners [21].

Figure 4.20 shows a dual-IMPATT integrated microstrip radiator, which is ideally suited for coupling to a rectangular waveguide. The configuration was used by Shillue et al. in a quasi-optical resonator in 1989 [22]. Improvements in oscillator spectrum was demonstrated at 56 GHz. In 1990, Frayne and Potter reported very good results at Ka-band [23]. Power was obtained in excess of 150 mW at 35.5 GHz for a three-element slot-coupled active patch antenna array.

Nakayama et al. demonstrated an 18-diode combiner and stabilizer at X-band in 1990 [24]. This work was later extended to millimeter-wave frequencies [25]. Figure 4.21 shows the open-resonator combiner configuration [24]. Sliding shorts are cleverly used to account for differences in device characteristics and impedance levels across the resonator beam waist. The

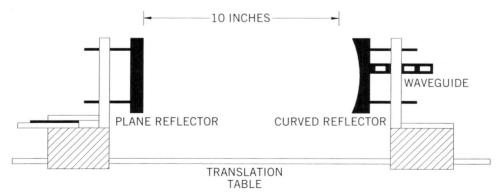

FIGURE 4.19. Adjustable quasi-optical cavity configuration.

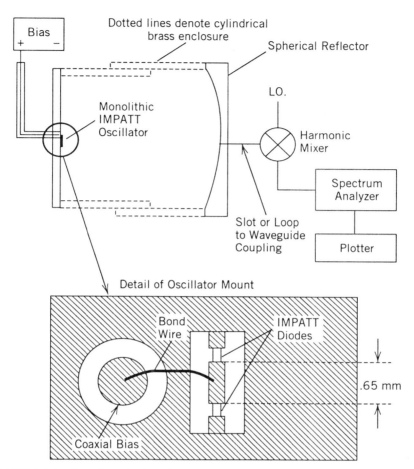

FIGURE 4.20. Quasi-optical cavity resonator containing monolithic IMPATT oscillator chip. (From Ref. 22 with permission from IEEE.)

18-diode oscillator and the six-FET combiner operated at X-band with very promising results. In 1991, Ge, Li, and Chen demonstrated 504 mW of power at 36 GHz with a 95% efficiency [26]. Very good spectrum and mechanical tuning range were shown for the combiner.

Many other investigations have been carried out for the open resonator [27–32]. Heron et al. have developed a circuit-level model for the open resonator [28]. Zeisberg et al. introduced a quasi-optical slab resonator [29]. The dyadic Green's function for a plano-concave open resonator and the impedance matrix were determined by Heron et al. [30, 31, 32].

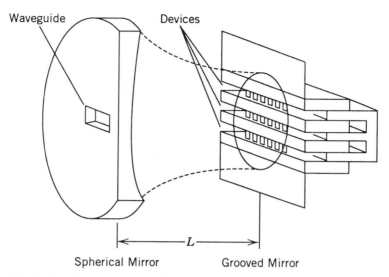

FIGURE 4.21. Cutaway view of quasi-optical combiner configuration. (From Ref. 24 with permission from IEEE.)

4.5 SPATIAL POWER COMBINING

Spatial power-combining methods also relax fabrication tolerances of cavity combiners and reduce the overall cost of millimeter- and sub-millimeter wave systems. Spatial power combiners radiate freely as large planar arrays of sources synchronously oscillating to produce a single larger power output at some point in space. Spatial power combining is not as limited by size, moding problems, or ohmic or dielectric losses, and it allows more active devices to be combined. It is further not limited by a specific area such as the beam waist in open resonators. Spatial power-combining methods enable arrays of active antennas to operate as one coherent transmitter with pre-dictable effective radiated power, beamwidth, and tuning bandwidth. The individual free-running low-power oscillators must be injection-locked via mutual coupling, external feedback, coupling networks, or an external source. Spatial power combiners have been demonstrated using distributed oscillators and amplifiers.

Spatial power combiners can be modeled directly with planar array theory. Most, if not all, equations given earlier for passive arrays apply to active spatial power combiners. Given that individual antenna free-running frequencies and power outputs are nearly identical and the entire array is locked to a common frequency, the radiation pattern will behave according to Equation (3.4). It is rewritten here for convenience:

$$E_{\text{total}} = I_1 f_1(\theta, \phi) \rho_1 \frac{e^{-j(k_0 r_1 + \beta_1)}}{4\pi r_1} + I_2 f_2(\theta, \phi) \rho_2 \frac{e^{-j(k_0 r_2 + \beta_2)}}{4\pi r_2}$$
$$+ \cdots + I_i f_i(\theta, \phi) \rho_i \frac{e^{-j(k_0 r_i + \beta_i)}}{4\pi r_i} + \cdots \tag{4.29}$$

The same far-field approximations apply for the phase and magnitude terms of the total electric field. Assuming identical elements also allows us to separate the element pattern from the array factor as given in Equation (3.6)

$$E_{\text{total}} = f(\theta, \phi) \frac{e^{-jk_0 r}}{4\pi r} \sum_{i=1}^{N} I_i e^{j(i-1)(k_0 d_z \cos\theta + \beta_z)}$$
$$= (\text{Element pattern})(\text{Array-Factor}) \tag{4.30}$$

The antenna element normally serves as a resonator, radiator, and dc bias pad. Biasing and device integration disturb the radiation patterns and disrupt polarization purity of the antenna. Also, individual device parameters vary, and tolerance errors during fabrication cause active antenna array sources to operate at different frequencies. Although the frequencies may be close enough and coupling between elements may be high enough to synchronize the entire array to a single frequency, the output power and combining efficiency suffers. Therefore, working on active antennas for power combining requires knowledge of a good combination of array fundamentals, coupling techniques and synchronization mechanisms.

Spatial power combiners are active radiating arrays with the potential to focus a pencil beam in a given direction. Phasing of spatial power-combining arrays has opened up a whole new area in active antenna technology. Similar to injection-locked amplifiers, the oscillation phase is a function of the self-oscillating frequency and the injection-locked combiner frequency. This phasing mechanism can result in a very inexpensive phased array which appears to be a contradiction in terms.

The use of distributed oscillators presents an interesting and potentially useful challenge. It can provide much needed system redundancy for graceful degradation as well as increased power at much lower costs. It also has potential for beam steering with relatively few controls.

REFERENCES

1. Michael T. Borkowski, 'Solid-State Transmitters', in *Radar Handbook* (M. Skolnik, ed.), 2nd Ed., McGraw-Hill, New York, 1990.

2. K. J. Russell, "Microwave Power Combining Techniques," *IEEE Transactions on Microwave Theory and Techniques*, Vol. 27, No. 5, pp. 472–478, May 1979.

3. K. Chang and C. Sun, "Millimeter-Wave Power-Combining Techniques," *IEEE Transactions on Microwave Theory and Techniques*, Vol. 31, No. 2, pp. 91–107, February 1983.

4. E. J. Wilkinson, "An N-way Hybrid Power Divider," *IRE Transactions on Microwave Theory and Techniques*, Vol. 8, No. 1, p. 116–118, January 1960.

5. I. J. Bahl, "Filters, Hybrids and Couplers, Power Combiners and Matching Networks," in *Handbook of Microwave and Optical Components*, Vol. 1 (K. Chang, ed.), Wiley, 1989.

6. N. Nagai, E. Maekawa, and K. Ono, "New N-Way Hybrid Power Dividers," *IEEE Transactions on Microwave Theory and Techniques*, Vol. 25, No. 12, pp. 1008–1012, December 1977.

7. R. Soares, J. Graffeuil, and J. Obregon, *Applications of GaAs MESFETs*, Artech House, Dedham, MA, 1983, Chapter 4.

8. A. A. M. Saleh, "Planar Electrically Symmetric N-Way Hybrid Power Dividers/Combiners," *IEEE Transactions on Microwave Theory and Techniques*, Vol. 28, No. 6, pp. 555–563, June 1980.

9. R. E. Collin, *Foundations for Microwave Engineering*, 2nd ed., McGraw-Hill, New York, 1992.

10. K. Chang, *Microwave Solid-State Circuits and Applications*, Wiley, New York, 1994.

11. D. M. Pozar, *Microwave Engineering*, Addison-Wesley, Reading, MA, 1990.

12. J. W. Mink, "Quasi-Optical Power Combining of Solid-State Millimeter-Wave Sources," *IEEE Transactions on Microwave Theory and Techniques*, Vol. 34, No. 2, pp. 273–279, February 1986.

13. K. Chang, K. A. Hummer, and J. L. Klein," Experiments on Injection Locking of Active Antenna Elements for Active Phased Arrays and Spatial Power Combiners," *IEEE Transactions on Microwave Theory and Techniques*, Vol. 37, No. 7, pp. 1078–1084, July 1989.

14. G. Goubau and F. Schwering, "On the Guided Propagation of Electromagnetic Wave Beams," *IRE Transactions on Antennas and Propagation*, Vol. 9, No. 3, pp. 248–256, May 1961.

15. H. Kogelnik and T. Li, "Laser Beams and Resonators," *Proceedings of the IEEE*, Vol. 54, No. 10, pp. 1312–1329, October 1966.

16. L. Wandinger and V. Nalbandian, "Millimeter-Wave Power Combiner Using Quasi-Optical Techniques," *IEEE Transactions on Microwave Theory and Techniques*, Vol. 31, No. 2, pp. 189–193, February 1983.

17. S. Kobayashi, R. Mittra, and R. Lampe," Dielectric Tapered Rod Antennas for Millimeter-Wave Applications," *IEEE Transactions on Antennas and Propagation*, Vol. 30, No. 1, pp. 54–58, January 1982.

18. G. Goubau and F. Schwering, "Free Space Beam Transmission," in *Microwave Power Engineering*, Vol. 1 (C. Okress, ed.), Academic Press, New York, pp. 241–255, 1968.

19. S. Young and K. D. Stephan, "Stabilization and Power Combining of Planar Microwave Oscillators with an Open Resonator," *IEEE MTT-S International Microwave Symposium Digest*, Las Vegas, Nevada, pp. 185–188, 1987.

20. K. J. Cogan, F. C. DeLucia and J. W. Mink, "Design of a Millimeter Wave Quasi-Optical Power Combiner for IMPATT Diodes," *Proceedings SPIE, Vol. 791, Millimeter Wave Technology IV and Radio Frequency Power Sources*, pp. 77–81, May 1987.

21. P. G. Frayne and C. J. Riddaway, "Efficient Power Combining Quasi-Optic Oscillator," *Electronics Letters*, Vol. 24, No. 16, pp. 1017–1018, August 1988.

22. W. P. Shillue, S. C. Wong, and K. D. Stephan, "Monolithic IMPATT Millimeter-Wave Oscillator Stabilized by Open-Cavity Resonator," *IEEE MTT-S International Microwave Symposium Digest*, pp. 739–740 (1989).

23. P. G. Frayne and J. Potter, "Efficient Power Transfer through Small Apertures," *Electronics Letters*, Vol. 26, No. 25, pp. 2070–2073, December 1990.

24. M. Nakayama, M. Heide, T. Tanaka, and K. Mizuno, "Millimeter and Sub-millimeter Wave Quasi-Optical Oscillator with Multi-elements," *IEEE MTT-S International Microwave Symposium Digest*, pp. 1209–1212 (1990).

25. J. Bae, Y. Aburakawa, H. Kondo, T. Tanaka, and K. Mizuno, "Millimeter and Submillimeter Wave Quasi-Optical Oscillator with Gunn Diodes," *IEEE Transactions on Microwave Theory and Techniques*, Vol. 41, No. 10, pp. 1851–1855, October 1993.

26. J. E. Ge, S. F. Li, and Y. Y. Chen, "Millimetre wave Quasi-Optical Power Combiner," *Electronics Letters*, Vol. 27, No. 10, pp. 880–882, May 1991.

27. H. M. Harris, A. Torabi, R. M. McMillan, C. J. Summers, J. C. Wiltse, S. M. Halpern, and D. W. Griffith, "Quasi-Optical Power Combining of Solid-State Sources in Ka-band," *IEEE MTT-S International Microwave Symposium Digest*, pp. 159–162 (1993).

28. P. L. Heron, G. P. Monahan, J. E. Byrd, M. B. Steer, F. W. Schwering, and J. W. Mink, "Circuit Level Modeling of Quasioptical Power Combining Open Cavities," *IEEE MTT-S International Microwave Symposium Digest*, pp. 433–436 (1993).

29. S. Zeisberg, A. Schuenemann, G. P. Monahan, P. L. Heron, M. B. Steer, J. W. Mink, and F. K. Schwering, "Experimental Investigation of a Quasi-Optical Slab Resonator," *IEEE Microwave and Guided Wave Letters*, Vol. 3, No. 8, pp. 253–255, August 1993.

30. P. L. Heron, F. K. Schwering, G. P. Monahan, J. W. Mink, and M. B. Steer, "A Dyadic Green's Function for the Plano-Concave Quasioptical Resonator," *IEEE Microwave and Guided Wave Letters*, Vol. 3, No. 8, pp. 256–258, August 1993.

31. P. L. Heron, G. P. Monahan, J. W. Mink, F. K. Schwering, and M. B. Steer, "Impedance Matrix of an Antenna Array in a Quasi-Optical Resonator," *IEEE Transactions on Microwave Theory and Techniques*, Vol. 41, No. 10, pp. 1816–1826, October 1993.

32. T. Matsui, K. Araki, and M. Kiyokawa, "Guassian-Beam Open Resonator with Highly Reflective Circular-Coupling Regions," *IEEE Transactions on Microwave Theory and Techniques*, Vol. 41, No. 10, pp. 1710–1714, October 1993.

CHAPTER FIVE

Integrated and Active Antenna Testing

5.1 ANTENNA TESTING CONSIDERATIONS

Accurate testing of antennas is a laborious task considering the effort, space, equipment setup, and massive amounts of data involved. Antennas require characterization of both circuit and radiation properties, and since electromagnetic (EM) fields couple to nearby objects, antennas under test (AUT) must be carefully isolated from the environment.

Antenna circuit properties (i.e., S_{11}, Z_{in}, etc.) are not as sensitive to the environment as are radiation characteristics. A network analyzer as shown in Figure 5.1 can be used to measure circuit properties of an AUT (i.e., input impedance, impedance bandwidth, VSWR, etc). Impedance measurements do not require an anechoic chamber, although, ideally, one would like to completely isolate the AUT. Figure 5.2 shows input return loss measurements of a microstrip antenna with and without an enclosure filled with radar-absorbing material (RAM). As shown, the enclosure reduces the effect of external objects on the input return loss of an antenna. Although small, the effects are noticeable. External objects play a bigger role when measuring radiation properties of antennas.

EM fields undergo noticeable changes as they emanate away from an antenna. These changes allow the volume surrounding a radiator to be classified into three regions: reactive near field, radiating near field, and far field. The boundaries of each region blend and the criteria for defining their separation are mostly defined by convention. Figure 5.3 shows the three regions with a small antenna denoted by the center dot. The radiating near field is the hatched region.

The reactive near field is the region immediately surrounding the antenna where the reactive field predominates. Any obstacles such as metals and dielectrics within this volume greatly affect both the circuit current distribution

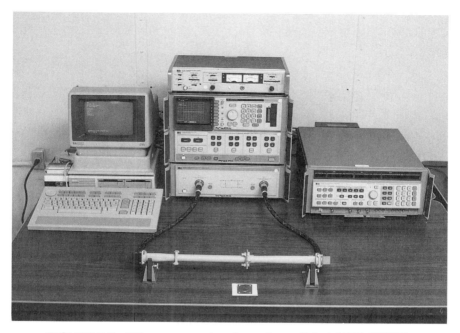

FIGURE 5.1. Microwave network analyzer (Model HP-8510B).

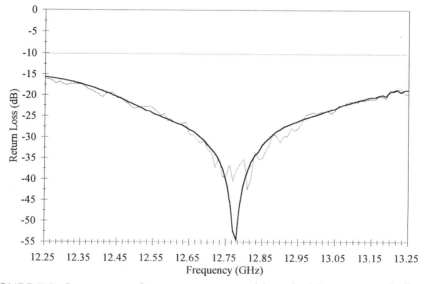

FIGURE 5.2. Input return loss measurement with and without external effects: (solid line) without external effects; (dotted line) with external effects.

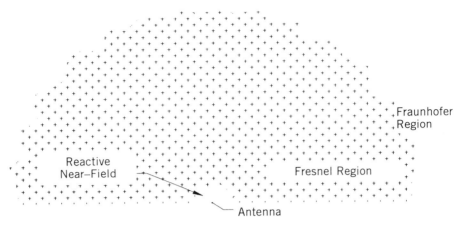

FIGURE 5.3. Field regions around an antenna.

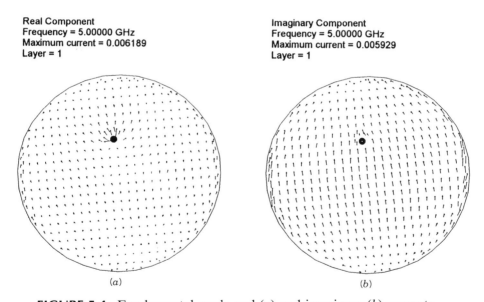

FIGURE 5.4. Fundamental mode real (*a*) and imaginary (*b*) currents.

and radiation characteristics of the AUT. The real and imaginary current distributions on the surface of a typical circular microstrip patch are shown in Figures 5.4*a* and *b*.

The radiating near field or Fresnel region is the volume between the reactive near field and the far field. In the Fresnel region, the radiation pattern still depends on the distance from the antenna. Foreign objects in this region can corrupt the antenna characteristics but will have little effect on the circuit characteristics. The boundary of this region is typically taken at

$$R \geqslant 0.62\sqrt{D^3/\lambda} \qquad (5.1)$$

where R is the separation between the AUT and the test source, D is the largest antenna dimension, and λ is the wavelength of operation.

Typically, antenna pattern measurements require larger separations to be in the Fraunhofer or far-field region. In the far field, the phase front across the main lobe is nearly spherical and the field distribution is independent of the distance from the antenna. If the dimension of the antenna is D and we require that the incident spherical wave differ from a uniform plane wave at the edges of the receiving antenna by some portion of a wavelength, $\Delta l = \lambda/k$, corresponding to a phase error $\Delta\phi = 360°/k$, where k is a constant, we obtain from Figure 5.5 that

$$R^2 = (R - \Delta l)^2 + \left(\frac{D}{2}\right)^2 = R^2 - 2R\Delta l + (\Delta l)^2 + \frac{D^2}{4} \qquad (5.2)$$

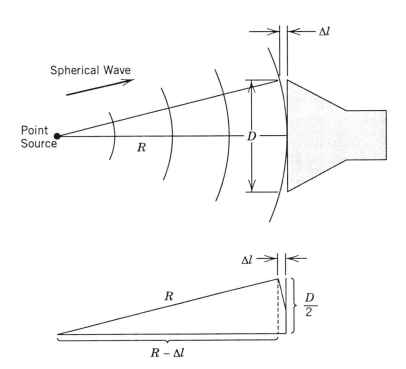

FIGURE 5.5. Derivation of the far-field criterion for surface-type antennas.

For $R \gg \Delta l$, a reasonable assumption, we obtain, from (5.2),

$$2R\Delta l \cong D^2/4 \tag{5.3}$$

Requiring Δl to be some portion of a wavelength such as $\Delta l = \lambda/k$, we obtain, from (5.3),

$$R = kD^2/8\lambda \tag{5.4}$$

The usual choice for k is $k = 16$ which gives $\Delta l = \lambda/16$ and $\Delta\phi = 22.5°$, whereby (5.4) becomes

$$R_{\text{far field}} = 2D^2/\lambda \tag{5.5}$$

The condition for far-field operation is

$$R \geqslant 2D^2/\lambda \tag{5.6}$$

Equation (5.6) is the standard distance used to designate the start of the far-field region. Longer ranges can decrease the phase error (i.e., $\Delta\phi \approx 11.25°$ for $R = 4D^2/\lambda$), while decreasing the range to $R = D^2/\lambda$ increases the phase error (i.e., $\Delta\phi \approx 45°$).

Phase errors due to finite range lengths have little effect on the main-beam amplitude, but they raise the level of nulls around the main beam. These errors become an important factor when testing arrays or larger apertures. Figure 5.6 shows the effect of measuring an aperture of dimension D at various R values [1]. As shown, phase errors of $\Delta\phi \approx 0°$, 2.813°, ..., 45° (which are beta shown in the figure) mainly affect the accuracy of measuring the first null and side lobe, which may not be a major concern for most antennas. However, this error should be recognized by the shoulders that it creates on the main beam.

The increase in cost is a major issue when decreasing the phase error $\Delta\phi$. Longer ranges require more space and better equipment to maintain similar operation conditions. An alternative solution is computer correction of the measured data for the phase error. Finite range lengths and other errors such as large source antennas or misalignment can cause variations of the amplitude distribution across the AUT, which will also corrupt the measured radiation pattern. The effects of errors from source illumination tapers, wide-angle scatters, and receiver nonlinearities are shown in Figure 5.7 [2].

To complicate matters further, complete characterization of an antenna requires a thorough three-dimensional spherical measurement of the copolarization and cross-polarization components of the EM fields. In most cases, however, a good estimate of the performance can be deduced from cuts along the elevation (E) and horizontal (H) planes. The copolarization and cross-polarization components of the fields along these planes are shown for a

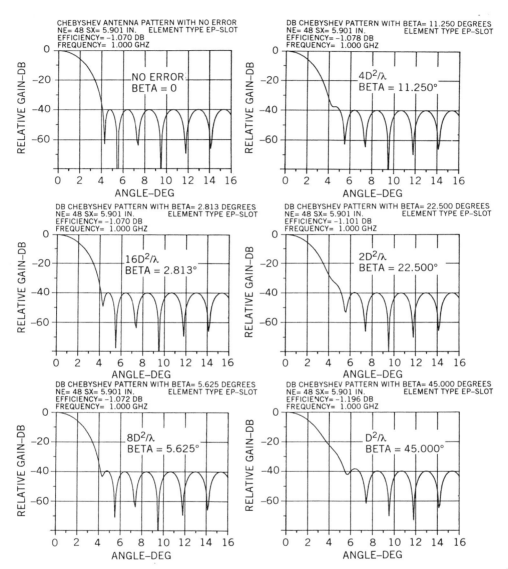

FIGURE 5.6. Finite range effect on pattern measurements. (From Ref. 1 with permission from IEEE.)

circular microstrip patch antenna in Figure 5.8. It is not unusual to take diagonal, great circle, or conical cuts to ensure correct performance at every important angle of coverage, especially on an irregular structure (i.e., airplane, missile, etc.). These cuts and the coordinate system used to test an antenna are shown in Figure 5.9.

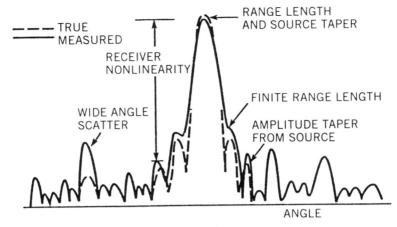

FIGURE 5.7. Typical range-induced errors on antenna patterns. (From Ref. 2 with permission from Artech House.)

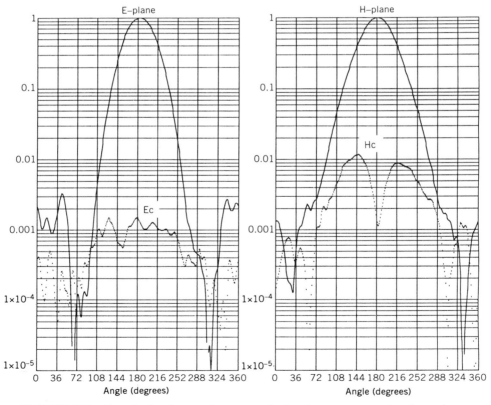

FIGURE 5.8. Copolarization and cross-polarization components of a microstrip patch antenna: (solid line) copolarization; (dotted line) cross-polarization.

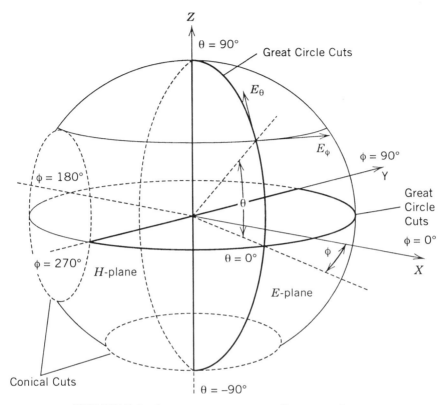

FIGURE 5.9. Antenna pattern coordinates and cuts.

Indoor or outdoor antenna ranges are set up to test antennas in the far field as defined by Equation (5.6). Testing the radiation lobes, half-power beamwidths, directivity, gain, and so on, of an antenna is entirely more complicated than testing the circuit properties. Figure 5.10 shows a basic configuration used to test antenna radiation patterns. In such a setup one is concerned with such things as outside interference, source amplitude illumination, phase error across the aperture, reflection paths, scatterers, dynamic range, and so on. Assuming that each of these concerns is within a specified level, carefully designed test fixtures and holders are necessary to test the antenna in the environment it is designed for. Scaled models are often used to facilitate testing of large structures such as airplanes and rockets. Analytic models such as infinite ground planes can only be approximated and are seldom seen in an actual application. Standard-gain horn antennas are normally used as a transmitter and the antenna under test as a receiver. The system is first calibrated by using two standard horns.

There are several types of antenna test ranges. Figures 5.11a–e show two outdoor and three indoor ranges. Figures 5.11a and b show an elevated and a

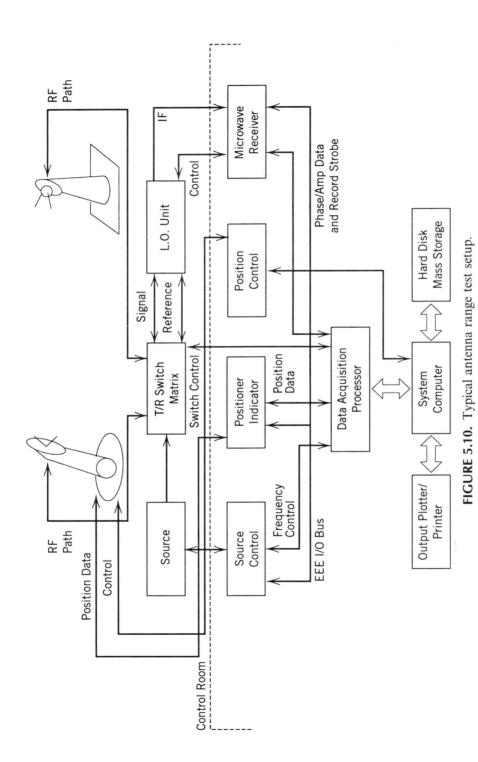

FIGURE 5.10. Typical antenna range test setup.

ground-reflection outdoor range. As shown, several precautions are taken to ensure that the measured pattern is not corrupted by reflections off the ground or external objects. External reflections may be due to local scatterers, a wide-source illumination beam, the AUT positioner, and other factors. A large antenna range free of interference and objects along or near the test path is sufficient to provide accurate antenna testing. When the necessary real estate is not available, antennas are tested within indoor anechoic ranges. Indoor chambers are often preferred due to the level of control one has over the test. Indoor anechoic chambers include near-field, compact, and standard indoor ranges. Figures 5.11c, d and e show a compact, a near-field and a standard indoor range. Indoor ranges rely on radar absorbing material to isolate an AUT from the building structure and simulate an unbounded medium. In most cases, external objects, finite ground planes, and other irregularities change the radiation patterns. In the end, one is also limited by the operation frequency and the test equipment. Receiver linearity–dynamic range, equipment noise as well as positioner and standard gain accuracies are some limiting factors. Typical trade-offs include cost budget, space allocation, frequency ranges, and types of applications.

An antenna range consists of a transmitter, transmit and receive antenas and a receiver. The antenna range simulates a free-space environment, and the energy traveling from the transmitter to the receiver is described through the Friis transmission equation [3]:

$$\frac{P_r}{P_t} = (1 - |\Gamma_r|)^2(1 - |\Gamma_t|)^2 \left(\frac{\lambda}{4\pi R}\right)^2 (e_{cd_r} D_r(\phi_r, \theta_r))(e_{cd_t} D_t(\phi_t, \theta_t))|\hat{\rho}_r \cdot \hat{\rho}_t^*|^2 \quad (5.7)$$

Equation (5.7) relates the received and transmitted power with both the circuit and radiation properties of the antenna. P_t and P_r are the power transmitted and the power received, respectively; Γ_t and Γ_r refer to the reflection coefficients of the transmit and receive antennas and express how much power reaches the antenna terminals. The wavelength of operation, λ, and the range length, R, are grouped to form the free-space loss factor, which accounts for spreading of the EM waves, D_i ($i = r$ or t) is the directive gain and is a function of ϕ and θ, while e_{cd_i} refers to the conductor and dielectric losses associated with each of the antennas, and ρ_i is the polarization vector of each antenna, and is used in the last term of the equation to determine the polarization loss factor.

Equation (5.7) can be reduced by several useful approximations, such as a good impedance match at each antenna ($|\Gamma_i| \approx 0$). Also, given a reasonable polarization, alignment of the transmit and receive antennas will remove most polarization losses (i.e., $|\hat{\rho}_r \cdot \hat{\rho}_t^*|^2 \approx 1$). Finally, accurate boresighting of the antennas allows the combination of the directive gain and efficiency into a single power gain term (i.e., $e_{cd_i} D_i(\phi_i, \theta_i) \equiv G_i$). These modifications reduce

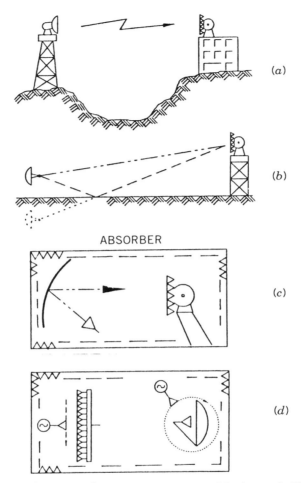

FIGURE 5.11. Basic types of pattern test ranges: (*a*) elevated, (*b*) ground, (*c*) compact, (*d*) near-field planar and cylindrical or spherical; (*e*) a standard indoor range.

Equation (5.7) to a more manageable form [4]:

$$\frac{P_r}{P_t} = \left(\frac{\lambda}{4\pi R}\right)^2 G_r G_t \tag{5.8}$$

The losses in the antenna range system are calibrated out by using the substitution method. Since we are seldom interested in the absolute power levels, the received power is calibrated with a standard-gain horn whose characteristics are well known. The horn is later replaced by the AUT and the measured data are compared to the standard horn. This calibration procedure

FIGURE 5.11. (*Continued*)

has been carried out for decades. Currently, the difference with past systems is in increased automation, sensitivity, and number-crunching capability. Otherwise, the method for testing passive antennas on far-field ranges has remained practically unchanged.

With the advent of integrated circuits and, later, integrated antennas, modifications to passive antenna testing were in order. Antennas integrated with passive devices (i.e., *pin*, varactor, transistors in passive operation, etc.) can be tested using standard ranges but require dc biasing lines for operation. Many modern radar systems use hundreds of integrated circuit modules to individually feed antennas in an array. These array are classified as active because of the distributed amplifying modules. Testing of these *active* array can be carried out in the transmit mode, which requires interchanging the transmit and receive cable routing of a conventional antenna test chamber. Typically, there is a low-power rf input and distribution network to the individual modules. Losses in the distribution network do little to the overall dc-to-rf conversion efficiency since they occur before the high-power amplification. The rf input also allows the use of a stable source (i.e., a synthesizer) to facilitate and ensure accurate testing of the antenna pattern. A similar situation occurs for active integrated antennas. As defined earlier, the term *active* refers to active circuit component functions, while *integrated* refers to components integrated directly at the antenna. For testing purposes, active integrated antennas used for receive amplification can be tested using a standard test range. Antennas used for amplification require rerouting of the transmit and receive sides of the test range. Since the antenna requires an rf input, a stable source can be used to ensure accurate results. For oscillating structures, the test setup changes. Since oscillating active antennas are inherently Doppler sensors, they sense movement and nearby objects. The "sensing" disturbs the oscillating frequency, which causes random errors in the pattern. Changes in the oscillating frequency can be overcome via injection locking, which can be accomplished with an rf input port on the antenna or externally. This creates added burden and cost for the test.

For oscillating active antennas, the same characteristics used to classify circuit oscillators are applicable. As radiating structures, they are further defined by their antenna properties. As free-running radiating sources, they cause difficulties during test.

5.2   ACTIVE ANTENNA OSCILLATOR TESTING

Oscillators are typically difficult to test accurately, although the procedures are well documented. As stated before antennas also cause considerable problems during tests. In addition, integrated antennas require dc lines to bias the solid-state devices. Most integrated antennas can be tested in a conventional antenna test range since they are usually reciprocal components. Active integrated antennas using an amplifier for reception are not reciprocal in operation but can also be tested in a conventional test range.

Amplifier integrated antennas used for transmitting and multiplying and those which generate their own signal (i.e., oscillators) require rerouting of the test range source and receive paths. Amplifer and multipliers have an rf input where a very high quality source can be used (i.e., synthesizer). This allows for more accurate tests. The free-running nature of oscillating active antennas causes problems which can be overcome if an rf input is added for injection locking. This may not always be feasible without disturbing the antenna structure and, consequently, the pattern.

Since the majority of range equipment test sets are dedicated to passive antenna testing and require considerable modifications to accomodate active antenna sources (i.e., cable rerouting, dc power supplies, etc.), most active antenna tests are carried out using awkward combinations of test equipment,

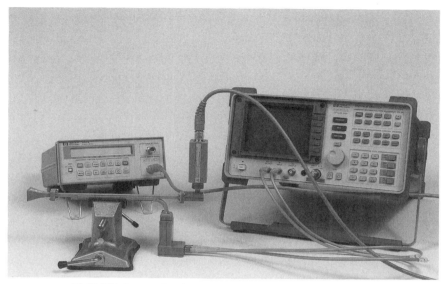

FIGURE 5.12. Equipment for active antenna testing.

power supplies, and positioners or makeshift ranges. Figure 5.12 shows typical test equipment used for active antenna tests: rf power meter, spectrum analyzer, external mixer, Ka-band coupler, and standard-gain horn.

Since an active antenna is a radiating oscillator, all circuit oscillator tests and characteristics are applicable. These include frequency, power, efficiency, stability, phase noise, bias-tuning range, load pulling, external quality factor, and injection-locking bandwidth, and gain. Unlike a circuit or waveguide type oscillator which has a well defined output port, an active antenna radiates directly to free space. We must therefore define how to obtain each of these measurements.

Figure 5.13a shows a test setup for injection-locking bandwidth and gain. The Q-factor of the circuit can be determined after the measurements of the locking bandwidth $2\Delta f_{max}$ and the locking gain. Equation (2.71b) can be rewritten as

$$Q = \frac{f_0}{\Delta f_{max}} \sqrt{\frac{P_i}{P_0}}$$ (5.9)

The locking gain in dB is defined as

$$G_L = 10 \log \frac{P_0}{P_i}$$ (5.10)

where

Q = external Q-factor

f_0 = operating frequency

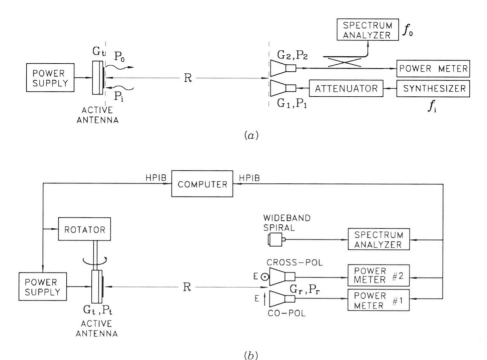

FIGURE 5.13. Active antenna test range setup: (*a*) injection-locking test setup; (*b*) active antenna test setup.

$\Delta f_{\max}$ = injection-locking bandwidth

P_i = injection-lock signal power

P_0 = free-running oscillator power

P_i and P_0 are calculated from the measured P_1 and P_2 using the Friis transmission equation

A basic active antenna test range is shown in Figure 5.13*b* complete with power supplies, spectrum analyzer, and rf power meters. The two power meters are attached to standard-gain horn antennas which detect the co- and cross-polarization components of the signal. The wideband spiral antenna detects harmonics and possible oscillator mode jumps which may occur during the bias ramping.

The oscillating active antenna can be seen as an oscillator which feeds the antenna directly, but unlike a circuit oscillator, it cannot be physically removed from the antenna to measure its oscillator characteristics. However, several methods have been set forth to tackle this problem. If we can assume that the active antenna's performance is similar to its passive counterpart, by using the passive antenna gain level, we can determine the oscillator power. From the co-polarization measurement in Figure 5.13b, the passive antenna gain level was first used by Chang et al. to deembed the oscillator power [5], using the Friis transmission equation:

$$P_t = P_r \left(\frac{4\pi R}{\lambda}\right)^2 \left(\frac{1}{G_r G_t}\right) \tag{5.11}$$

Comparing the performances of passive and active antennas is reasonable when the integrated devices are relatively small. This approximation is a very conservative estimate of the power level of the oscillator since bias lines typically increase cross-polarization components, thereby reducing copolarization efficiency. For most active antennas described in the literature, the cross-polarization levels are comparable to the copolarization component. In such situations, the active devices have truly changed the operation of the antenna, and comparing it to its passive counterpart becomes less justifiable.

If one is unable to accurately approximate the gain of the passive antenna, we can lump the antenna and oscillator properties together into an effective isotropic radiated power (EIRP) level. This measure does not separate the antenna properties from the oscillator characteristics. It is simply defined as [6]

$$\mathrm{EIRP} \equiv P_t G_t = \frac{P_r}{G_r} \left(\frac{4\pi R}{\lambda}\right)^2 \tag{5.12}$$

The EIRP is the amount of power which would be transmitted by an isotropic radiator given the measured received power. Although the inflated power value is, in itself, not a useful quantity, it provides a measure by which others can compare their results. It becomes a valuable tool for testing, since little need be known of the antenna or the active devices involved. An effective transmitted power has been proposed which provides another figure by which to compare or gauge the performance of an active antenna [7]:

$$P_{\mathrm{eff}} \equiv \frac{\mathrm{EIRP}}{D_t} = \frac{P_t G_t}{D_t} = P_t \eta_t = \frac{P_r}{D_t G_r} \left(\frac{4\pi R}{\lambda}\right)^2 \tag{5.13}$$

where D_t is the directivity and η_t is the efficiency of the antenna integrated with the oscillator.

5.3   BIAS TUNING

A spectrum analyzer, power meter, standard-gain horn, and directional coupler are used along with a variable power supply. A typical Gunn integrated active patch antenna oscillator frequency and power versus bias-tuning level is shown in Figure 5.14. The data gathered with an automated range shown in Figure 5.13*b* show turn-on threshold, frequency tuning range, and power response for the co- and cross-polarization components. Similar results are shown in Figure 5.15 for a FET integrated patch antenna.

The wideband spiral is used to detect the fundamental mode, harmonics and possible oscillator mode jumps which may occur during the bias ramping. Power levels of harmonics can also be used to test active antenna patterns for other oscillator modes.

5.4   SPECTRUM QUALITY, STABILITY, AND NOISE

The test set in Figure 5.13*a* or *b* is useful in determining the spectral qualities, stability, and noise level for an active antenna. A modern spectrum analyzer has most of these functions built in. The FM and AM noise levels can be determined. A modern spectrum analyzer can also be used to measure the noise figure [8].

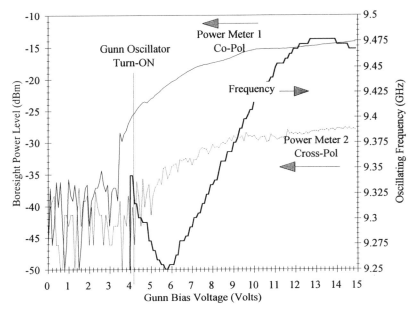

FIGURE 5.14. Active Gunn-integrated antenna frequency and power versus Gunn bias voltage.

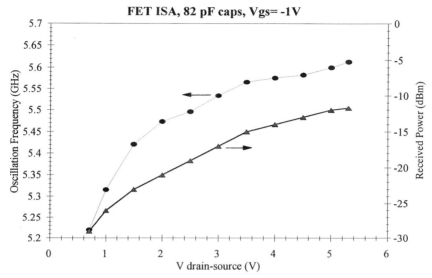

FIGURE 5.15. Active FET integrated patch bias tuning results.

5.5 LOCKING GAIN AND LOCKING BANDWIDTH

The test setup in Figure 5.13a allows the determination of locking gain and locking bandwidth of an active oscillator. For this case the analyzer confirms that the free-running active antenna is or is not synchronized to the external source. Figure 5.16 shows typical injection-locking results for an active antenna. The plot shows the increase of locking bandwidth with an increase in injection-locking signal.

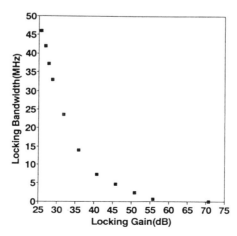

FIGURE 5.16. Locking gain vs. locking bandwidth for an active antenna.

The locking-gain versus locking-bandwidth graph is especially important for active antennas because they are easily affected by the environment. In power-combining applications, the locking gain and locking bandwidth determine how much power is required to keep the distributed sources synchronized to a master source.

5.6 ACTIVE ANTENNA PATTERN MEASUREMENTS

Active antenna pattern measurements require the same precautions stated previously for oscillators and antennas. However, active antennas are "directly" connected to free space, and they are inherently good Doppler sensors. Therefore, any disturbance or quick movements by the rotator are translated into frequency shifts of the active antenna. Although this is a great quality for automatic door openers and car alarms, it adds more uncertainties during a pattern test.

With a passive antenna or other integrated antennas, the test frequency is determined by stable oscillators and receivers. This is not the case for an active antenna. Methods to ensure that the operating frequency of an active antenna remains constant are limited because the oscillator and antenna are "connected" to free space. A ferrite isolator would work nicely, but it would have to come between the active antenna and free space. High-Q cavities or resonators would also change the configuration. External injection locking of the active antenna to a stable internal or external source is viable but may interfere with the antenna pattern.

Figure 4.8c shows several stabilization methods used for active antennas. Each method has its share of advantages and disadvantages. Of these methods, injection locking with external circuitry appears to have the most promise. A locking signal through an external circuit can be used to maintain a constant frequency during a pattern measurement.

5.7 CONVERSION EFFICIENCY

The dc-to-rf conversion efficiency is, perhaps, the most important parameter in active antennas. It depends on the solid-state device used, and it explains the evolution of active patch antennas from diodes to transistors. The mere conversion of power from dc to rf is a bit more complicated for an active antenna than a conventional oscillator.

If the rf power conversion is very high but is going to the antenna's cross-polarization component, it is essentially wasted. In this sense it is more important to speak of the rf power converted to the copolarization component for active antennas. The actual oscillator conversion is more difficult to ascertain than in a conventional oscillator since the antenna is physically inseparable from the device. However, the effective power and EIRP are figures of merit which can be used for cataloging and comparing different active antennas.

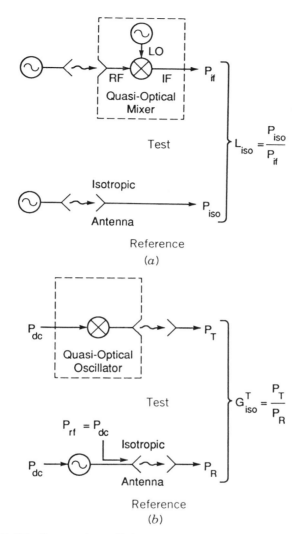

FIGURE 5.17. Conversion efficiency. (*a*) Isotropic conversion loss L_{iso} for a quasi-optical mixer. (*b*) Isotropic conversion gain G_{iso}^T for an active antenna. (From Ref. 9 with permission from IEEE.)

For active integrated antennas used for a quasi-optical mixer, the antenna test is more like a conventional antenna test. The determination of rf to if conversion loss requires some definitions as shown in Figure 5.17 [9]. Specifically, the isotropic conversion loss is defined by a reference power level divided by the if power from the mixer:

$$L_{iso} = P_{iso}/P_{if} \tag{5.14}$$

where P_{iso} is the rf power that would be received if the receiver circuit were replaced by a fictitious isotropic antenna under the same measurement conditions. L_{iso} accounts for the conversion loss and antenna gain.

Also in the figure is a similar test approach for an active antenna. The isotropic conversion gain (G_{iso}^T) for an active antenna is the ratio of the transmitted power (P_T) to received power (P_R):

$$G_{iso}^T = P_T/P_R \qquad (5.15)$$

where P_R is obtained from a fictitious 100% efficient source driving an isotropic antenna. The gain G_{iso}^T can exceed unity for an aperture with considerable gain even though the device efficiency is below 100%.

REFERENCES

1. P. S. Hacker and H. E. Schrank, "Range Distance Requirements for Measuring Low and Ultra-low Sidelobe Antenna Patterns," *IEEE Transactions on Antennas and Propagation*, Vol. 30, No. 5, pp. 956–966, September 1982.

2. Gary E. Evans, *Antenna Measurement Techniques*, Artech House, 1990.

3. H. T. Friis, "A Note on a Simple Transmission Formula," *Proceedings of the IRE on Waves and Electrons*, Vol. 34, No. 5, pp. 254–256, May 1946.

4. C. A. Balanis, *Antenna Theory-Analysis and Design*, Wiley, New York, 1982.

5. K. Chang, K. A. Hummer, and J. L. Klein, "Experiments on Injection Locking of Active Antenna Elements for Active Phased Arrays and Spatial Power Combiners," *IEEE Transactions on Microwave Theory and Techniques*, Vol. 37, No. 7, pp. 1078–1084, July 1989.

6. W. A. Morgan, Jr. and K. D. Stephan, "An X-Band Experimental Model of a Millimeter-Wave Interinjection-Locked Phased Array System," *IEEE Transactions on Antennas and Propagation*, Vol. 36, No. 11, pp. 1641–1645, November 1988.

7. M. Gouker, "Toward Standard Figures-of-Merit for Spatial and Quasi-optical Power-Combined Arrays," *IEEE Transactions on Microwave Theory and Techniques*, Vol. 43, No. 7, pp. 1614–1617, July 1995.

8. C. Slater, "Spectrum-Analyzer-Based System Simplifies Noise Figure Measurement," *RF Design*, pp. 24–32, December 1993.

9. K. D. Stephan and T. Itoh, "Recent Efforts on Planar Components for Active Quasi-Optical Applications," *IEEE MTT-S International Microwave Symposium Digest*, pp. 1205–1208 (1990).

CHAPTER SIX

Active Antennas: Work before 1987

6.1 CONVENTIONAL "ACTIVE" ANTENNA ARRAYS

Traditionally, functional components have remained within guided-wave structures far away from the antenna terminals. These structures can shield circuits from each other, thereby reducing mutual coupling or crosstalk. Also, within circuit layouts, there is an obvious progression from various inputs to various outputs. Antennas, on the other hand, radiate directly into free space where fields are not so easily controlled. Devices at the terminals of an antenna can disturb radiation patterns, which, in turn, disrupt the component's performance. Isolation between circuits and the environment also becomes an important issue. In this chapter we review the conventional approach to transmitting and receiving systems, define the concept of integrated and active integrated antennas, and briefly review the early history of integrated and active antennas.

The conventional approach is shown in Figure 6.1. It uses separately designed components and antennas interconnected via transmission lines. The interconnections are made up of at least one transmission line and two transitions. One transition is needed from the circuit to the transmission line, another from the transmission line to the antenna. Transitions may be in the form of a probe, loop, or an aperture, while transmission lines use waveguides, coaxial cables, twin lines, wires, and so on. Components are designed by using variations of waveguide and coaxial circuits, while antennas use some derivative of a horn or a dipole.

In the conventional approach there is freedom to optimize the performance of components and antennas independently because there is an obvious distinction between the circuits and the radiating structure. This allows one to obtain the optimal performance from the individual parts of the system and later match them together. One drawback of this approach has been the size

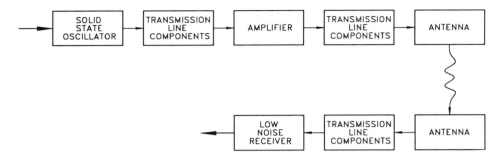

FIGURE 6.1. Typical microwave system.

and weight of these systems; they are often bulky, expensive to manufacture, and difficult to reproduce.

A modified conventional approach replaces waveguides, coaxial circuits, and interconnections with integrated circuits (i.e., MICs, MMICs), thereby reducing the overall size and weight of the system. Monolithic phase shifters, switches, oscillators, mixers, and amplifiers are fabricated simultaneously with transmission lines and distribution networks. These are later coupled to antennas through apertures, loops, or probes. In this approach there remains a well-defined boundary between where the guided-wave structure ends and the antenna begins.

Many modern array implementations fall into this category. Major issues in the design of these systems include cost, weight, manufacturability, and reliability. In a complete transceiver, integration of multiple layers becomes critical. These layers combine active devices and circuits, dc power and control lines, rf distribution networks, and antennas in relatively small volumes. Figure 6.2 shows a Ka-band array built for the NASA–Lewis Research Center using this integration approach and based on the aperture-coupled microstrip antenna [1, 2]. The concept is not unlike that shown by Staiman et al. [3] in 1968, describing the use of distributed amplifiers to overcome power deficiencies of single solid-state devices. It provided an alternative to circuit-level cavity combiners, but it was not used by others in the field for several years. Over a decade later, a paper by Durkin et al. [4] in 1981 discussed an IMPATT oscillator power combiner using a similar approach. These implementations are illustrated in Figures 6.3a and b.

6.2 THE ACTIVE INTEGRATED ANTENNA APPROACH

The integrated antenna approach differs from other more conventional approaches in that there is no obvious distinction or boundary between the component and the antenna. Packaged devices lie within the volume normally associated with the radiating structure. Component functions are imbedded directly at the antenna terminals, and the antenna is an integral part of the component. The antenna serves as a load for the component and as the radiator and/or receiver for the system. To avoid losses and transitions, transmission lines are not used. These configurations are made up of at least two components: a solid-state device and an antenna. The type of device and antenna combination determines the final microwave component function and its classification. To date, most integrated antenna combinations have concentrated on single solid-state devices. However, more complicated functions will require combinations of several devices and/or MICs. In this fashion, all possible rf component functions can be localized with the antenna to provide the smallest possible system.

The smallest possible system, however, is not without its share of problems. Due to the strong coupling between the MICs and the antenna, there is considerable degradation of both component and antenna performance. The combination of antennas and circuits on a single substrate forces a significant trade-off in performance. Substrate properties cannot be separately optimized for the circuit nor the antenna. MICs and devices also require dc biasing lines near the antenna, which can disturb currents flowing on exposed surfaces. Similarly, the antenna may have to be modified to accommodate these devices, which often degrades radiation characteristics. Many configurations have been devised to address these issues.

6.3   ACTIVE ANTENNAS: THE EARLY HISTORY

The implementation of the integrated antenna concept may be in its infancy, but the idea is not new. It could be argued that Hertz' introduced the concept with his end-loaded dipole transmitter and resonant square-loop antenna receiver [5]. Neither the transmitter nor the receiver used any matching networks between the circuit and antenna terminals. The dipole served as the tank circuit for the transmitter, and the resonant square loop was the only frequency filter for the receiver. However, this system offers little control over the stability and unwanted harmonics. In fact, its use would cause widespread interference.

Spark-gap transmitters operate along these lines and thus were banned from commercial applications by the Federal Communications Commission (FCC) in 1934. But control over harmonics and isolation from the environment can be easily accomplished within the circuit away from the antenna. As technology advanced, the oscillator, amplifier, mixer, and filter functions became part of the circuit (not the antenna), and what remained we refer to as the conventional approach to transmitting and receiving systems.

The status quo remained for nearly three decades until the integrated antenna concept surfaced again in the early- and mid-1960s [6]. Several investigators in Ohio State University, funded primarily by the United States

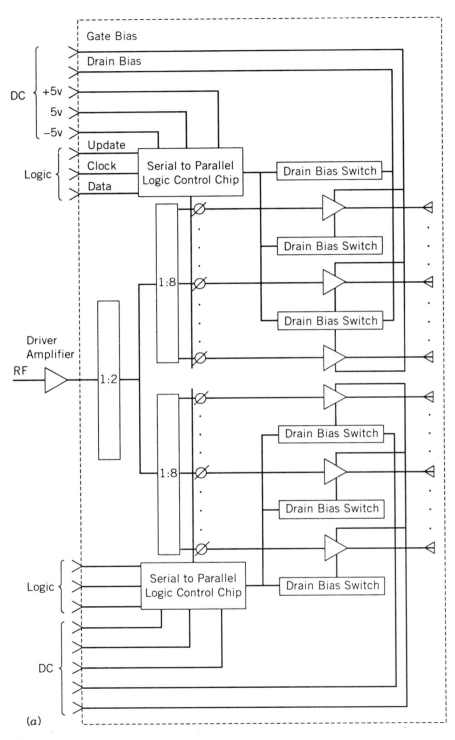

(a)

124

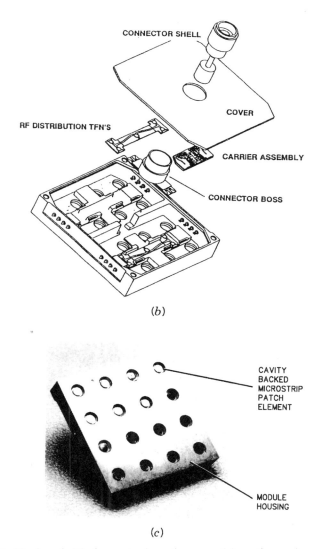

(b)

(c)

FIGURE 6.2. Ka-band 16-element phased array: (a) active subarray module block diagram; (b) module assembly; (c) active 16-element subarray module antenna configuration. (From Refs. 1 and 2 with permission from IEEE.)

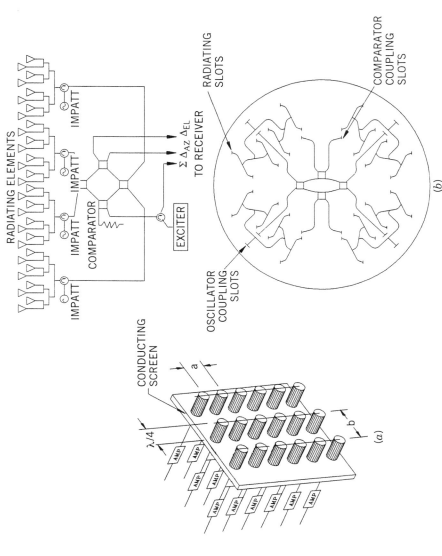

FIGURE 6.3. Spatial power combining using distributive sources (*a*) rf power combination in free space using an array of individually fed, closely spaced dipoles; (*b*) 35-GHz spatial combiner block diagram and antenna array layout. (From Refs. 3 and 4 with permission from IEEE.)

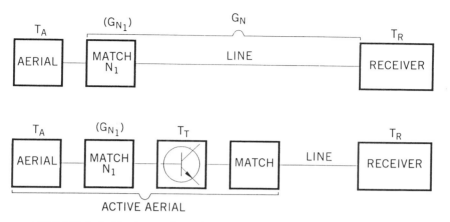

FIGURE 6.4. Early work on integrating devices and antennas [8].

Air Force, demonstrated both diode- and transistor-integrated antennas. Using a tunnel diode and a spiral antenna, Copeland and Robertson demonstrated a mixer-integrated antenna which they describe as an 'antennaverter'. They also used a traveling wave antenna, together with tunnel diodes, to operate as a traveling wave amplifier which they called an 'antennafier'. In 1961, Pedinoff demonstrated a slot amplifier. A thorough description of a tunnel-diode integrated dipole antenna amplifier was later shown by Fujimoto. A single 'antennafier' demonstrated a gain of 10 dB and 6 dB noise figure at 420 MHz. During the IEEE Antennas and Propagation Conference of 1968 held in Boston, Massachusetts, Meinke and Landstorfer [7] described the mating of a FET transistor to the terminals of a dipole to serve as a VHF amplifier for reception at 700 MHz. The following year, at the European Microwave Conference, Landstorfer presented a paper entitled "Applications and Limitations of Active Aerials at Microwave Frequencies" [8]. This early work illustrated in Figure 6.4 uses matching networks near the antenna to optimize performance, but it shows remarkable foresight in its concept of integration as well as the idea of varactor tuning of the antenna. In her comments [9] the section chair, E. A. Killick, stated:

> The term *active antenna* could be taken to mean any antenna in which the power source or head amplifier is very closely associated with the radiating or receiving element. The term seems most appropriate when the active component is coupled directly to the element without an intervening match (or mismatch) to any sort of transmission line.

Our definition for an active antenna follows this statement closely; however in much of the literature the definition of the active antenna is broadened to

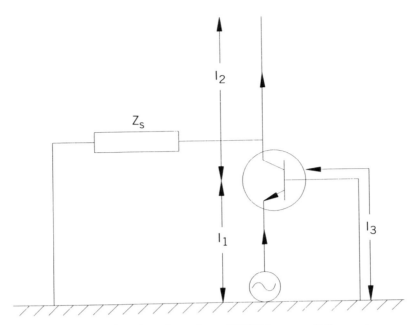

FIGURE 6.5. Fed-emitter base loop (FEBL) transmitting antenna.

include the integration of any solid-state device. Strictly speaking, we should distinguish passive integrated antennas and active integrated antennas:

1. *Passive integrated antennas* are antennas incorporating one or more passive solid-state devices and circuits for switching, tuning, modulating, mixing, multiplying, detecting, or any other functions.

2. *Active integrated antennas* are antennas incorporating one or more active solid-state devices and circuits to amplify or generate rf frequencies. This category usually involves a negative resistance for rf power generation or amplification. In some cases, this distinction is difficult since both types of solid-state devices are involved in the integration. For example, a transistor oscillator and a mixer can be integrated with an antenna to form a transceiver.

Following the work of Meinke and Landstorfer, Ramsdale and Maclean [10] used BJTs and dipoles for transmitting applications in 1971. They demonstrated large height reductions in 1974 [11], and later in 1975 [12] by using the active antenna concept. Their configuration is shown in Figure 6.5. Meanwhile, in 1974 a concept using injection-locked distributed oscillators demonstrated beam steering for active antenna arrays [13]. Figure 6.6 shows the beam-steerable active array configuration. Phase shifts are induced through

RADIATING ELEMENTS

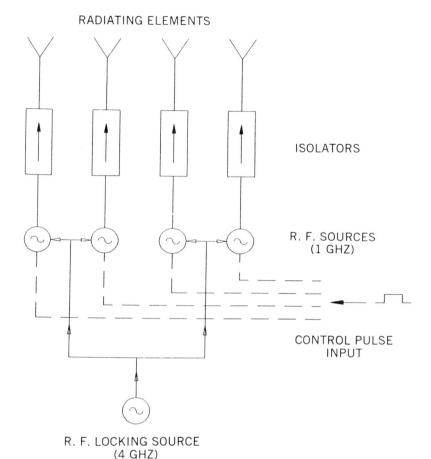

ISOLATORS

R. F. SOURCES
(1 GHZ)

CONTROL PULSE
INPUT

R. F. LOCKING SOURCE
(4 GHZ)

FIGURE 6.6. The beam steerable active array. (From Ref. 13 with permission from IEEE.)

the use of an external injection-locking pulse. This idea found very little use until the mid-1980s when active antennas and spatial power combining were widely sought to solve power deficiencies of solid-state devices in millimeter-wave bands.

6.4   ACTIVE MICROSTRIP ANTENNAS

It took several years for active integrated antennas to return to the literature after the initial transistor integrated aerials of the early 1970s. Perhaps the usefulness of the integration was in the microwave and millimeter-wave bands,

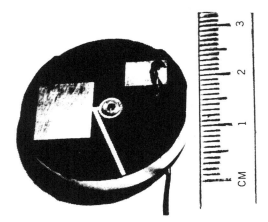

FIGURE 6.7. BARITT-diode microstrip Doppler sensor. (From Ref. 14 with permission from IEEE.)

while device technology had not advanced enough to take advantage of it. Following the work of Maclean and Ramsdale, Armstrong et al. demonstrated the use of a single active antenna for Doppler sensing applications in 1980 [14]. As a radiating oscillator coupled directly to free space, an active antenna is very sensitive to changes in its immediate vicinity. Variations in its environment (i.e., temperature, relative motion) are translated into shifts in the operating frequency. These shifts make an active antenna ideal for Doppler sensing. A BARITT diode was integrated with a microstrip patch antenna and used as a self-oscillating mixer. This integration represents the first active antenna demonstration using microstrip antenna technology. Its use as a proximity detector highlights a practical application for individual active antennas. As shown in Figure 6.7, it is an ideal active antenna product — small, compact, and inexpensive. The proximity detector application does not require excessive amounts of power, and it can be used in automatic door openers, burglar alarms, and other commercial applications.

During the 1980s, improvements in solid-state devices and integrated circuit techniques made millimeter-wave operating frequencies more accessible. Integrated circuits increased in its uses and functionality. Monolithic circuits began to replace hybrid integrated circuits in some applications. Improvements in packaging technology made antenna integrations possible by reducing their effects on the antenna structure. Early in the decade, investigators began to miniaturize circuits and bring them closer to the antenna.

In 1982, the varactor-tunable passive microstrip patch antenna was introduced by Bhartia and Bahl [15]. The simple integration in Figure 6.8 turns a narrowband antenna ($<2\%$) into a fairly wide band element ($\sim 30\%$). The

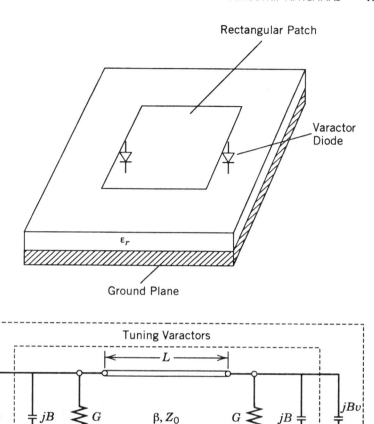

FIGURE 6.8. Varactor-tuned rectangular patch antenna and its equivalent circuit. (From Ref. 15 with permission from *Microwave Journal*.)

quick-tuning ability of the solid-state varactors makes it an ideal preselector for receiving systems. The loading effect on the antenna input impedance gives it potential to remove scan blindness in phased arrays. This successful integration opened possibilities for other devices and different functions. But this valuable contribution stirred very little interest because, at the time, more emphasis was placed on circuit-level integration.

Even improved solid-state devices were fundamentally limited to low output powers which could not meet most commercial or military applications. Low

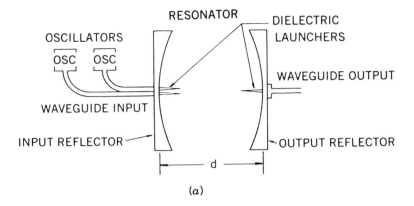

(a)

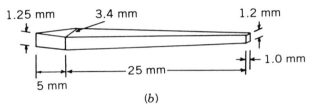

(b)

FIGURE 6.9. Quasi-optical millimeter-wave power combiner schematic: (a) complete setup; (b) tapered dielectric launcher (polyethylene). (From Ref. 18 with permission from IEEE.)

output power had been a problem since the 1960s, and investigators have continually sought new power-combining techniques [16, 17] to overcome power deficiencies. In Reference 17, power-combining techniques are divided into chip, circuit, and spatial methods. A variation on spatial combining— quasi-optical power combining—would prove to be instrumental in the development of active antennas.

In 1983, Waldinger and Nalbandian [18] demonstrated the concept of an open-cavity resonator for power combining. This combining method is referred to as a quasi-optical power combiner due to its similarities to laser cavities. Two oscillators couple directly to the modes of the open resonator defined by two reflecting surfaces as in Figure 6.9. This power-combining scheme is ideally suited for active antennas. It requires radiating oscillators, which is in fact defined as an active antenna. However, it took a few years for the connection between active antennas and quasi-optical resonator to take place.

In 1984, quite by accident, the first active endfire antenna [19] was developed, as shown in Figure 6.10. It was ahead of its time due to its monolithically integrated Gunn diode and its use of a uniplanar CPW

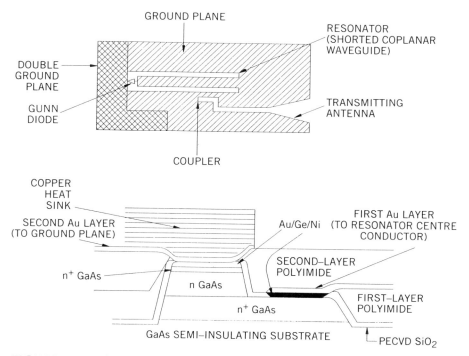

FIGURE 6.10. A monolithic Gunn oscillator chip and the cross-sectional drawing of a Gunn diode. (From Ref. 19 with permission from *Electronics Letters.*)

resonator. The oscillator was coupled to a tapered-slot antenna for measurement. It is seldom regarded as an active antenna because the tapered-slot radiator was not an integral part of the Gunn oscillator design. It was a means of avoiding the use of a coaxial connector for direct measurement. The clever technique allowed measurement of the oscillating frequency without direct contact and only required an antenna and spectrum analyzer.

What is generally accepted as the first modern active antenna was developed and published by Thomas et al. on February 1985. It was a Gunn integrated rectangular microstrip patch antenna operating at X-band frequencies [20]. Coincidentally, one of the authors, G. Morris, previously worked with T. S. M. Maclean on active integrated dipoles in the mid-1970s. The active microstrip patch was a compact, inexpensive microwave source. It can be used for Doppler sensing or spatial power-combining applications. It provided a means for external injection locking and stabilization. However, due to package disturbance of the antenna cavity, the radiating performance was very poor, with a +2-dB cross-polarization component. The following year an IMPATT integrated circular microstrip patch was demonstrated by Perkins [21]. The design had a much-improved cross-polarization level (-8 dB) and better overall radiation performance. The circuit design provided an external port for injection locking and stabilization. These first two active antenna designs, shown in Figure 6.11, provided a means of developing very inexpensive rf sources for power combining and many other applications.

6.5 SPATIAL AND QUASI-OPTICAL POWER COMBINING

These developments set the stage for what would become a very competitive and intriguing research area. It was not until an article by James Mink [22] and the subsequent support by the Army Research Office that active antennas for quasi-optical power combining began in earnest. Mink describes the use of an array of currents within a quasi-optical cavity. The large resonator dimensions provide the array of sources with a very-high-Q cavity for stability. The array spacing and number of distributed oscillators determine the efficiency in which the energy couples to the main resonator mode. The technique provides

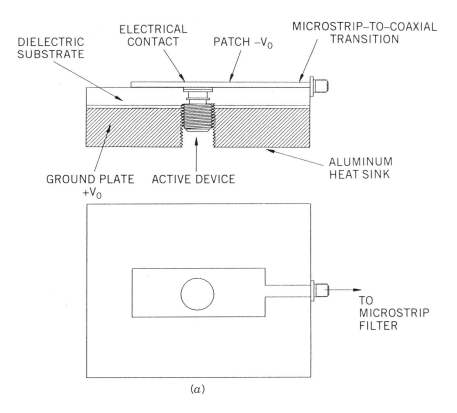

(a)

(*b*)

FIGURE 6.11. (*a*) The active patch under investigation used a conventionally encapsulated Gunn diode and 0.8-mm-thick RT-Duroid 5870 with an ε_r of 2.33 [20]. (*b*) Active microstrip circular patch antenna integrated with an IMPATT diode [21].

a viable alternative for power combining at very high millimeter-wave frequencies. The large resonator dimensions and ease of integration have potential to overcome low solid-state power levels. The technique is not as costly as cavity combiners nor is it as limited by multimoding effects, tight machining tolerances, or conductor losses.

In 1986 Stephan [23] simulated the use of injection-locked oscillators for spatial power combining and even proposed a beam steering method. An X-band experimental model was later built for an injection-locked phased-array system [24]. Shown in Figures 6.12*a* and *b* are block diagrams for a conventional phased array and the proposed method [24]. Figure 6.12*c* shows the actual system implementation. In 1987, Young and Stephan [25] used the open-resonator approach to stabilize the free-running microwave sources. Similarly, Cogan et al. used a quasi-optical cavity for power combining in 1987 [26]. Figure 6.13 shows the resonator implementations.

This was the start of active antenna spatial and quasi-optical power combining. From 1987 to the present, active antenna articles and applications

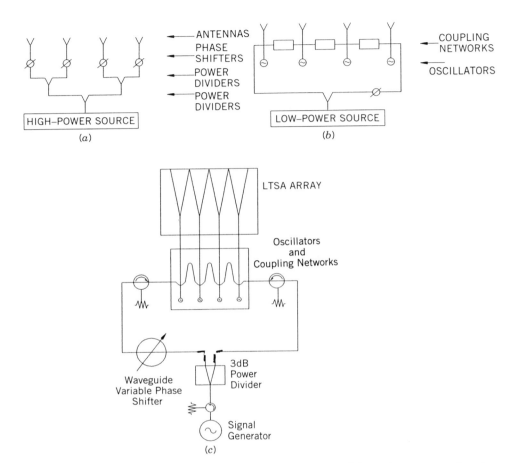

FIGURE 6.12. (*a*) One-dimensional phased array driven by conventional phase control network. (*b*) Phased array driven by interinjection-locked oscillators. (*c*) Block diagram of phased-array system. (From Ref. 24 with permission from IEEE.)

have been growing continually. Although, the concept was originally intended to meet power deficiencies of single solid-state devices at millimeter-wave frequencies, it has since expanded to include many other commercial and military applications. Many different configurations have been shown and various operation frequencies have been demonstrated. It has presented many investigators with a strong and difficult challenge to overcome. It has the potential of making rf and millimeter-wave systems a commodity for everyday uses.

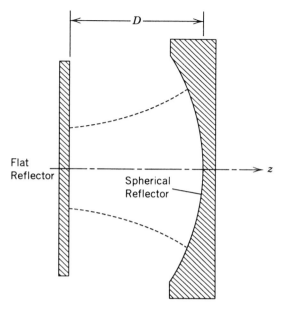

FIGURE 6.13. Quasi-optical cavity.

REFERENCES

1. S. Sanzgiri, W. Pottenger, D. Bostrom, D. Denniston, and R. Q. Lee, "Active Subarray Module Development for Ka-Band Satellite Communication Systems," *IEEE Antennas and Propagation Symposium*, Vol. 2, pp. 860–863, Seattle, Washington, 1994.

2. J. A. Navarro, K. Chang, J. Tolleson, S. Sanzgiri, and R. Q. Lee, "A 29.3 GHz Cavity-Enclosed Aperture-Coupled Circular-Patch Antenna for Microwave Circuit Integration," *IEEE Microwave and Guided Wave Letters*, Vol. 1, No. 7, pp. 170–171, July 1991.

3. D. Staiman, M. E. Breese, and W. T. Patton, "A New Technique for Combining Solid-State Sources," *IEEE Journal of Solid-State Circuits*, Vol. SC-3, No. 3, pp. 238–243, September 1968.

4. M. F. Durkin, R. J. Eckstein, M. D. Mills, M. S. Stringfellow, and R. A. Neidhard, "35 GHz Active Aperture," *IEEE MTT-S International Microwave Symposium*, pp. 425–427, June 1981.

5. R. Hertz, *Electric Waves*, Macmillan, 1983.

6. (a) A. D. Frost, "Parametric Amplifier Antennas," *Proc. IRE*, Vol. 48, p. 1163, June 1960.

ACTIVE ANTENNAS: WORK BEFORE 1987

(b) J. R. Copeland and W. J. Robertson, "Antenna-verters and antennafiers," *Electronics*, pp. 68–71, October 1961.

(c) M. E. Pedinoff, "The Negative Conductance Slot Amplifier," *IRE Transactions on Microwave Theory and Techniques*, Vol. 9, pp. 557–566, November 1961.

(d) W. J. Robertson, J. R. Copeland and R. G. Verstraete, "Antennafier Arrays," *IEEE Trans. on Antennas and Propagation*, Vol. 2, pp. 227–233, March 1964.

(e) K. Fujimoto, "Active Antennas: Tunnel-Diode-Loaded Dipole," *Proceedings of the IEEE*, Vol. 53, p. 174, February 1965.

(f) H. H. Meinke, "Tunnel Diodes Integrated with Microwave Antenna Systems," *Radio and Electronic Engineer*, Vol. 31, No. 2, pp. 76–80, February 1966.

7. H. Meinke and F. M. Landstorfer, "Noise and Bandwidth Limitations with Transistorized Antennas," *IEEE Antennas and Propagation Symposium*, Boston, September 1968.

8. F. M. Landstorfer, "Applications and Limitations of Active Aerials at Microwave Frequencies," *European Microwave Conference*, pp. 141–144, September 1969.

9. E. A. Killick, "Scanning and Active Antennas," *European Microwave Conference*, pp. 122–123, September 1969.

10. P. A. Ramsdale and T. S. M. Maclean, "Active Loop-Dipole Aerials," *Proceedings of the IEE*, Vol. 118, No. 12, pp. 1698–1710, December 1971.

11. T. S. M. Maclean and P. A. Ramsdale, "Short Active Aerials for Transmission," *International Journal of Electronics*, Vol. 36, No. 2, pp. 261–269, February 1974.

12. T. S. M. Maclean and G. Morris, "Short Range Active Transmitting Antenna with Very Large Height Reduction," *IEEE Transactions on Antennas and Propagation*, Vol. 23, No. 3, pp. 286–287, March 1975.

13. A. H. Al-Ani, A. L. Cullen, and J. R. Forrest, "A Phase-Locking Method for Beam Steering in Active Array Antennas," *IEEE Transactions on Microwave Theory and Techniques*, Vol. 22, No. 6, pp. 698–703, June 1974.

14. B. M. Armstrong, R. Brown, F. Rix, and J. A. C. Stewart, "Use of Microstrip Impedance-Measurement Technique in the Design of a BARITT Diplex Doppler Sensor," *IEEE Transactions on Microwave Theory and Techniques*, Vol. 28, No. 12, pp. 1437–1442, December 1980.

15. P. Bhartia and I. J. Bahl, "Frequency Agile Microstrip Antennas," *Microwave Journal*, Vol. 25, No. 10, pp. 67–70, October 1982.

16. K. J. Russell, "Microwave Power Combining Techniques," *IEEE Transactions on Microwave Theory and Techniques*, Vol. MTT-27, No. 5, pp. 472–478, May 1979.

17. K. Chang and C. Sun, "Millimeter-Wave Power-Combining Techniques, *IEEE Transactions on Microwave Theory and Techniques*, Vol. MTT-31, No. 2, pp. 91–107, February 1983.

18. L. Wandinger and V. Nalbandian, "Millimeter-Wave Power Combiner Using Quasi-Optical Techniques," *IEEE Transactions on Microwave Theory and Techniques*, Vol. MTT-31, No. 2, pp. 189–193, February 1983.

19. N. Wang and S. E. Schwarz, "Monolithically Integrated Gunn Oscillator at 35 GHz," *Electronics Letters*, Vol. 20, No. 14, pp. 603–604, 5th July 1984.

20. H. J. Thomas, D. L. Fudge, and G. Morris, "Gunn Source Integrated with Microstrip Patch," *Microwaves and RF*, Vol. 24, No. 2, pp. 87–91, February 1985.

21. T. O. Perkins III, "Active Microstrip Circular Patch Antenna" *Microwave Journal*, Vol. 30, No. 3, pp. 109–117, March 1987.

22. J. W. Mink, "Quasi-Optical Power Combining of Solid-State Millimeter-Wave Sources," *IEEE Transactions on Microwave Theory and Techniques*, Vol. MTT-34, No. 2, pp. 273–279, February 1986.

23. K. D. Stephan, "Inter-Injection-Locked Oscillators for Power Combining and Phased Arrays," *IEEE Transactions on Microwave Theory and Techniques*, Vol. MTT-34, No. 10, pp. 1017–1025, October 1986.

24. W. A. Morgan Jr. and K. D. Stephan, "An X-Band Experimental Model of a Millimeter-Wave Interinjection-Locked Phased Array System," *IEEE Transactions on Antennas and Propagation*, Vol. 36, No. 11, pp. 1641–1645, November 1988.

25. S. L. Young and K. D. Stephan, "Stabilization and Power Combining of Planar Microwave Oscillators with an Open Resonator," *IEEE MTT-S Int. Microwave Conference Digest*, pp. 185–188 (1987).

26. K. J. Cogan, F. C. De Lucia, and J. W. Mink, "Design of a Millimeter-Wave Quasi-Optical Power Combiner for IMPATT Diodes," in *Millimeter Wave Technology IV and Radio Frequency Power Sources*, SPIE, Vol. 791, pp. 77–81, May 1987.

CHAPTER SEVEN

Active Microstrip Patch Antennas and Power Combining

7.1 INTRODUCTION

Over the past decade there have been countless configurations of integrated and active integrated antennas. In fact, the number of publications increased dramatically in the early 1990s. The majority of these designs uses some variation of a microstrip patch antenna. This chapter reviews active microstrip patch antennas, beginning with diode integrations and later transistor integrations. As work in this area matured, publications emphasized the power-combining aspect of the designs over the single element.

7.2 DIODE INTEGRATED ACTIVE MICROSTRIP PATCH ANTENNAS

Many different solid-state devices have been integrated with antennas to provide several functions. By far, the most popular antenna for integration has been the microstrip patch antenna. When integrated with active solid-state devices, microstrip patch antennas can generate rf power, amplify an incoming or outgoing signal, and serve as multipliers, frequency translators, or mixers. Single active antennas, when used as self-oscillating mixers, make ideal, compact, low-cost Doppler sensors. In the late 1970s, Kwok and Weller showed the use of BARITT diodes for Doppler sensing [1], which led Armstrong et al. to an active BARITT integrated microstrip patch antenna Doppler sensor in 1980 [2]. Figure 7.1 shows that the diode was integrated on the same substrate using a patch antenna for radiation and an open microstrip stub for impedance matching. However, the use of the BARITT diode was

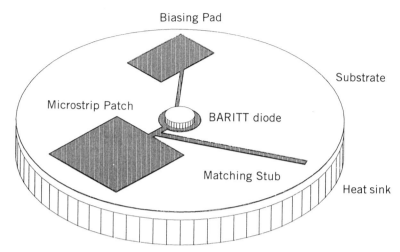

FIGURE 7.1. BARITT-diode microstrip Doppler sensor. (From Ref. 2 with permission from IEEE.)

short-lived, and this type of integrated antenna approach was filed away for several years.

The need for inexpensive rf sources for quasi-optical power combiners brought the active integrated antenna concept back in the mid-1980s. G. Morris, who contributed to the active aerials of the early 1970s (see Ref. 12 of Chap. 6), demonstrated one of the first active microstrip patch antennas in 1984 [3]. The design consists of a Gunn diode and a microstrip antenna. The antenna serves as a resonator and load for the radiating oscillator. The active device is biased to oscillate at 10.891 GHz [4]. This hybrid integration, shown in Figure 7.2a, unlike the BARITT Doppler sensor, does not use any external rf matching networks. The diode can be variously positioned along the antenna, and thus the device can be matched for efficient rf power generation. A high-impedance line at a low-impedance point on the patch is used to provide dc bias to the active device. To find the best point of insertion for the active device, one must determine the fields within the patch resonator.

For a microstrip patch antenna with length L and width W as shown in Figure 7.2, the fields under the patch antenna are given by

$$E_z = \frac{V_0}{h} \cos\left(\frac{m\pi}{W} x\right) \cos\left(\frac{n\pi}{L} y\right)$$

$$H_x = -\frac{j\omega\varepsilon_0}{k_0^2} \frac{n\pi}{L} \frac{V_0}{h} \cos\left(\frac{m\pi}{W} x\right) \sin\left(\frac{n\pi}{L} y\right) \qquad (7.1)$$

$$H_y = \frac{j\omega\varepsilon_0}{k_0^2} \frac{m\pi}{W} \frac{V_0}{h} \sin\left(\frac{m\pi}{W} x\right) \cos\left(\frac{n\pi}{L} y\right)$$

where V_0 is the voltage at a patch corner, h is the height of patch substrate over the ground plane, and m and n depict the mode of operation. The resonant frequency for the lowest-order mode ($m = 0$, $n = 1$) is

$$f_0 = \frac{1}{2\sqrt{\mu\varepsilon}} \sqrt{\left(\frac{m}{W}\right)^2 + \left(\frac{n}{L}\right)^2} = \frac{c}{2\sqrt{\varepsilon_r}} \frac{1}{L} \qquad (7.2)$$

where c is the speed of light and ε_r is the relative dielectric constant of the substrate material.

For the dominant mode the real part of the input impedance is a minimum at the center of the patch (i.e., $L/2$) and is a maximum near the end of the patch. The impedances at different points along the patch are needed to match the active device and ensure that oscillations are sustained.

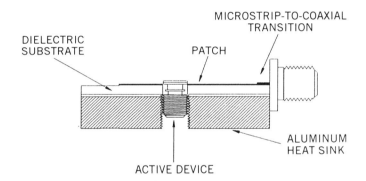

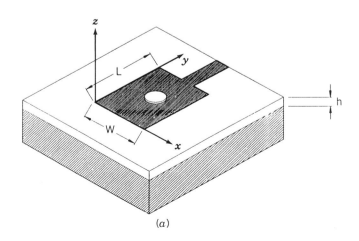

(a)

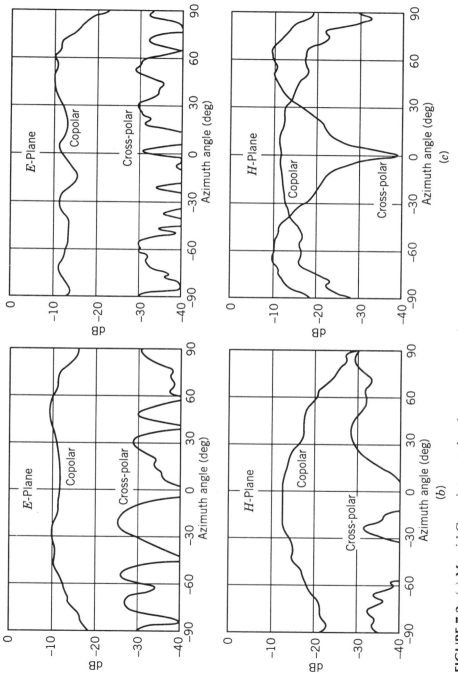

FIGURE 7.2. (*a*) Morris' Gunn integrated active antenna configuration. (*b*) Passive antenna radiation patterns. (*c*) Active antenna radiation patterns. (From Ref. 4 with permission from *Microwave and RF.*)

The real part of the input impedance seen at a position y is

$$R_{in} = R_0 \left(\cos\left(\frac{\pi}{L} y\right) \right)^2 \tag{7.3}$$

where R_0 is the radiation resistance at the edge of the patch. If we neglect discontinuities and perturbations, the solid-state device can be matched by solving Equation (7.3) for the device position:

$$y = \frac{L}{\pi} \cos^{-1}(\sqrt{R_d(G_r - G_m)}) \tag{7.4}$$

where R_d is the real part of the device impedance $(-R_d + jX_d)$, G_r is the patch radiation conductance, and G_m is the mutual conductance of the two radiating edges. The active antenna integration will begin to oscillate if R_d is approximately 20% greater than the input resistance (R_{in}) seen by the device at the frequency where the reactances cancel. The conditions for oscillation are repeated here for convenience:

$$R_d \geqslant R_{in}, \qquad X_d = -X_{in} \tag{7.5}$$

A packaged device is typically large with respect to the antenna, which will disturb the fields within the antenna. If the device could be integrated without disturbing the antenna (i.e., unpackaged, monolithically), the far-field patterns of the patch radiator would be closer to ideal. The far fields for an undisturbed patch antenna are approximately given by

$$E(\theta, \phi) =$$

$$\left| \cos(k_0\sqrt{\varepsilon_r} h \cos(\theta)) \; \frac{\sin\left(\dfrac{k_0 W}{2} \sin(\theta)\sin(\phi)\right)}{\dfrac{k_0 W}{2} \sin(\theta)\sin(\phi)} \; \cos\left(\frac{k_0 L}{2} \sin(\theta)\cos(\phi)\right) \cos(\phi) \right| \tag{7.6a}$$

$$H(\theta, \phi) = \left| \cos(k_0\sqrt{\varepsilon_r} h \cos(\theta)) \; \frac{\sin\left(\dfrac{k_0 W}{2} \sin(\theta)\sin(\phi)\right)}{\dfrac{k_0 W}{2} \sin(\theta)\sin(\phi)} \right.$$

$$\left. \times \cos\left(\frac{k_0 L}{2} \sin(\theta)\cos(\theta)\right) \cos(\theta) \sin(\phi) \right| \tag{7.6b}$$

where E and H refer to the elevation and horizontal planes for a linearly polarized microstrip patch antenna etched on a substrate with height h and dielectric constant ε_r.

The insertion of solid-state devices does, however, disturb fields which exist within the antenna cavity and currents which flow along the surfaces. These disturbances often cause large deviations from the calculated radiation pattern. Active antennas often exhibit irregular patterns and high cross-polarization levels (CPL), and the need for biasing lines further degrades the radiation performance. As shown in Figure 7.2a, Morris' active patch antenna also includes an rf input line for external injection locking or synchronization. Figure 7.2b shows measurements of a passive microstrip patch antenna, while Figure 7.2c shows the Gunn integrated patch antenna patterns. As expected, the relative size of the diode package to the antenna causes a very high H-plane cross-polarization component ($+2\,\mathrm{dB}$ higher than co-polarization component).

In 1986, Dydik [5] proposed an IMPATT diode integrated on a circular patch antenna. Similar to the rectangular patch antenna, the fields within a circular patch (Fig. 7.3) on a thin substrate can be expressed as

$$
E_z = E_0 J_1(k\rho) \cos(\phi)
$$

$$
H_\phi = -\frac{jk}{\omega\mu_0} E_0 J'_1(k\rho) \cos(\phi)
\tag{7.7}
$$

where ρ and ϕ are the radial and angular coordinates for the circular configuration. J_1 is the Bessel function of order 1 and the prime sign signifies differentiation with respect to the argument.

The diameter D for a particular frequency of operation is determined for the fundamental mode from the equation

$$
D = \frac{1.8411c}{\pi f_0 \sqrt{\varepsilon_r}}
\tag{7.8}
$$

where f_0 is the fundamental operating frequency and c is the speed of light on a substrate with dielectric constant ε_r.

Similar to the rectangular configuration, the real part of the input impedance is needed to match an active device. This is given by

$$
R_{\mathrm{in}} = R_0 \left[\frac{J_1^2(k\rho)}{J_1^2(kD/2)} \right]
\tag{7.9}
$$

where ρ is the device position on a patch of diameter D, and R_0 is the radiation resistance at $\rho = D/2$.

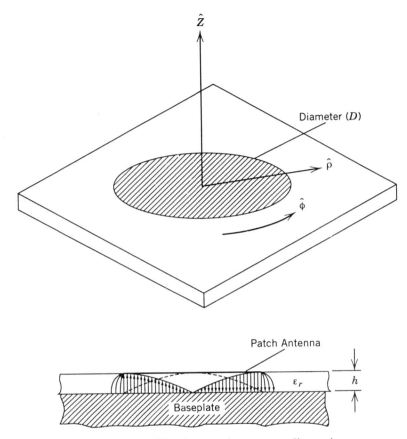

FIGURE 7.3. Circular patch antenna dimensions.

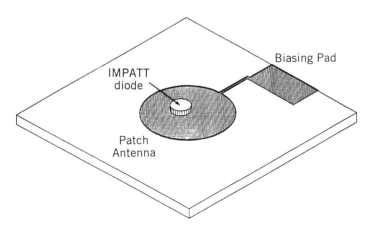

FIGURE 7.4. IMPATT integrated circular patch antenna.

The IMPATT integrated circular patch configuration was demonstrated by Perkins in 1986 [6, 7]. Figure 7.4 shows Perkins' active circular patch antenna, which operates at 6.8 GHz with an effective radiated power of 41.5 dBm. The radiation patterns show that the principal plane patterns are smooth and symmetric and the cross-polarization component is 8 dB below the maximum of the co-polarization. To date, this early diode integration remains a standard for other modified configurations. The lower operating frequency and a better biasing network improve this active antenna's radiation performance.

In 1986, Stephan envisioned a spatial power combiner which could not only overcome low power levels of solid-state devices but also potentially provide

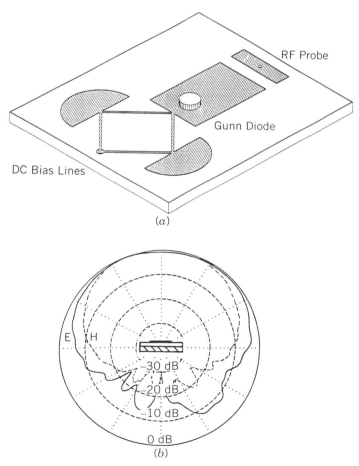

FIGURE 7.5. Gunn integrated microstrip patch antenna: (*a*) configuration; (*b*) radiation patterns. (From Ref. 9 with permission from SPIE.)

beam steering [8]. Although the original work used VHF transistor oscillators, he later investigated the concept with active microstrip antennas, as shown in Figure 7.5a [9]. The active microstrip patch antenna uses a symmetrical dc bias low-pass filter and a probe-fed patch edge coupled section. The configuration exhibits smooth principal plane patterns as shown in Figure 7.5b, but no cross-polarization levels are mentioned. The probe feed can be used to externally tune, monitor, or injection-lock the active antenna.

In 1986, the emphasis shifted from single active antennas to arrays of distributed sources for power combining [10]. Stephan used an X-band model to demonstrate the distributed oscillator concept for power combining and beam steering [11]. His theoretical predictions concentrated on the phase dynamics of an array of distributed sources. Other investigations which followed attempted to improve the active integrated antenna in size, conversion efficiency, oscillator quality, and radiation performance. Since individual active antennas cannot meet many system-level power requirements, a large number are needed. Also, since active devices are the cost drivers in current designs, it is essential to mass produce active antennas at extremely low costs. Given a low-cost, reliable active antenna, large arrays of distributed radiating oscillators can be developed to provide high power with electronic beam steering. The active array system will provide graceful degradation during device failures and will avoid complex, lossy rf distribution networks. The high-risk areas of the array will be in heat dissipation, stability, integration, and dc control.

An obvious path toward a compact, inexpensive design (in mass production) is to use a monolithic active antenna. Typically, the high-dielectric-constant monolithic substrates using GaAs are ideal for devices, but they reduce the radiation efficiency of the antenna. However, devices are directly fabricated with the antenna without package parasitics or integration discontinuities. Such a monolithic active antenna chip would be ideal for developing a large array of distributed sources. One such integration, shown in Figure 7.6a, uses two IMPATT diodes which feed the edges of a microstrip patch antenna and have a dc biasing network as well as an open stub for frequency trimming. The monolithic active antenna operates at 43.3 GHz with an output power of 27 mW at 7.2% dc-to-rf conversion efficiency [12, 13]. Figure 7.6b shows very good principal plane patterns, but no cross-polarization levels are mentioned. Half-power beamwidths in the H-plane are approximately 80° and over 120° in the E-plane. Figure 7.7 shows the output power and conversion efficiency versus biasing current of the active antenna at 43.3 GHz. This early monolithic integration power combines two and three elements in the H-plane. Injection locking is ensured by using 50-Ω microstrip lines between the active antennas. Figure 7.8a demonstrates in-phase and out-of-phase coupling (i.e., phasing between the two active antenna elements in an array), while Figure 7.8b shows H-plane beam sharpening versus number of elements in the spatial power-combining arrays.

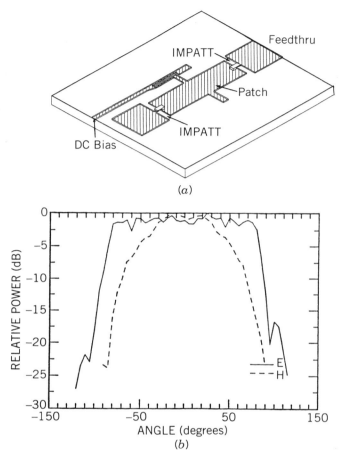

FIGURE 7.6. Dual-IMPATT integrated microstrip patch antenna. (From Ref. 13 with permission from IEEE.)

The IMPATT integrated monolithic configuration is ideally suited for coupling to a rectangular waveguide or for a large planar array. The configuration was used by Shillue et al. in a spatial array within a quasi-optical resonator in 1989 [14]. A single antenna was stabilized by the resonator at 56.097 GHz, improving the spectral purity of the signal. Similarly, the quasi-optical resonator, as described by Mink, can synchronize a large array of the individual sources to operate at a single frequency. If the monolithic active antenna can provide sufficient yield, a large active array aperture can be cost effective.

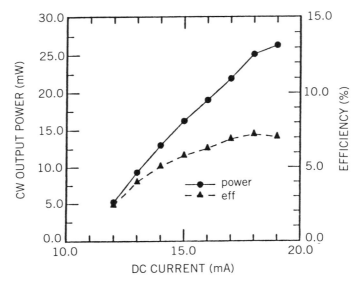

FIGURE 7.7. Monolithic IMPATT power output and efficiency vs. bias current. (From Ref. 13 with permission from IEEE.)

Gunn integrated microstrip patch antennas [15] have demonstrated good power-combining efficiency and fairly good radiation patterns. Early in 1988, Hummer and Chang investigated Gunn integrated microstrip antennas for spatial power combining [16, 17]. Both the single patch antenna and the two-element array were constructed on a Duroid 5870 substrate 1.524 mm thick. The circuit configurations are shown in Figures 7.9a and 7.11a.

The antenna dimensions were determined by the equation given by James et al. [18]. The antenna length was chosen to be

$$L = \frac{\lambda_g}{2} - 2\Delta l_{e0} \tag{7.10}$$

and the antenna width is

$$W = 0.3\lambda_0 \tag{7.11}$$

where

λ_g = guided wavelength

Δl_{e0} = equivalent length to account for open-end fringing capacitance

λ_0 = free-space wavelength

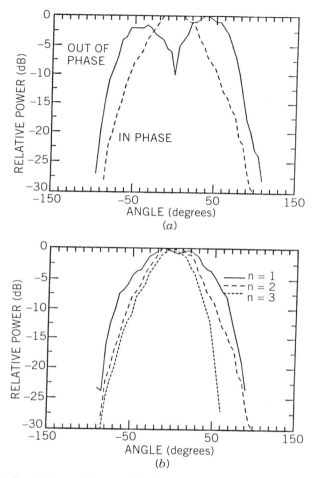

FIGURE 7.8. Spatial combiner radiation patterns. (*a*) *H*-plane patterns of two-element array injection-locked inphase and out of phase. (*b*) *H*-plane patterns of one-, two-, and three-element array patterns injection locked inphase. (From Ref. 13 with permission from IEEE.)

The placement of the active device was chosen such that the device impedance was matched to the input impedance of the patch. The diode placement location D is given by

$$D = \frac{L}{\pi} \cos^{-1}\left[Z_{\text{in}} G_r\left(1 - \frac{G_m}{G_r}\right)\right]^{1/2} \tag{7.12}$$

where

G_r = radiation conductance
G_m = mutual conductance of the two edges of the antenna
D = distance from either antenna edge to the feed position
Z_{in} = input impedance to the antenna at the diode location

For a rectangular patch of $W = 0.3\lambda_0$, G_r and G_m can be found from James et al. [18]:

$$G_r = W^2/90\lambda_0^2 \tag{7.13}$$

$$G_m/G_r = 0.32 \tag{7.14}$$

For impedance matching, the input impedance Z_{in} is set equal to the active device resistance, which is assumed to be 8 ohms. A single active antenna shown in Fig. 7.9a exhibited 15 mW at 10.1 GHz. A 3-dB tuning range of 839 MHz was achieved from 9.278 to 10.117 GHz (i.e., 9%). The power and frequency response versus dc biasing voltage for a single active antenna are shown in Figure 7.9b along with E-plane pattern dependency for various voltages in Figure 7.9c. As shown, the HPBW in the E-plane of 90° remains constant for various voltages (i.e., over the frequency tuning range). A free-running and an injection-locked signal at 9.8 GHz is shown in Figures 7.10a and b. Figure 7.10c shows a plot of injection-locking bandwidth versus locking gain, which can be used to determine the active antenna's external Q-factor through the relation

$$Q_{ext} = \frac{f_0}{\Delta f} \sqrt{\frac{P_{inj}}{P_o}} \tag{7.15}$$

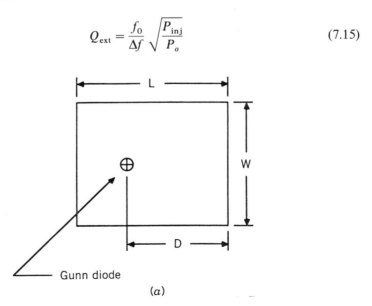

(a)

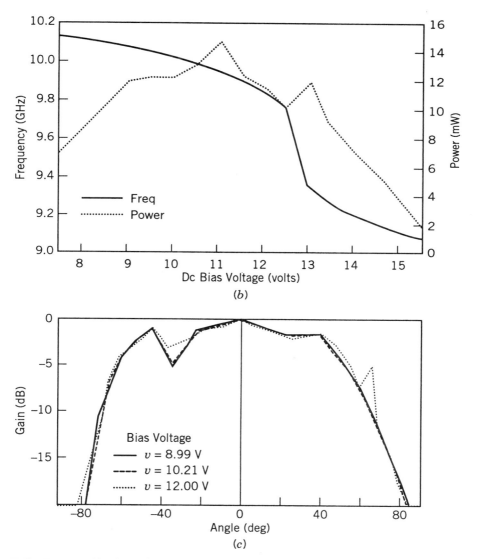

FIGURE 7.9. Single active antenna configuration and performance. (*a*) Single active patch antenna. (*b*) output power as a function of frequency and bias voltage for a single active antenna. (*c*) *E*-plane antenna patterns for several different bias voltage levels for a single active antenna. (From Ref. 17 with permission from IEEE.)

where Δf is the single-side locking bandwidth for an injected signal of power P_{inj}, and the active antenna oscillates at f_0 with an output power of P_o. The active antenna was externally injection-locked with a stable source connected to a standard-gain horn antenna. Through the Friis transmission equation both the injected power and the oscillator power can be calculated (see Chapter 5).

A two-element E-plane power combiner injection-locked via mutual coupling exhibited relatively smooth patterns, as shown in Figure 7.11b. The combiner's output power was 30 mW at 10.42 GHz, which was twice that seen for a single patch [19]. The combining power and frequency versus biasing voltage showed a mode jump at a bias voltage of 15.45 V. The jump causes an out-of-phase power combination between the two elements, verifying results by Camierelly and Bayraktarouglu. Figure 7.11c shows in-phase and out-of-phase combiner patterns as well as performance. This phase inversion would later be exploited to steer a spatial power combiner's beam merely by altering the individual oscillator frequencies.

A Gunn integrated active patch antenna was used by York, Compton, and others to investigate synchronization of weakly coupled distributed oscillators [20, 21, 22]. The active antenna used is shown in Figure 7.12 along with its radiation patterns at 7.6 GHz. A 4×4 array of Gunn integrated microstrip patches shown in Figure 7.13a exhibits a 22-W EIRP with nearly 1% efficiency. Figure 7.13b shows the E-plane patterns and the H-plane patterns of the 16-element power combiner at 9.6 GHz. There is relatively good agreement with calculated results. Possible discrepancies include variations in individual element amplitude and phase. The elements were synchronized through an external partially reflecting dielectric surface. The dielectric layer in front of the array ensures stable injection locking for all elements in the array.

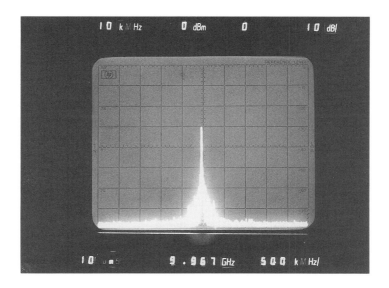

(a)

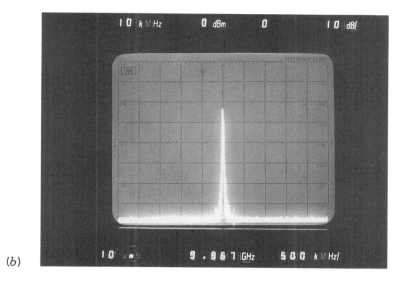

(b)

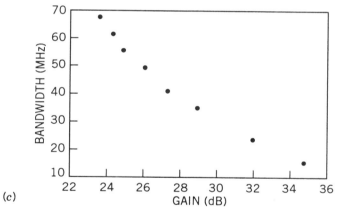

(c)

FIGURE 7.10. Single active antenna injection locking performance: (*a*) free running (horizontal, 500 kHz/div, vertical, 10 dB/div); (*b*) injection locked (horizontal, 500 kHz/div, vertical, 10 dB/div); (*c*) injection-locking bandwidth as a function of locking gain. (From Ref. 17 with permission from IEEE.)

As shown in Figure 7.12, the individual active antennas exhibit high cross-polarization levels, which lower an active antenna's radiation efficiency and usefulness. York and Compton use a dual Gunn integrated antenna (as shown Fig. 7.14) to provide more power and lower cross-polarization [22]. For a patch designed to resonate at 10 GHz, a Gunn integrated active antenna operated at 10.4 GHz with an output power of 26 mW. Placing a second Gunn

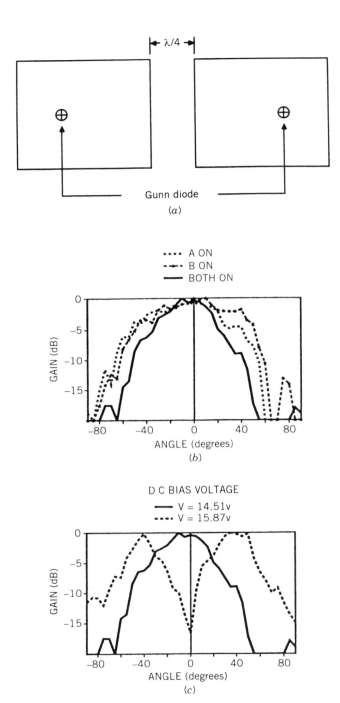

FIGURE 7.11. Two-element combiner configurating and performance: (*a*) two-element array; (*b*) measured *E*-plane power pattern for a two-element array; (*c*) pattern broken up above 15.45 V. (From Ref. 19 with permission from IEEE.)

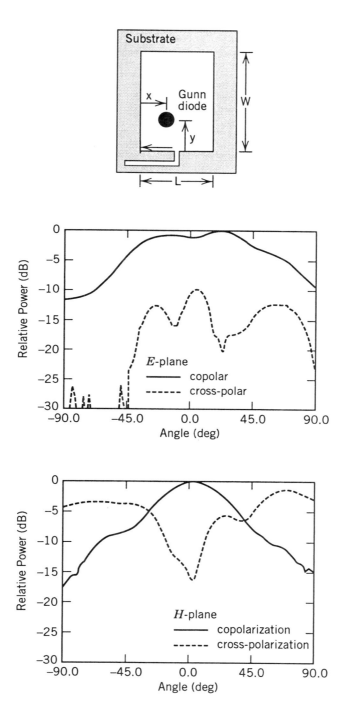

FIGURE 7.12. Gunn integrated active antenna patterns. (From Ref. 21 with permission from IEEE.)

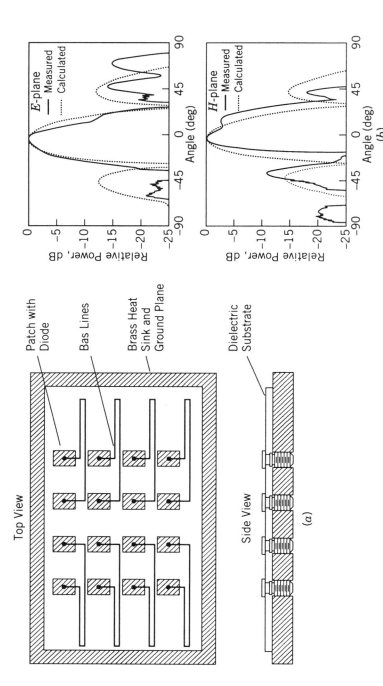

FIGURE 7.13. Gunn integrated 4×4 spatial combiner: (*a*) 4×4 array configuration. (*b*) *E*-plane and *H*-plane patterns at 9.6 GHz for the active array. The theoretical results are calculated by combining the pattern of a single patch with a 4×4 array factor. The dielectric slab above the array ($\varepsilon_r = 4$) has a small effect on the patterns. Good qualitative agreement between the measured and calculated curves indicates that the elements are nearly in phase with similar amplitudes. (From Ref. 21 with permission from IEEE.)

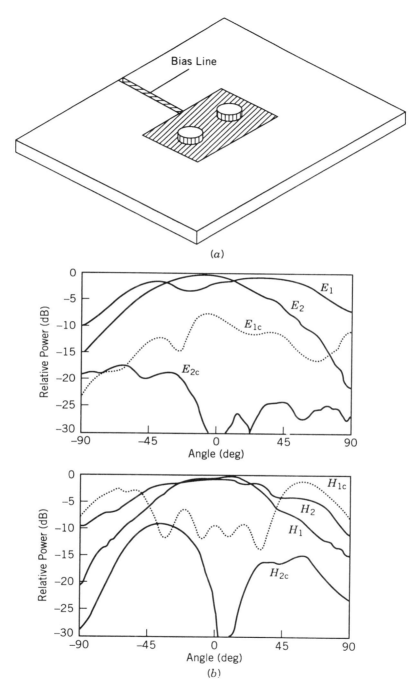

FIGURE 7.14. Dual Gunn integrated microstrip patch. (*a*) configuration; (*b*) radiation patterns. E_1, E_{1c}, H_1, H_{1c}: single Gunn, $F_0 = 10.4$ GHz; E_2, E_{2c}, H_2, H_{2c}: dual Gunn, $F_0 = 10.7$ GHz. (From Ref. 22 with permission from *Electronics Letters.*)

diode diametrically opposite of the first, as shown in Figure 7.14*a*, nearly doubles the power at an operating frequency of 10.7 GHz. The configuration's symmetry helps to improve the overall radiation pattern and lower the CPL from 2 to 10 dB below the maximum. Figure 7.14*b* shows the radiation performance for single and dual Gunn integrated antennas. As shown, there is a clear improvement over a single Gunn integrated antenna. *E*-plane and *H*-plane cross-polarization levels are greatly reduced, while the *E*-plane is sharpened and the *H*-plane is broadened. The approach also demonstrates the ability to combine chip, circuit, and spatial power combining simultaneously in a single patch. Another method used in improving an active antenna's radiation performance is to use the concept of aperture coupling.

Although aperture coupling removes active devices away from the antenna, it solves many of the radiation problems which occur due to device integration

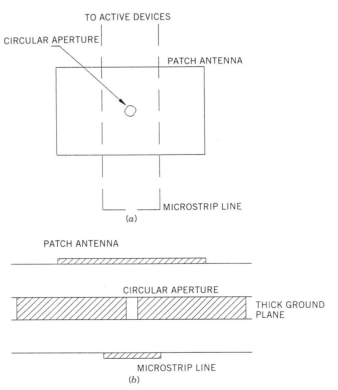

FIGURE 7.15. Aperture-coupled microstrip to patch antenna circuit for active array. (*a*) Top view. (*b*) Cross-sectional view. (From Ref. 23 with permission from IEEE.)

and dc biasing. The use of an aperture-coupled patch for hybrid and mono-lithic circuit integration was proposed by Gao and Chang in 1988 [23]. As shown in Figures 7.15*a* and *b*, aperture coupling solves many radiation problems associated with an active integrated antenna. Since the circuit is isolated from the antenna by a metallic ground plane and coupling occurs through an aperture, spurious feed line and device radiation are minimized. Active devices are biased on the backside away from the antenna, and the two substrates can be optimized separately. Unlike the original aperture-coupled patch antennas introduced by Pozar in 1985 [24], this active configuration uses a circular aperture on a thick metal support layer. The thick metal layer increases heat-sinking volume and provides structural support. Figures 7.16*a* and *b* show the *H*- and *E*-plane patterns of two-element *E*-plane array of

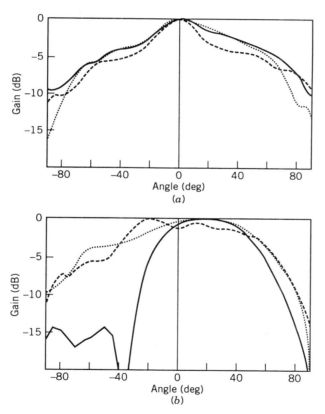

FIGURE 7.16. Aperture-coupled spatial power combiner: (*a*) *H*-plane pattern of a two-element aperture-coupled active array; (*b*) *E*-plane power pattern of a two-element aperture-coupled active array. (– – –) A on, (- - -) B on, (———) both A and B on. (From Ref. 19 with permission from IEEE.)

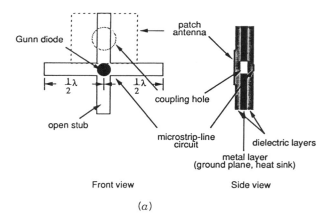

Front view   Side view

(a)

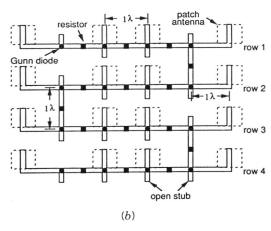

(b)

FIGURE 7.17. Aperture-coupled power combiner. *(a)* Single oscillator unit of two-dimensional array in multilayer structure. *(b)* Circuit layout of the 4×4 array in multilayer structure. (From Ref. 25 with permission from IEEE.)

aperture-coupled active antennas [19]. As expected, patterns are smooth and symmetric, and a 25° beam shift along the *E*-plane demonstrates the ability to steer the beam as proposed by Stephan.

The concept of aperture-coupled antennas for power combiners has recently been shown by Lin and Itoh in a 4×4 array (Fig. 7.17) [25]. The array uses strongly coupled Gunn oscillators on one substrate aperture-coupled to patch antennas through a circular hole in a thick metal ground plate. Interconnecting lines between the Gunn oscillators ensure strong coupling for injection locking,

TABLE 7.1. Array Data as Given in Reference 25

	Columns								Linear Array		Planar Arrays			
	1		2		3		4		1 × 4		2 × 4		4 × 4	
Rows	GHz	dBm	GHz	dBm	GHz	dBm	GHz	dBm	GHz	dBm	GHz	dBm	GHz	dBm
1	11.813	2.5	11.819	1.5	11.767	1.5	11.802	1.5	11.834	12.5				
2	11.778	−0.5	11.789	5.5	11.790	5.5	11.803	−4.5	11.826	11.5	11.784	16.5		
3	11.816	−4.5	11.800	5.5	11.785	3.5	11.791	−4.5	11.815	9.5				
4	11.791	1.5	11.795	3.5	11.825	1.5	11.824	1.5	11.850	11.5	11.724	16.5	11.750	21.5

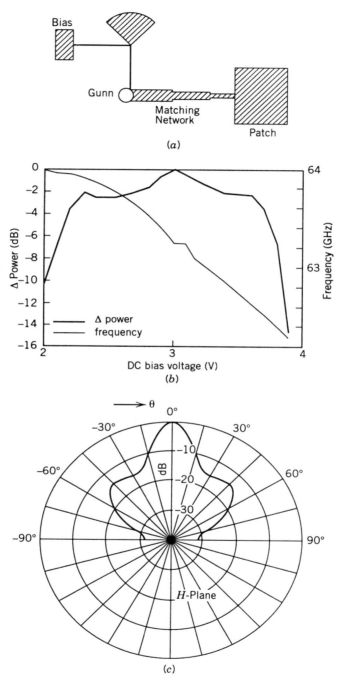

FIGURE 7.18. 60-GHz Gunn integrated microstrip patch: (*a*) configuration; (*b*) power and frequency vs. voltage; (*c*) radiation pattern. (From Ref. 31 with permission from *Electronics Letters.*)

while chip resistors suppress undesired modes. Individual oscillators provide an average of 1 dBm of power at 11.80 ± 0.05 GHz. Table 7.1 lists the frequency and power outputs of the individual oscillators, 1×4, 2×4 and 4×4 arrays. The first four data columns list the individual measured oscillating frequencies and output powers. The column labeled 1×4 array depicts power combining between four antennas for each row of the planar array. Similarly, the 2×4 column lists the output frequency and power for an 8-element power combiner. Finally, the entire array is used as a power combiner with an output power of 21.5 dBm and an operating frequency of 11.75 GHz.

The table shows that the EIRP is proportional to the square of the number of elements. The E-plane pattern of the array is as predicted, but the H-plane nulls are not as expected. A single element is shown in Figure 7.17a. Figure 7.17b shows the spatial power-combining array. The configuration moves active devices away from the antenna, allowing separate optimization of the circuit and the antenna. Other investigators have demonstrated this concept with transistors [26, 27] and monolithic chips [28, 29].

A V-band power combiner using 60-GHz pulsed IMPATT oscillators was shown by Davidson et al. [30]. Due to the high operating frequency, the package and antenna combine to affect the frequency of oscillations. Eight of these active antennas combined to produce over 2 W of CW power. For the 61 W of dc power required to drive the array, the conversion efficiency is approximately 3.3%. Pulsing was accomplished using a 2-μs, low-duty-cycle, 4-kHz bias.

An active V-band antenna implementation is shown in Figure 7.18a along with bias-tuning characteristics and radiation patterns in Figures 7.18b and c, respectively [31]. It uses a packaged Gunn diode connected to a patch antenna through a microstrip transformer. The Gunn is biased to provide 9.26 mW at 63.24 GHz. It has over 1 GHz of bias-tuning bandwidth ($\sim 1.6\%$) without mode jumps. Although the E-plane patterns or cross-polarization levels are not reported, the H-plane pattern seems overly directive (30° for 10-dB beamwidth).

Although diodes have shown higher operating frequencies and higher output power levels, most investigators have also used transistors in their spatial power-combining schemes. Transistors are lower priced, monolithically compatible, and provide higher dc-to-rf conversion efficiencies at lower operating voltages. Smaller operating currents translate to less heat-sinking requirements. Furthermore, transistors perform a variety of functions using a single technology, thereby allowing greater flexibility in a multifunction design. As described earlier, the original transistor integrated aerials and loops occurred in the mid-1960s and 1970s (see Refs. 6–11 in Chap. 6). For whatever reason, the approach did not stir much interest and it remained a laboratory curiosity for over a decade.

7.3   TRANSISTOR INTEGRATED ACTIVE MICROSTRIP PATCH ANTENNAS

The integration of transistors has been essential for the development of active antennas. The use of three-terminal devices has brought on a wide range of integrated antennas for oscillators, amplifiers, multipliers, and other components.

The FET integrated microstrip patch antenna was first demonstrated by Chang et al. [32] in 1988. Figure 7.19a shows that the patch antenna serves as a feedback element for the FET oscillator circuit and a radiator. The C-band radiating oscillator circuit operates at 5.7 GHz with a power output of 17 mW. Figure 7.19b shows the principal plane patterns of the FET integrated active antenna. The 3-dB beamwidth is 56° in the E-plane and 32° in the H-plane with high cross-polarization levels. The erratic radiation patterns are due to exposed microstrip lines, device interconnections, impedance mismatches, and bias lines. A similar configuration was analyzed by Fusco and Burns in 1990 [33]. The FET integration operated at 10.2 GHz with output power of 30 mW. Figures 7.20a and b show the configuration and schematic for the feedback active patch antenna oscillator. The procedure to synthesize the active antenna design requires the following relationships:

$$B_2 = \frac{C_1 + C_3}{V_{gi}} \tag{7.16a}$$

$$B_1 = C_2 + C_4 + V_{gr}B_2 \tag{7.16b}$$

$$B_3 = \frac{V_{gi}C_1 + (1 - V_{gr})[-C_4 - V_{gr}B_2]}{(1 - V_{gr})^2 + V_{gi}^2} \tag{7.16c}$$

$$G_3 = \frac{C_1 - V_{gi}B_3}{1 - V_{gr}} \tag{7.16d}$$

$$C_1 = \mathrm{Re}\left(\frac{I_1}{V_1}\right), \quad C_2 = \mathrm{Im}\left(\frac{I_1}{V_1}\right), \quad C_3 = -\mathrm{Re}\left(\frac{I_2}{V_1}\right), \quad C_4 = -\mathrm{Im}\left(\frac{I_2}{V_1}\right),$$

$$V_g = \frac{V_2}{V_1} = V_{gr} + jV_{gi} \tag{7.16e}$$

where stubs are used to realize B_1 and B_2 and G_3 and B_3 are realized by locating on the patch the position where G_3 is satisfied, using [32]

$$Y_{in}(z) = G_3 + jB_3 = 2G/\cos^2(\beta z) \tag{7.17}$$

where G is the radiating conductance, B is the propagation constant, and z is the location of the feed.

Some transistor integrated antenna designs have been introduced by Birkeland and Itoh [34, 35]. Figure 7.21a shows a linear microstrip array active

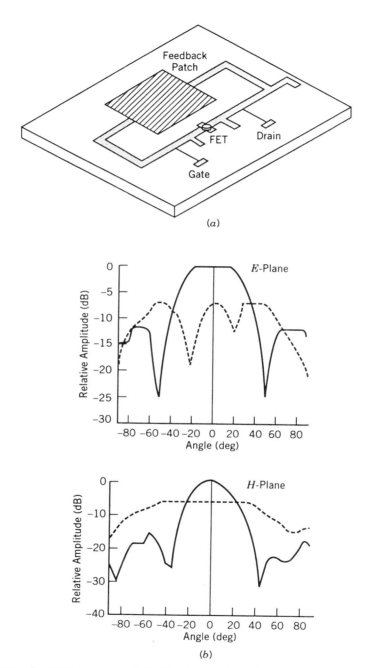

FIGURE 7.19. FET integrated feedback patch active antenna: (*a*) configuration; (*b*) radiation patterns. (From Ref. 32 with permission from *Electronics Letters.*)

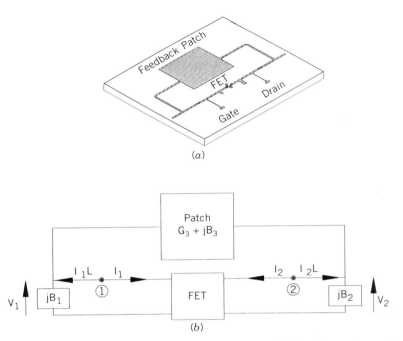

FIGURE 7.20. FET integrated active patch antenna. (a) Configuration and (b) schematic diagram. (From Ref. 33 with permission from *Electronic Letters.*)

antenna, and Figure 7.21b shows the transceiver configuration. The active antenna provides frequency tuning from 9.50 to 9.54 GHz through a 1.5-V change in the gate voltage (V_{gs}). Figures 7.22a and b show the active antenna array frequency and power performance and radiation patterns, respectively. The antenna uses a 17-antenna series-fed microstrip array to provide a very good fan beam active antenna.

Birkeland and Itoh introduced an approach to edge-couple FETs to microstrip patches in 1990 [36]. The integrations operate at 6 GHz with typical EIRPs of 19 dBm. Two and four FETs were integrated with a single patch antenna as shown in Figure 7.23. Two-, three-, and four-element E-plane arrays were demonstrated with EIRPs of 22.9, 28.9, and 31.7 dBm at 10 GHz, respectively. This compact integrated active patch antenna appears ideal for the power combiner, which includes chip, circuit, and spatial power combining. Figures 7.24a and b show the configuration and radiation patterns of the three-element spatial power combiner. Asymmetry in the combiner patterns is due to phase and power deviations of individual active antennas in the array. A combining efficiency of 98% with EIRP of 28.9 dBm was achieved.

Fusco also investigated an edge-fed transistor integrated active antenna as a more compact design over the feedback amplifier design in 1992 [37]. The

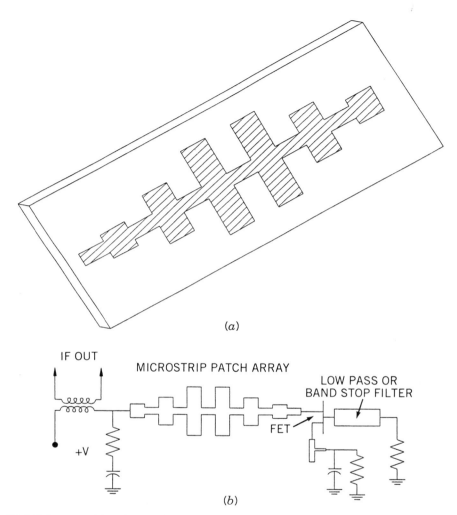

FIGURE 7.21. (*a*) Microstrip patch linear array. (*b*) Schematic view of single-device transceiver circuit. (From Ref. 35 with permission from IEEE.)

series feedback active antenna design uses the edge-fed microstrip patch antenna. The schematic of Figure 7.25*b* shows the gate load (jX_1), source load (jX_2), and drain load $G_3 + jX_3$ definitions for the configuration in Figure 7.25*a*. The values required to synthesize the design are

$$X_1 = \frac{-\operatorname{Re}(V_1(I_1 + I_2))}{\operatorname{Im}(I_1(I_1 + I_2)^*)} \tag{7.18a}$$

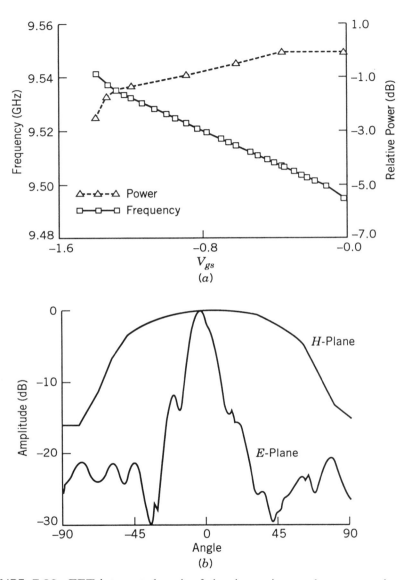

FIGURE 7.22. FET-integrated series-fed microstrip patch array active antenna. (*a*) Oscillation frequency and relative output power for the single-device oscillator. (*b*) Radiation patterns for single-device oscillator with 17-element antenna. (From Ref. 35 with permission from IEEE.)

$$X_2 = \frac{\text{Re}(V_1 I_1)}{\text{Im}((I_1 + I_2)I_1^*)} \qquad (7.18b)$$

$$X_3 = \frac{-\text{Im}(V_2 I_2^*)}{|I_2|^2} - \frac{X_2\text{Re}((I_1 + I_2)I_2^*)}{|I_2|^2} \qquad (7.18c)$$

$$G_3 = \left[-jX_3 - \frac{V_2}{I_2} - jX_2\frac{I_1 + I_2}{I_2} \right]^{-1} \qquad (7.18d)$$

where I_1, I_2, V_1, V_2, are defined in Figure 7.25b. Figure 7.25c shows the frequency and power performance versus biasing voltage of the active antenna.

The FET active antenna provides 8 mW at 9.625 GHz (10% dc-to-rf conversion efficiency) with 90 MHz of bias tuning. The radiation performance of the configuration is affected by the patch as well as the open stub used for matching, which causes some asymmetry [38]. Fusco has investigated the FET integrated patch antenna for Doppler sensing [39]. The FET self-oscillating mixer provides an efficient, compact, low-cost alternative for inexpensive proximity detectors.

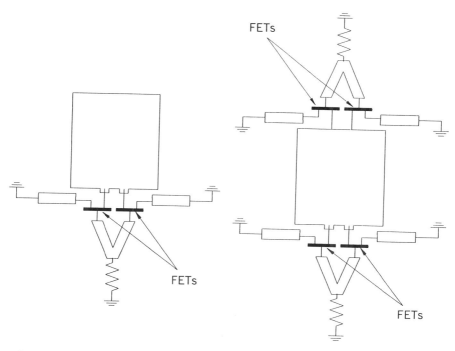

FIGURE 7.23. Two- and four-FET active patch antenna. (From Ref. 36 with permission from IEEE.)

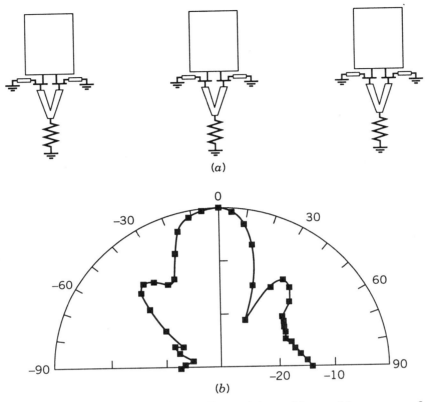

(a)

(b)

FIGURE 7.24. Dual-FET Edge-FED spatial combiners: (a) array configuration; (b) radiation pattern for a three-element array. (From Ref. 36 with permission from IEEE.)

(a)

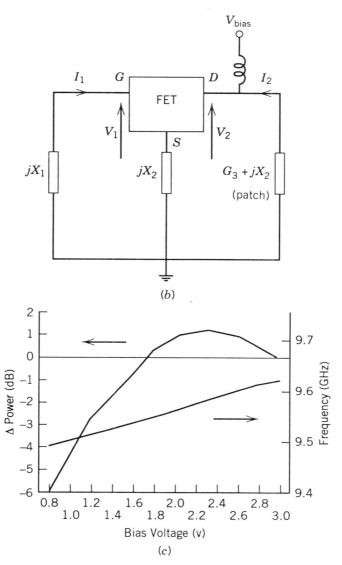

FIGURE 7.25. FET series feedback active patch antenna: (*a*) active radiating element; (*b*) series feedback arrangement; (*c*) frequency pushing characteristics. (From Ref. 37 with permission from IEEE.)

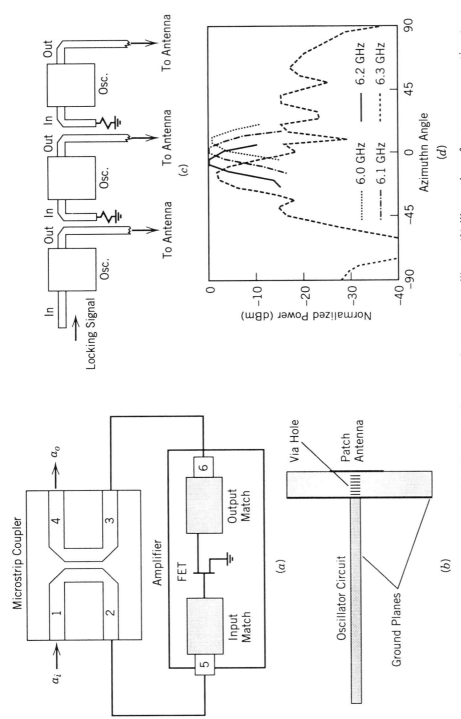

FIGURE 7.26. Two-port active antenna: (*a*) schematic diagram of two port oscillator; (*b*) illustration of antenna connection to oscillator circuit; (*c*) series array of injection-locked elements; (*d*) antenna pattern for five-element oscillator array. (From Ref. 42 with permission from IEEE).

Another edge-fed dual-FET integrated microstrip patch antenna oscillator was demonstrated by Hall and Haskins for beam-scanning applications in 1992 [40] and 1994 [41]. An operating frequency of 2.28 GHz and an external injection-locking signal allows over 40° of beam steering for a four-element *H*-plane array. The external probe feed at each patch for the injection-locking signal may prove costly in a large array, but it provides an accurate and stable injection-locking signal for beam steering. Array patterns are smooth with beam steering of 40°.

Figures 7.26*a*, *b*, and *c* show another approach introduced by Birkeland and Itoh in 1991. It uses a directional microstrip coupler to develop two-port FET oscillators [42]. The port can be used for external injection locking. It uses probe coupling to connect to the patch antennas for radiation. Figure 7.26*d* shows the radiation pattern for the five-element array. Note the frequency beam steering characteristic. The design has an increase in injection-locking bandwidth, which is essential for an array of oscillators (500 MHz at 6 GHz). For a 16-element array of these elements, a locking bandwidth of 453 MHz at 6 GHz (7.5%) was demonstrated for an injected power of 10.3 dBm [43]. The array showed an EIRP of 28.2 W CW with an isotropic conversion gain of 9.9 dB at 6 GHz. The radiation patterns are very good with very low cross-polarization levels. Relatively stable gain over bandwidth and EIRP with frequency are demonstrated. The same configuration was also modified with an input and output patch antenna separated by a ground plane [44].

A similar amplification concept is shown by Mader et al. [45]. A linearly polarized wave is received by a rectangular patch, amplified by a MESFET,

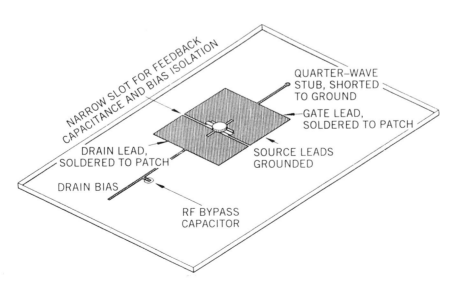

FIGURE 7.27. Compact FET integrated active antenna configuration. (From Ref. 48 with permission from *Electronics Letters.*)

and retransmitted by another patch. A single-amplifier combination had a 7.1-dB gain at 10 GHz with over 5 dB of gain from 9 to 11 GHz. A 4 × 6 array showed 21-dB gain enhancement over a single amplifier. The design is also flexible in receiving one type of polarization and transmitting another. A single-amplifier module demonstrated a circular output from a linear input at 10 GHz with 6.8-dB gain and 0.3-dB axial ratio. Tsai and York developed a reflection amplifier module for quasi-optical arrays, which differs from the arrays shown earlier [46]. Two orthogonally polarized patches use a resistive

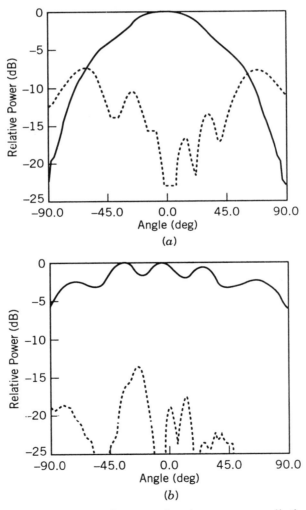

FIGURE 7.28. Compact FET integrated active antenna radiation patterns. (*a*) *H*-plane. (*b*) *E*-plane. (From Ref. 48 with permission from *Electronics Letters.*)

feedback amplifier module to receive an incoming signal and retransmit the signal with orthogonal polarization. An isotropic power gain of 17 dB at 4.2 GHz with 1% bandwidth was measured. Another quasi-optical amplifier was shown by Sheth et al. [47]. It is a nine-element HEMT amplifier array with a gain of 5.5 dB at 10.9 GHz and a 3-dB bandwidth of 1 GHz. The E-plane pattern shown agrees with the calculated pattern, but the H-plane pattern and CPL are not mentioned.

In 1990, York et al. introduced the compact FET active integrated antenna shown in Figure 7.27 [48]. Unlike previous designs the FET is integrated within the antenna structure. A narrow slit splits the patch to provide dc isolation between the drain and gate during biasing. The active antenna requires vias to connect the FET source leads and a quarter-wave stub on the gate side of the antenna to ground. The drain terminal is biased through a low-pass filter using a capacitor and inductive line. Typical operating voltage for maximum power output was 6 V at 40 mA. The active antenna oscillates at 8.2 GHz with about 250 MHz of bias-tuning range. As shown in Figure 7.28, the principal radiation patterns are relatively smooth with a cross-polarization component 8 dB below the maximum. The EIRP is about 40 mW with a dc-to-rf conversion efficiency of 5%. The compact active antenna is ideally suited for quasi-optical power-combining arrays. Figures 7.29a and b show a 4 × 4 array of FET integrated microstrip patches along with radiation patterns. The array operates at 8.27 GHz with good H-plane patterns and cross-polarization levels.

A dual push-pull FET integrated patch was introduced by Wu et al. in 1992 [49]. The configuration allows separate loading of the drain and gate ports of the FET, which simplifies the design. The dual-FET split patch antenna configuration shown in Figure 7.30a generated 47 mW at 8.1 GHz. Figure 7.30b shows very smooth, symmetrical principal plane patterns with a CPL of −12 dB. The dual device reduces the cross-polarization level and increases the power output from a single patch.

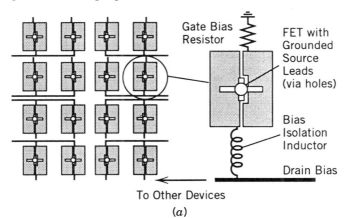

FIGURE 7.29. (*Continues on next page*)

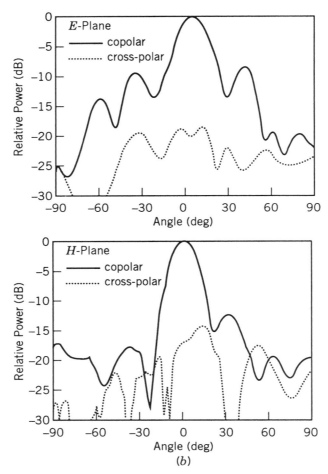

FIGURE 7.29. FET integrated 4×4 spatial power combiner. (*a*) Sketch of the array which uses Fujitsu FSX 02 MESFETs, showing bias arrangement and individual element design. Elements measure 11 mm by 15 mm, and the spacing of the elements is $0.67\lambda_0$ between centers. The bias inductor reduces element interactions along the bias line. (*b*) E-plane and H-plane patterns for the array. The measurements were made at 8.27 GHz, using a flat 2.5-cm-thick dielectric reflector with a dielectric constant of 4. The good patterns indicate in-phase operation. (From Ref. 21 with permission from IEEE.)

Wu and Chang introduced another active antenna in 1994 [50]. It uses one transistor and two circular, instead of rectangular, patches to load the transistor terminals, as shown in Figure 7.31 along with its radiation patterns. Single active antenna modules exhibited an EIRP of 120 mW at 8.4 GHz. The

CPL was relatively high at $-6\,dB$. A two-element array has an EIRP of 520 mW at 8.4 GHz with a CPL at least 13 dB below the maximum. A four-element linear array had an EIRP of 1.3 W at 8.4 GHz with a CPL of $-16\,dB$. A 2×2 array achieved an EIRP of 1.5 W at 8.4 GHz with a CPL of $-16\,dB$. A 2×4 array showed an EIRP of 3.8 W at 8.99 GHz. Patterns for these arrays are relatively well behaved as shown in Figures 7.32a–e. A dielectric slab was placed in front of the array to increase the feedback for injection locking.

An aperture-coupled FET integrated microstrip patch antenna was shown by Shen et al. [26]. Figure 7.33 shows the prototype circuit of the FET active patch antenna with dimensions. The maximum power is 4.81 mW at 4.893 GHz with 50 MHz tuning. The radiation patterns show half-power beamwidths of $40°$ and $52°$ in the E- and H-plane, respectively. Another aperture-coupled design was shown by Simons and Lee [27]. They introduced in 1992 a dielectric resonator-stabilized HEMT oscillator active antenna using a CPW aperture-coupled microstrip patch antenna. It operates at 7.6 GHz with approximately 1.1 mW of output power. The CPL is $-20\,dB$.

Superconducting Y-Ba-Cu-O/GaAs hydbrid oscillator coupled to circular patch at 10 GHz was reported in 1993 [51]. The reflection-mode oscillator uses a circular patch antenna proximity-coupled to a microstrip line connected to a Y-Ba-Cu-O/GaAs hybrid oscillator. Although the E-plane is disturbed, the H-plane patterns are reasonable and there is a noticeable increase in efficiency with decreasing temperature.

An active patch antenna with very good oscillating characteristics was demonstrated by Martinez and Compton in 1994 [52]. Figures 7.34a and b show the active integration which uses coupled feedback from a patch nonradiating edge to maintain oscillations and the equivalent circuit. The FET

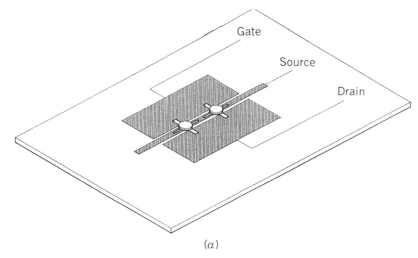

(a)

FIGURE 7.30. (*Continues on next page*)

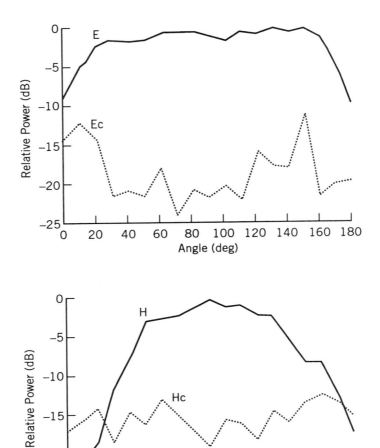

FIGURE 7.30. Dual-FET integrated split patch antenna: (*a*) configuration; (*b*) radiation patterns. (From Ref. 49 with permission from *Electronics Letters.*)

patch antenna oscillator operates at 9.84 GHz with 25-MHz tuning range and a ±1-dB power deviation. The EIRP is 360 mW with a dc-to-rf conversion efficiency of 58%. Radiation patterns, however, were irregular with a high cross-polarization level, as shown in Figure 7.34*c* for the *E*-plane and Figure 7.34*d* for the *H*-plane.

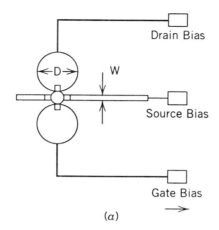

(a)

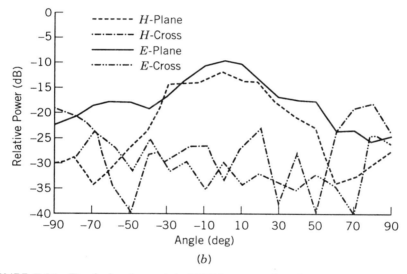

(b)

FIGURE 7.31. Dual circular patch FET integrated active antenna: (a) circuit configuration of a FET active circular patch antenna; $W = 0.5$ mm, $D = 6.5$ mm. (b) E- and H-plane patterns of the single-element active antenna. (From Ref. 50 with permission from IEEE.)

The poor radiation performance of the original circuit was improved with the symmetrical circuit in Figure 7.35a. The modified circuit had an EIRP of 410 mW with a 44% conversion efficiency. Figures 7.35b and c show that the E- and H-plane patterns are smooth and relatively symmetric with a CPL of −12 dB. The E-plane pattern has ∼30° squint off broadside. The configuration

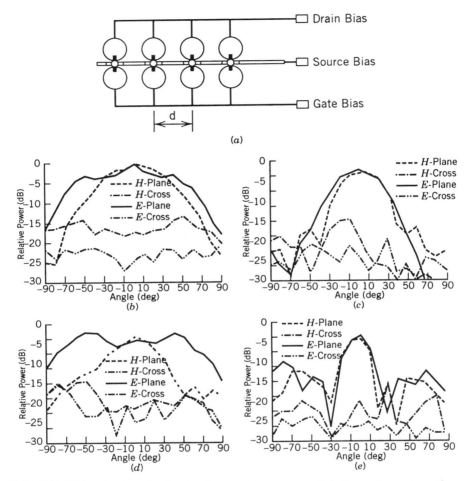

FIGURE 7.32. Active dual circular patch arrays: (*a*) four-element active antenna linear array; (*b*) *E*- and *H*-plane patterns of a 1×2 element linear array; (*c*) *E*- and *H*-plane patterns of the 2×2 active antenna array; (*d*) *E*- and *H*-plane patterns of the 1×4 element linear array; (*e*) *E*- and *H*-plane patterns of the 2×4 active antenna array. (From Ref. 50 with permission from IEEE.)

is readily integrated as an AM active antenna [53]. The 10-GHz active antenna oscillator/modulator has 3-dB modulation bandwidth of 1.4 GHz. The spectral oscillator jitter is below 50 kHz and the single-sideband noise is $-65 \, \text{dBc/Hz}$ at 10 kHz from the carrier. The intermodulation distortion is less than $-20 \, \text{dBc}$ for all modulation powers.

A voltage controlled transistor oscillator using a varactor diode is shown in Figure 7.36*a*. This multilayer design, whose layout is shown in Figure 7.36*b*,

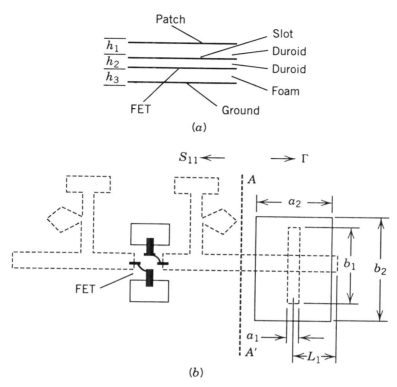

FIGURE 7.33. Active multilayer slot coupling patch antenna: (*a*) side view; (*b*) top view; $h_1 = 3.83$ mm, $h_2 = 1.67$ mm, $h_3 = 10.96$ mm, $b_1 = 19.5$ mm, $b_2 = 22.2$ mm, $a_1 = 1$ mm, $a_2 = 15.3$ mm, $L_1 = 10.4$ mm. (From Ref. 26 with permission from *Electronics Letters*.)

was introduced by Haskins, Hall, and Dahele in 1991 [54, 55]. It uses a BAR-28 diode to provide 100 MHz of tuning range at 2.2 GHz for the active antenna. Figure 7.36*c* shows the circuit resonance versus dc bias. Figure 7.36*d* shows excellent radiation performance with very smooth principal plane patterns and low cross-polarization levels. Another VCO active antenna was shown by Liao and York and used in a 1×10 element array [56]. The configuration (Fig. 7.37) is similar to that shown by Martinez and Compton [52] in that the patch serves as feedback from the drain to the gate through a coupled line section. The transistor requires vias to ground for the source leads and a quarter-wave stub on the gate terminal. The varactor has a maximum to minimum capacitance ratio of 3. The VCO operates at 8.45 GHz with 150 MHz of tuning bandwidth (1.8%). The VCO phase noise is exceptional at -90 dBc/Hz at 100 kHz from the carrier. The EIRP of the 10-element array was 10.5 W at 8.43 GHz. The VCO allows the spatial combiner to steer the beam from $-10°$

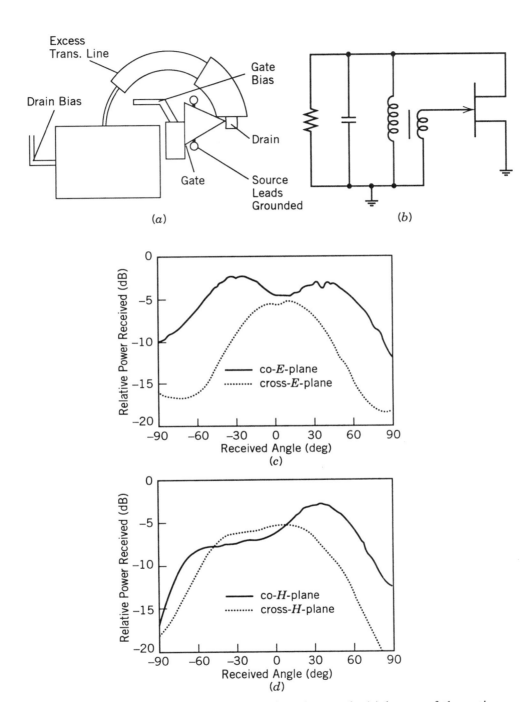

FIGURE 7.34. FET integrated feedback active patch: (*a*) layout of the entire microstrip-patch oscillator; (*b*) equivalent lumped-element circuit, showing feedback applied to the gate; (*c*) *E*-plane radiation pattern; (*d*) *H*-plane radiation pattern. (From Ref. 52 with permission from IEEE.)

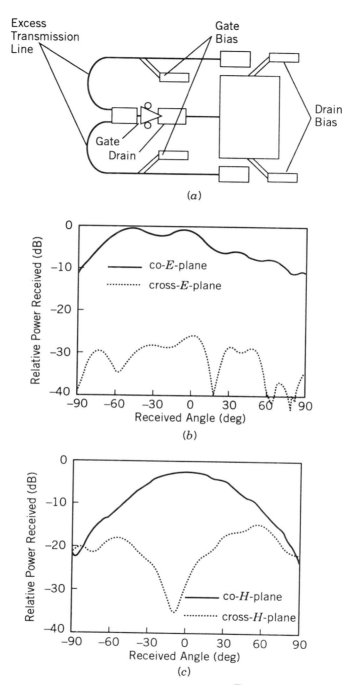

FIGURE 7.35. FET integrated symmetrically fedback patch: (*a*) A linearly polarized radiating oscillator, two balanced pickups quench unwanted cross-pol radiation; (*b*) *E*-plane pattern; (*c*) *H*-plane pattern. (From Ref. 52 with permission from IEEE.)

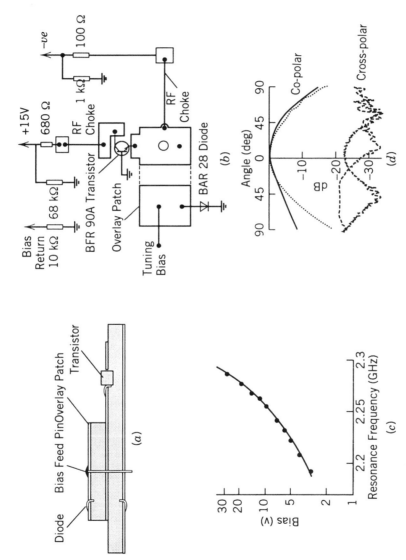

FIGURE 7.36. Multilayer transistor integrated microstrip patch: (*a*) longitudinal cross section of active patch; (*b*) bias circuitry; (*c*) bias voltage to resonant frequency; (*d*) radiation pattern: (———) *E*-plane, (----) *H*-plane. (From Ref. 54 with permission from *Electronics Letters*.)

to 20° by adjusting the free-running frequencies. The VCO active antenna has since been improved to provide a much wider tuning bandwidth [57]. A varactor with a maximum to minimum capacitance ratio of 3.4 is located on a radiating edge of the patch. Its capacitance varies from 1.7 to 0.5 pF and allows the active antenna to electronically tune from 6.8 to 8.1 GHz with a 3-dB variation in output power. This represents a very useful 17% electronic tuning bandwidth for beam steering power combiners. The EIRP is 29.5 mW with a dc-to-rf conversion efficiency of 11%. The radiation patterns are smooth with a CPL of -10 dB.

An 8- and a 16-element MESFET array combiner with 5.54 and 17.38 W of EIRP at 10 GHz was demonstrated by Balasubramaniyan and Mortazawi [58]. The arrays exhibit excellent radiation patterns in both principal planes which agree well with the theoretical prediction. Figure 7.38 shows the configuration of the 16-element array. Measurements for one and two active antennas are 0.058 W and 0.266 W EIRP, respectively. Mortazawi and De Loach have analyzed spatial power combiners, using an extended resonance technique [59]. Very good combining results have been demonstrated for various planar spatial power combiners.

An amplifying active antenna array using conventional microstrip patches with an integrated amplifier at its terminals has shown excellent radiation patterns and very good efficiency [60]. Since the distribution network carries very little power, microstrip losses encountered are relatively small compared with the amplified output of the array. Very smooth symmetric patterns are achieved for a single transmitter, which is shown to have 12 dB of gain.

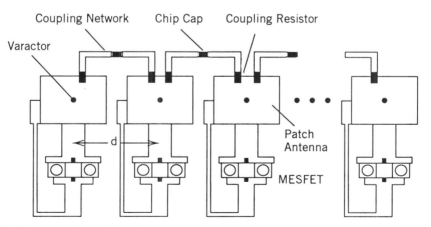

FIGURE 7.37. Coupled VCO array. Coupling between oscillators is accomplished with a resistively loaded transmission line. The chip capacitor is included as a dc block. Not shown are the bias lines to the varactors, which are connected via the coupling network. (From Ref. 56 with permission from IEEE.)

Some of the most promising active antenna investigations deal with electo-optical applications. Transistors are ideal candidates for millimeter-wave power generation through optical interaction. A 60-GHz CW generator through optical mixing in FET integrated antennas has been shown by Plant et al. [61]. Three-terminal devices such as the high-electron-mobility transistors (HEMTs) [62] and heterojunction bipolar transistors (HBTs) [63] can also be used. Optically driven solid-state devices can generate microwaves by mixing two continuous-wave lasers or by using a mode-locked laser [64]. Using fiber-optic lines throughout a complex circuit and later converting it to microwaves at an antenna appears to be an attractive alternative to current approaches.

Many other oscillators [65], amplifiers [66, 67], receivers [68–71], and transponders [72] have been shown in the literature. Recent papers have embarked on full 3D field simulations of active antenna structures, including

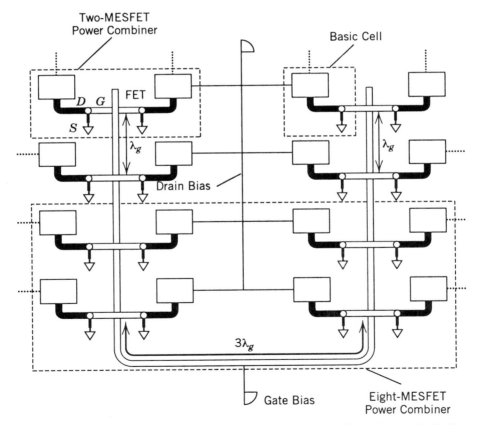

FIGURE 7.38. A 16-MESFET periodic spatial power combiner. (From Ref. 58 with permission from IEEE.)

solid-state devices [73, 74]. Such an undertaking will allow accurate analysis to understand and correct many radiation problems often encountered with active antennas. These transistor integrated antennas are essential for multi-function integration since they allow a variety of different operations. They make ideal oscillators and transceivers which can be applied to single or large arrays. The increased dc-to-rf efficiency and lower operating levels are ideal for distributed oscillators in a power-combining array. The innovations provide a wide range of options to successfully develop such a power-combining array.

REFERENCES

1. S. P. Kwok and K. P. Weller, "Low Cost X-band MIC BARITT Doppler Sensor," *IEEE Transactions on Microwave Theory and Techniques*, Vol. 27, No. 10, pp. 844–847, October 1979.

2. B. M. Armstrong, R. Brown, F. Rix, and J. A. C. Stewart, "Use of Microstrip Impedance-Measurement Technique in the Design of a BARITT Diplex Doppler Sensor," *IEEE Transactions on Microwave Theory and Techniques*, Vol. 28, No. 12, pp. 1437–1442, December 1980.

3. G. Morris, H. J. Thomas, and D. L. Fudge, "*Active Patch Antennas,*" *1984 Military Microwave Conference*, London, pp. 245–249, September 1984.

4. H. J. Thomas, D. L. Fudge, and G. Morris, "Gunn Source Integrated With Microstrip Patch," *Microwaves and RF*, Vol. 24, No. 2, pp. 87–91, February 1985.

5. M. Dydyk, "Planar Radial Resonator Oscillator," *IEEE MTT-S International Microwave Symposium*, pp. 167–168 (1986).

6. T. O. Perkins III, "Microstrip Patch Antenna with Embedded IMPATT Oscillator," *IEEE AP Symposium Digest*, pp. 447–450 (1986).

7. T. O. Perkins III, "Active Microstrip Circular Patch Antenna," *Microwave Journal*, Vol. 30, No. 3, pp. 109–117, March 1987.

8. K. D. Stephan, "Inter-Injection-Locked Oscillators with Applications to Spatial Power Combining and Phased Arrays," *IEEE MTT-S International Microwave Symposium Digest*, pp. 159–162 (1986).

9. S. L. Young and K. D. Stephan, "Radiation Coupling of Inter-Injection-Locked Oscillators," SPIE, Vol. 791, *Millimeter Wave Technology IV and Radio Frequency Power Sources*, pp. 69–76, May 1987.

10. J. W. Mink, "Quasi-Optical Power Combining of Solid-State Millimeter-Wave Sources," *IEEE Transactions on Microwave Theory and Techniques*, Vol. MTT-34, No. 2, pp. 273–279, February 1986.

11. W. A. Morgan and K. D. Stephan, "An X-Band Experimental Model of a Millimeter-Wave Inter-Injection-Locked Phased Array System," *IEEE Transactions on Antennas and Propagation*, Vol. 36, No. 11, pp. 1641–1645, November (1988).

12. N. Camirelli and B. Bayraktarouglu, "Monolithic Millimeter Wave IMPATT Oscillator and Active Antenna," *IEEE MTT-S International Microwave Symposium Digest*, pp. 955–958 (1988).

13. N. Camirelli and B. Bayraktarouglu, "Monolithic Millimeter Wave IMPATT Oscillator and Active Antenna," *IEEE Transactions on Microwave Theory and Techniques*, Vol. 36, No. 12, pp. 1670–1676, December 1988.

14. W. P. Shillue, S. C. Wong, and K. D. Stephan, "Monolithic IMPATT Millimeter-Wave Oscillator Stabilized by Open Cavity Resonator," *IEEE MTT-S International Microwave Symposium Digest*, pp. 739–740 (1989).

15. J. A Navarro, K. A. Hummer, and K. Chang, "Active Integrated Antenna Elements," *Microwave Journal*, Vol. 35, No. 1, pp. 115–126, January 1991.

16. K. A. Hummer and K. Chang, "Spatial Power Combining Using Active Microstrip Antennas," *Microwave and Optical Technology Letters*, Vol. 1, No. 1, pp. 8–9, March 1988.

17. K. A. Hummer and K. Chang, "Microstrip Active Antennas and Arrays," *IEEE MTT-S International Microwave Symposium Digest*, pp. 963–966 (1988).

18. J. R. James, P. S. Hall, and D. Wood, *Microstrip Antenna: Theory and Design*, Peregrinus, London, 1981, Ch. 4.

19. K. Chang, K. A. Hummer, and J. L. Klein, "Experiments on Injection Locking of Active Antenna Elements for Active Phased Arrays and Spatial Power Combiners," *IEEE Transactions on Microwave Theory and Techniques*, Vol. 37, No. 7, pp. 1078–1084, July 1989.

20. D. B. Rutledge, Z. B. Popovic, R. M. Weikle, M. Kim, K. A. Potter, R. Compton, and R. A. York, "Quasi-Optical Power Combining Arrays," *IEEE MTT-S International Microwave Symposium Digest*, pp. 1201–1204, May 1990.

21. R. A. York and R. C. Compton, "Quasi-Optical Power Combining Using Mutually Synchronized Oscillator Arrays," *IEEE Transactions on Microwave Theory and Techniques*, Vol. 39, No. 6, pp. 1000–1009, June 1991.

22. R. A. York and R. C. Compton, "Dual-Device Active Patch Antenna with Improved Radiation Characteristics," *Electronics Letters*, Vol. 28, No. 11, pp. 1019–1021, May 1992.

23. X. Gao and K. Chang, "Network Modeling of an Aperture Coupling between Microstrip Line and Patch Antenna for Active Array Applications," *IEEE Transactions on Microwave Theory and Techniques*, Vol. 36, No. 3, pp. 505–513, March 1988.

24. D. M. Pozar, "A Microstrip Antenna Aperture-Coupled to a Microstrip Line," *Electronics Letters*, Vol. 21, No. 2, pp. 49–50, January 1985.

25. J. Lin and T. Itoh, "Two-Dimensional Quasi-Optical Power Combining Arrays Using Strongly Coupled Oscillators," *IEEE Transactions on Microwave Theory and Techniques*, Vol. 42, No. 4, pp. 734–741, April 1994.

26. Y. Shen, R. Fralich, C. Wu, and J. Litva, "Active Radiating Oscillator using a Reflection Amplifier Module," *Electronics Letters*, Vol. 28, No. 11, pp. 991–992, May 1992.

27. R. N. Simons and R. Q. Lee, "Planar Dielectric Resonator Stabilized HEMT Oscillator Integrated with CPW/Aperture Coupled Patch Antenna," *IEEE MTT-S International Microwave Symposium*, pp. 433–436 (1992).

28. J. A. Navarro, K. Chang, J. Tolleson, S. Sanzgiri, and R. Q. Lee, "A 29.3 GHz Cavity Enclosed Aperture-Coupled Circular Patch Antenna for Microwave Circuit Integration," *IEEE Microwave and Guided Wave Letters*, Vol. 1, No. 7, pp. 170–171, July 1991.

29. S. Sanzgiri, W. Pottenger, D. Bostrom, D. Denniston, and R. Q. Lee, "Active Subarray Module Development for Ka-Band Satellite Communication Systems," *IEEE AP Symposium Digest*, Seattle, Washington, pp. 860–863 (1994).

30. A. C. Davidson, F. W. Wise, and R. C. Compton, "A 60 GHz IMPATT Oscillator Array with Pulsed Operation," *IEEE Transactions on Microwave Theory and Techniques*, Vol. 41, No. 10, pp. 1845–1850, October 1993.

31. D. S. Hernandez and I. Robertson, "60 GHz-Band Active Microstrip Patch Antenna for Future Mobile Systems Applications," *Electronics Letters*, Vol. 30, No. 9, pp. 677–678, April 1994.

32. K. Chang, K. A. Hummer, and G. Gopalakrishnan, "Active Radiating Element Using FET Source Integrated with Microstrip Patch Antenna," *Electronic Letters*, Vol. 24, No. 21, pp. 1347–1348, October 1988.

33. V. F. Fusco and H. O. Burns, "Synthesis Procedure for Active Integrated Radiating Elements," *Electronic Letters*, Vol. 26, No. 4, pp. 263–264, February 1990.

34. J. Birkeland and T. Itoh, "Quasi-Optical Planar FET Transceiver Modules," *IEEE MTT-S International Microwave Symposium*, pp. 119–122, 1989.

35. J. Birkeland and T. Itoh, "FET-Based Planar Circuits for Quasi-Optical Sources and Transceivers," *IEEE Microwave Theory and Techniques*, Vol. 37, No. 9, pp. 1452–1459, September 1989.

36. J. Birkeland and T. Itoh, "Spatial Power Combining using Push-Pull FET Oscillators with Microstrip Patch Resonators," *IEEE MTT-S International Microwave Symposium*, pp. 1217–1220 (1990).

37. V. F. Fusco, "Series Feedback Integrated Active Microstrip Antenna Synthesis and Characterization," *Electronics Letters*, Vol. 28, No. 1, pp. 89–91, January 1992.

38. D. S. McDowall and V. F. Fusco, "Electromagnetic Far-Field Pattern Modeling for an Active Microstrip Antenna Module," *Microwave and Optical Technology Letters*, Vol. 6, No. 5, pp. 275–277, April 1993.

39. V. F. Fusco, "Self-Detection Performance of Active Microstrip Antennas," *Electronics Letters*, Vol. 28, No. 14, pp. 1362–1363, July 1992.

40. P. S. Hall and P. M. Haskins, "Microstrip Active Patch Array with Beam Scanning," *Electronics Letters*, Vol. 28, No. 22, pp. 2056–2057, October 1992.

41. P. S. Hall, I. L. Morrow, P. M. Haskins, and J. S. Dahele, "Phase Control in Injection-Locked Microstrip Antennas," *IEEE MTT-S International Microwave Symposium Digest*, pp. 1227–1230 (1994).

42. J. Birkeland and T. Itoh, "Two-Port FET Oscillators with Applications to Active Arrays," *IEEE Microwave and Guided Wave Letters*, Vol. 1, No. 5, pp. 112–113, May 1991.

43. J. Birkeland and T. Itoh, "A 16 Element Quasi-Optical FET Oscillator Power Combining Array with External Injection Locking," *IEEE Transactions on Microwave Theory and Techniques*, Vol. 40, No. 3, pp. 475–481, March 1992.

44. J. Birkeland and T. Itoh, "An FET Oscillator Element for Spatially Injection Locked Arrays," *IEEE MTT-S International Microwave Symposium*, pp. 1535–1538 (1992).

45. T. Mader, J. Schoenberg, L. Harmon, and Z. B. Popovic, "Planar MESFET Transmission Wave Amplifier," *Electronics Letters*, Vol. 29, No. 19, pp. 1699–1701, September 1993.

46. H. S. Tsai and R. A. York, "Polarization-Rotation Quasioptical Reflection Amplifier Cell," *Electronics Letters*, Vol. 29, No. 24, pp. 2125–2127, November 1993.

47. N. Sheth, T. Ivanov, A. Balasubramaniyan, and A. Mortazawi," A Nine HEMT Spatial Amplifier," *IEEE MTT-S International Microwave Symposium*, pp. 1239–1242 (1994).

48. R. A. York, R. D. Martinez, and R. C. Compton, "Active Patch Antenna Element for Array Applications," *Electronics Letters*, pp. 494–495, Vol. 26, No. 7, March 1990.

49. X. D. Wu, K. Leverich, and K. Chang, "Novel FET Active Patch Antenna," *Electronics Letters*, Vol. 28, No. 20, pp. 1853–1854, September 1992.

50. X. D. Wu and K. Chang, "Novel Active FET Circular Patch Antenna Arrays for Quasi-optical Power Combining," *IEEE Transactions on Microwave Theory and Techniques*, Vol. 42, No. 5, pp. 766–771, May 1994.

51. N. J. Rohrer, M. A. Richard, G. J. Valco, and K. B. Bhasin, "A 10 GHz Y-Ba-Cu-O/GaAs Hybrid Oscillator Proximity Coupled to a Circular Microstrip Patch Antenna," *IEEE Transactions on Applied Superconductivity*, Vol. 3, No. 1, pp. 23–27, March 1993.

52. R. D. Martinez and R. C. Compton, "High-Efficiency FET/Microstrip-Patch Oscillators," *IEEE Antennas and Propagation Magazine*, Vol. 36, No. 1, pp. 16–19, February 1994.

53. R. D. Martinez and R. C. Compton, "A Quasi-Optical Oscillator/Modulator for Wireless Transmission," *IEEE MTT-S International Microwave Symposium Digest*, pp. 839–842 (1994).

54. P. M. Haskins, P. S. Hall, and J. S. Dahele, "Active Patch Antenna Element with Diode Tuning," *Electronics Letters*, Vol. 27, No. 20, pp. 1846–1847, September 1991.

55. P. S. Hall, "Analysis of Radiation from Active Microstrip Antennas," *Electronics Letters*, Vol. 29, No. 1, pp. 127–129, January 1993.

56. P. Liao and R. A. York, "A 1 Watt X-band Power Combining Array Using Coupled VCOs," *IEEE MTT-S International Microwave Symposium Digest*, pp. 1235–1238 (1994).

57. P. Liao and R. A. York, "A Varactor-Tuned Patch Oscillator for Active Arrays," *IEEE Microwave and Guided Wave Letters*, Vol. 4, No. 10, pp. 335–337, October 1994.

58. A. Balasubramaniyan and A. Mortazawi, "Two-Dimensional MESFET-Based Spatial Power Combiners," *IEEE Microwave and Guided Wave Letters*, Vol. 3, No. 10, pp. 366–368, October 1993.

59. A. Mortazawi and B. C. De Loach, "Spatial Power Combining Oscillators Based on an Extended Resonance Technique, "*IEEE Transactions on Microwave Theory and Techniques*, Vol. 42, No. 12, pp. 2222–2228, December 1994.

60. B. Robert, T. Razban, and A. Papiernik, "Compact Active Patch Antenna," *IEE 8th Int. Conf. On Antennas and Propagation*, Edinburgh, UK, pp. 307–310, April 1993.

61. D. V. Plant, D. C. Scott, D. C. Ni, and H. R. Fetterman, "Generation of Millimeter-Wave Radiation by Optical Mixing in FET's Integrated with Printed Circuit Antennas," *IEEE Microwave and Guided Wave Letters*, Vol. 1, No. 6, pp. 132–134, June 1991.

62. D. V. Plant, D. C. Scott, D. C. Ni, and H. R. Fetterman, "Optically-Generated 60 GHz mm-Waves Using AlGaAs/InGaAs HEMTs, Integrated with Both Quasi-optical Antenna Circuits and MMICs," *IEEE Photon. Tech. Letters*, Vol. 4, pp. 102–105 (1992).

63. D. C. Scott, D. V. Plant, and H. R. Fetterman, "60 GHz Sources Using Optically Driven Heterojunction Bipolar Transistors," *Applied Physics Letters*, Vol. 61, pp. 1–3 (1992).

64. D. V. Plant, D. C. Scott, and H. R. Fetterman, "Optoelectronic mm-Wave Sources," *Microwave Journal*, Vol. 36, No. 4, pp. 62–72, April 1993.

65. T. Razban, M. Nannini, and A. Papiernik, "Integration of Oscillators with Patch Antennas," *Microwave Journal*, Vol. 36, No. 1, pp. 104–110, January 1993.

66. G. F. Avitabile, S. Maci, G. Biffi Gentili, F. Ceccuti, and G. F. Manes, "A Basic Module for Active Antenna Applications," *IEE 8th Int. Conf. On Antennas and Propagation*, Edinburgh, UK, pp. 303–306, April 1993.

67. H. An, B. K. J. C. Nauwelaers, G. A. E. Vandenbosch, and A. R. Van de Capelle, "Active Antenna Uses Semi-Balanced Amplifier Structure," *Microwaves and RF*, Vol. 33, No. 13, pp. 153–156, December 1994.

68. R. Gillard, H. Legay, J. M. Floch, and J. Citerne, "Rigorous Modeling of Receiving Active Microstrip Antenna," *Electronics Letters*, Vol. 27, No. 25, pp. 2357–2359, 5th December 1991.

69. H. An, B. Nauwelaers, and A. Van de Capelle, "Noise Figure Measurement of Receiving Active Microstrip Antennas," *Electronics Letters*, Vol. 29, No. 18, pp. 1594–1596, September 1993.

70. J. J. Lee, "G/T and Noise Figure of Active Array Antennas," *IEEE Transactions on Antennas and Propagation*, Vol. AP-41, No. 2, pp. 241–244, February 1993.

71. U. Dahlgren, J. Svedin, H. Johansson, O. J. Hagel, H. Zirath, C. Karlsson, and N. Rorsman, "An Integrated Millimeterwave BCB Patch Antenna HEMT Receiver," *IEEE MTT-S International Microwave Symposium Digest*, pp. 661–664 (1994).

72. K. Cha, S. Kawasaki, and T. Itoh, "Transponder Using Self-Oscillating Mixer and Active Antenna," *IEEE MTT Symposium Digest*, pp. 425–428 (1994).

73. V. A. Thomas, K. M. Ling, M. E. Jones, B. Toland, J. Lin, and T. Itoh, "FDTD Analysis of an Active Antenna," *IEEE Microwave and Guided Wave Letters*, Vol. 4, No. 9, pp. 296–298, September 1994.

74. B. Toland, J. Lin, B. Houshmand, and T. Itoh, "Electromagnetic Simulation of Mode Control of a Two Element Active Antenna," *IEEE MTT-S International Microwave Symposium Digest*, pp. 883–886 (1994).

CHAPTER EIGHT

Integrated and Active Grids

8.1 INTRODUCTION

The term *grid* has been applied to a style of wire antennas, periodic frequency-selective surfaces (FSS), and, more recently, to a method for active and integrated antennas and spatial power combiners. Large grids used for antenna arrays have been shown by Kraus' endfire antenna [1] in 1964 and the chain antenna by Tiuri et al. in 1974 [2]. In 1981 Conti et al. [3] showed the wire grid microstrip antenna for array applications. Furthermore, passive grids have been used as frequency-selective surfaces to manipulate space waves through transmission, reflection, or depolarization. This was shown by Chen [4], Arnaud and Pelow [5], Anderson [6], and Watanabe [7].

The use of solid-state devices within a grid style structure was reported as early as 1961 [8]. A grid array of diodes was demonstrated by Sabbagh and George of Purdue University [9]. The diode grids were used to convert rf energy to dc power in a microwave wireless power transmission link. Figure 8.1a shows the 680 diode rectenna grid array. It operates at 2.45 GHz, and, unlike a typical detector, each diode operates at very large power densities. Figure 8.1b shows the rectification efficiency versus dc power output for a 25-Ω output load. The power converted by the rectifying antenna, or *rectenna*, was used to power a helicopter-type platform [10, 11]. A review of microwave wireless power transmission is given by W. C. Brown [12].

In 1972, Lee and Fong proposed the integration of active devices within a periodic corrugated surface [13]. The diodes within the corrugations serve as localized amplifiers for the impinging space wave. A large array of these diodes functions together as a quasi-optical reflection amplifier. This served as one of the original active quasi-optical amplifier integrations; however, work in this area did not flourish till the late 1980s.

In the early 1980s, Rutledge and Muha used a large array of diodes for millimeter-wave and far-infrared imaging [14]. The use of antennas integrated with diodes (i.e., Schottky, superconducting tunnel junctions, microbolometers)

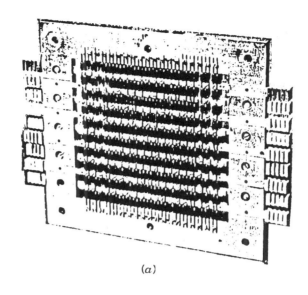

(a)

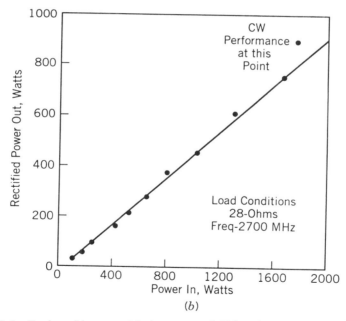

(b)

FIGURE 8.1. Early grid array. (*a*) An array of 680 point-contact semiconductor diodes arranged in a series-parallel arrangement. This array was inserted into an enlarged waveguide with a reflecting plate approximately 1/4 wavelength back of the device. (*b*) Relationship of rectified dc output power to rf input power in the QR1222. A 25-ohm load was used. Data is pulse data with exception of the highest power point on the graph, which was continuous output [8, 9].

195

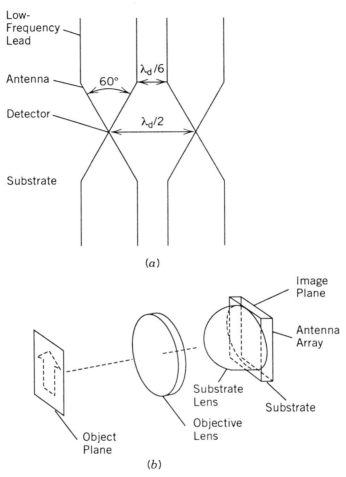

FIGURE 8.2. Diode integrated bow tie imaging array. (*a*) Modified bow-tie for an imaging antenna array. (*b*) "Reverse microscope" configuration. A substrate and substrate lens have been added to the system in (*a*). (From Ref. 14 with permission from IEEE.)

for the large array distributes the detector function throughout the array, thus increasing redundancy, reducing losses, and improving system performance. The use of integrated devices allows the reproducibility of a system well into the millimeter-wave region. Figure 8.2*a* shows the modified bow tie antennas. Individual detector diodes are integrated at each antenna location. Figure 8.2*b* shows the configuration of the imaging array. The concept is ideal for detectors and other passive device applications, and it was later used by Rutledge and colleagues at Caltech for many spatial oscillators, amplifiers, and mixers.

The method described as a "grid" is well suited for quasi-optical components, and it has been used to develop quasi-optical oscillators, amplifiers, multipliers, mixers, phase shifters, and switches. It is readily accessible to monolithic fabrication techniques and can be used in the millimeter-wave range. This range has been sparsely used due to lack of power, device inefficiencies, and tolerance difficulties. The grid approach provides a method which may allow us to overcome some of these difficulties. The grid approach differs from other spatial-combining approaches in that the integrated grid is made up of devices which are relatively close together with respect to wavelength. Very strong coupling and the collective interaction of all the individual elements create a quasi-optical component.

8.2   GRID EQUIVALENT CIRCUITS

At millimeter and submillimeter wavelengths, large combinations of synchronized devices are essential to overcome low power and tight tolerances. For a large array of devices, characterization of the electromagnetic properties of individual components becomes paramount. Rigorous analysis of periodic grid structures has been carried out by many different approaches, including method of moments [15], finite elements, and conjugate gradient methods [16]. These methods are often time consuming and require large amounts of computing power. Equivalent circuit models have proven to be reasonably accurate and require little computing power.

The equivalent circuit of many periodic grids has been developed through the use of the induced EMF method. The EMF method, introduced by Brillouin [17] and later developed by Carter [18], was used to calculate the self-impedance of various antennas. The method uses the Poynting's theorem to find an expression for the driving-point impedance of an antenna. This was demonstrated by Tai [19] and Elliot [20] for biconical and dipole antennas, respectively.

The EMF method requires that the current distribution across the antenna aperture be known to some degree, which results in its use for relatively simple radiating structures. The inhomogeneous wave equation relates the impressed current ($\mathbf{J}(\mathbf{r})$) to the resulting electric field ($\mathbf{E}(\mathbf{r})$):

$$\nabla \times \nabla \times \mathbf{E}(\mathbf{r}) - k^2 \mathbf{E}(\mathbf{r}) = -j\omega\mu\mathbf{J}(\mathbf{r}) \qquad (8.1)$$

where $k = \omega\sqrt{\mu\varepsilon}$. The solution to the equation typically involves the dyadic Green's function:

$$\mathbf{E}(\mathbf{r}) = -j\omega\mu \int_V \bar{G}(\mathbf{r}\,|\,\mathbf{r}') \cdot \mathbf{J}(\mathbf{r}')\,dV' \qquad (8.2)$$

where the primed coordinates denote the region containing the impressed current distribution. The Poynting theorem is then used to determine the radiated power from the assumed current distribution:

$$\oint_S \mathbf{E} \times \mathbf{H}^* \cdot d\mathbf{s} = -\int_V \mathbf{E} \cdot \mathbf{J}^* \, dV - j\omega \int_V (\mathbf{H}^* \cdot \mathbf{B} - \mathbf{E} \cdot \mathbf{D}^*) \, dV \qquad (8.3)$$

where the quantities in the equation are rms values. The LHS is the average power crossing the closed surface S. The RHS is the power loss due to electric currents minus the reactive power stored within the volume enclosed by S.

When the EMF method is applied, the terms in Equation (8.3) are rearranged so that the total complex power is equated to the average power radiated across the surface plus the reactive power stored within the volume:

$$\int_V \mathbf{E} \cdot \mathbf{J}^* \, dV = -\oint_S \mathbf{E} \times \mathbf{H}^* \cdot d\mathbf{s} - j\omega \int_V (\mathbf{H}^* \cdot \mathbf{B} - \mathbf{E} \cdot \mathbf{D}^*) \, dV \qquad (8.4)$$

Evaluating the LHS of the equation over the assumed current distribution ($|I|^2$), one can find the driving-point impedance Z:

$$Z = -\frac{1}{|I|^2} \int_V \mathbf{E}(\mathbf{r}) \cdot \mathbf{J}^*(\mathbf{r}) \, dV \qquad (8.5)$$

The driving-point impedance for a given grid requires excessive computation. There are some problems associated with the determination of the driving-point impedance in a finite grid. Each element in the grid couples to every other element, and the uncertainty of the fields at the edges of the grid also causes problems. In order to simplify this calculation, the grid is assumed to be infinite, thereby ignoring edge effects. Furthermore, grids are made up of periodic unit cells. The EMF calculation is then carried out over the span of a unit cell with the appropriate electric or magnetic walls at the planes of symmetry. This results in the driving-point impedance for the grid unit cell.

This impedance can then be used in conjunction with transmission line models. Quasi-optical combiners often use large mirrors and dielectric layers on either side of the grid to provide feedback and tuning. In the combiner, air regions are modeled by transmission lines of characteristic impedance $Z_0 = 377\,\Omega$, and dielectric regions use $Z_d = 377/\sqrt{\varepsilon_r}\,\Omega$. The metal reflector is modeled by a short circuit, while the unit cell is made up of lumped elements derived from the driving-point impedance calculations. The following sections describe oscillators, multipliers, mixers, phase shifters, and amplifiers using these techniques.

8.3   GRID OSCILLATORS

The first grid oscillator used in quasi-optical power combining was introduced by Popovic and Rutledge in June 1988 [21]. Figures 8.3a and b show a front view of the grid and the quasi-optical oscillator configuration, respectively. The separation between sources (i.e., unit cells) is approximately $\lambda/6$, which allows the grid to be treated as a uniform active sheet. The grid oscillator can be modeled using a simple transmission line equivalent circuit as shown in Figure 8.3c. The metallic ground reflector is treated as a short circuit separated from the current sheet by a length of air transmission line with a characteristic impedance of $Z_0 = 377\,\Omega$. The dielectric layer is a length of transmission line with a characteristic impedance of $Z_d = 377/\sqrt{\varepsilon_r}\,\Omega$. The metallic ground reflector, diode, and dielectrics collectively form a quasi-optical oscillator. The

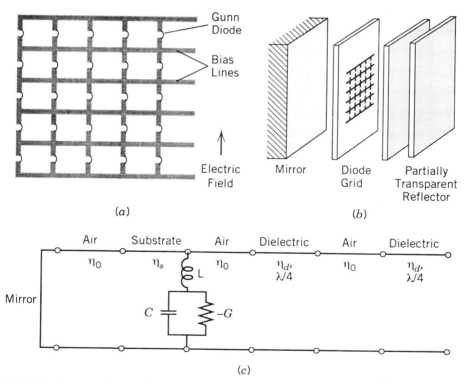

FIGURE 8.3. (a) Gunn-diode loaded grid; (b) power-combiner configuration; (c) transmission-line equivalent circuit. (From Ref. 21 with permission from IEEE.)

diode grid oscillator operated at 10 GHz. Problems encountered during this initial array included heat-sinking deficiencies, injection locking, and mode jumping. However, the concept for the grid oscillator was demonstrated, and methods to correct these deficiencies began.

To increase the device efficiency, diodes can be replaced by transistors. A 25-element MESFET grid was demonstrated by Popovic et al. in July 1988 [22]. The grid operates at 9.7 GHz with an EIRP of 20.7 W and a conversion efficiency of 15%. As stated earlier, there are differences between a grid oscillator and an active antenna spatial power-combining array. To reiterate, an active antenna combiner uses many individual oscillators which are synchronized through mutual coupling (weak coupling) or an external source. The grid, on the other hand, oscillates due to the collective interaction (strong coupling) of the entire grid. The grid oscillation power and frequency depend on the active device used, dimensions of the unit cell, and external reflectors.

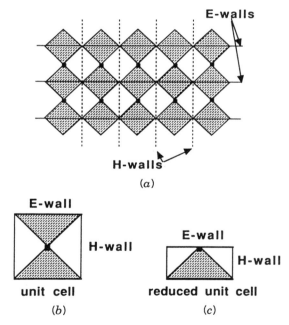

FIGURE 8.4. (a) Locations of symmetry planes determine magnetic and electric walls in infinite 2D array. (b) The cross section of the equivalent 0D array. The unit cell consists of a single antenna in a waveguide with electric walls top and bottom and magnetic walls at the sides. (c) The reduced unit cell constructed using the horizontal symmetry plane of the unit cell. (From Ref. 24 with permission from IEEE.)

The grid unit cell is the key component, and grids are typically large enough so that an infinite array approximation can be employed. Many planar elements have been analyzed under these assumptions. Sjogren and Luhmann use a method of moments (MOM) solution in an equivalent waveguide to obtain a useful circuit impedance model for a quasi-optical diode array [23].

Bow tie and double-V arrays are considered in the same family of the grid array. Figure 8.4 shows a bow-tie and double-vee unit cells analyzed and measured by Pance and Wengler [24]. A dielectric waveguide measurement method is used to obtain the driving point impedance of a single unit cell. Dipoles, bow ties, double-Vs, and slot antennas were analyzed and measured. Each of these is useful in the creation of a large grid. Bundy and Popovic have also analyzed the geometries shown in Figure 8.5, using MOM [25].

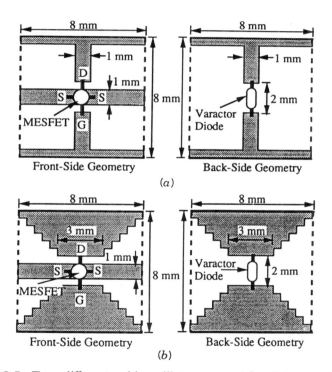

FIGURE 8.5. Two different grid oscillator geometries, (*a*) one with narrow-strip radiating elements and (*b*) another with approximated short bow-tie radiating elements. Both structures are printed on 1.5-mm-thick substrates with $\varepsilon_r = 2.2$. The MESFET metalization is on one side and the diode metalization is on the other side of the dielectric substrate. (From Ref. 25 with permission from IEEE.)

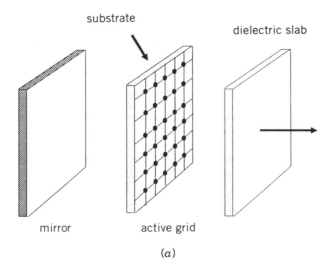

substrate

dielectric slab

mirror   active grid

(a)

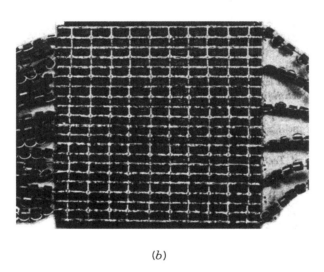

(b)

FIGURE 8.6. A 10×10 FET grid oscillator: (a) Quasi-optical power-combining array configuration. The active devices are placed on a substrate inside a Fabry-Perot cavity. (b) Photograph of the planar MESFET grid oscillator. The transistors are Fujitsu low-noise MESFETs. (From Ref. 26 with permission from IEEE.)

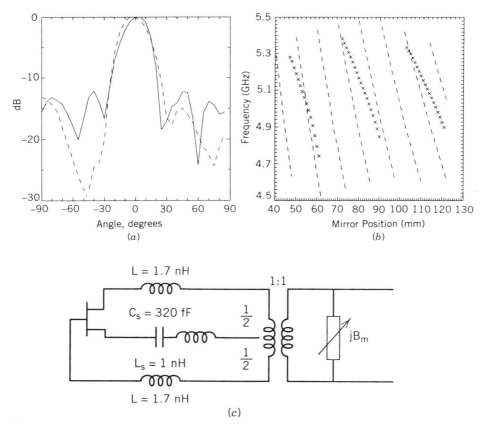

FIGURE 8.7. A 10×10 FET grid oscillator circuit and performance: (*a*) (—) *E*-plane and (- - -) *H*-plane antenna patterns of the MESFET grid oscillator. (*b*) Theoretical (—) and measured ($\times$) mirror tuning curves for the planar MESFET grid. The theoretical curve was obtained by including the MESFET in the model as a two-port with measured small-signal *s*-parameters. (*c*) Transmission line model for the planar MESFET grid. The values of the lumped elements are calculated from an EMF analysis of the grid. (From Ref. 26 with permission from IEEE.)

A description of several early power-combining arrays is described by Rutledge et al. in 1990 [26]. Of special interest is the 10×10 MESFET grid oscillator. Unlike an active antenna combiner, transistor spacing was only 0.133 wavelength. For each device the impedance presented to its terminals can be described by using the symmetry of the grid to develop an equivalent waveguide. EMF analysis is then used to determine the impedance matrix. Figures 8.6*a* and *b* show the Fabry-Perot cavity configuration and a 10×10 grid. A back plate and dielectric layer used to provide feedback to control the oscillation frequency are modeled using shunt reactances. Coupling of the transistor gate, drain, and source to various waveguide modes are modeled using a center-tapped transformer. Figures 8.7*a* and *b* show the far-field radiation patterns and the mirror-tuning characteristics for the MESFET grid oscillator, respectively. It operates at 5 GHz with an EIRP of 25 W and dc input power of 3 W. Actual radiated power of 0.625 W for the grid oscillator points to a 20% dc-to-rf conversion efficiency. The theoretical results shown in Figure 8.7*b* are based on the equivalent circuit in Figure 8.7*c*.

An EIRP of 28 W at 9.21 GHz for a 16-element MESFET array was shown by Rutledge et al. in 1992 [27]. The actual radiated power of 2 W is used to calculate a conversion efficiency of 28%.

Bundy et al. used the grid oscillator configuration and added a grid of varactors to create a grid VCO [28]. Figure 8.8 shows the grid VCO configuration, and Figure 8.9 shows two realizations of the approach. One unit cell consists of dipoles and the other uses bow tie antennas. The 7×7 grid of dipoles oscillated at 2.5 GHz, while the bow tie grid operates at 3.5 GHz. The grids can be tuned by using either the gate bias or the varactor bias level. One obtains 80- and 20-MHz tuning ranges for the dipole and bow tie grids, respectively. A maximum of 10% tuning was obtained for the bow tie grid by varactor tuning. There was less than 2 dB in power variation. The radiation

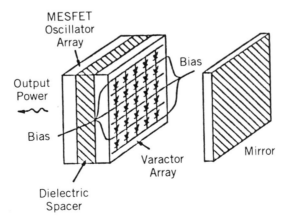

FIGURE 8.8. A quasi-optical VCO consists of a grid oscillator and a grid tuner backed by a mirror. The radiated frequency can be tuned by changing the bias of the varactor array. (From Ref. 28 with permission from IEEE.)

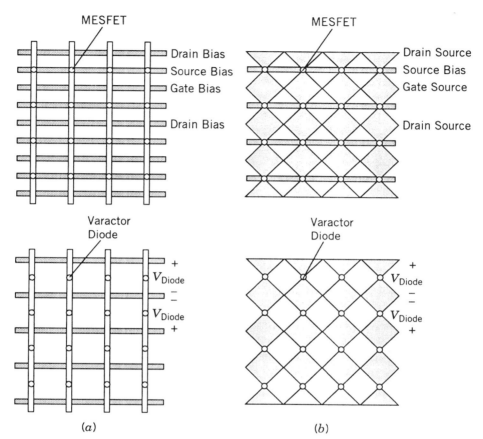

FIGURE 8.9. Grid VCO configurations: (*a*) The geometries of the dipole MESFET oscillator and varactor tuner arrays. The arrays consist of 49 elements each. (*b*) The geometries of the bow-tie MESFET oscillator and varactor tuner arrays with 49 elements each. (From Ref. 28 with permission from IEEE.)

pattern of the grid VCOs and gate bias-tuning results are shown in a later paper [29].

A 100-element MESFET grid oscillator was shown again by Popovic et al. [30]. The array oscillates at 5 GHz with an output power of 600 mW and conversion efficiency of 20%. The transistors' source leads are oriented horizontally with the drain and gate aligned vertically. For vertical polarization, the gate lead radiation causes strong gate coupling to the oscillation mode. The grid will then oscillate at lower frequencies where the transistors have higher

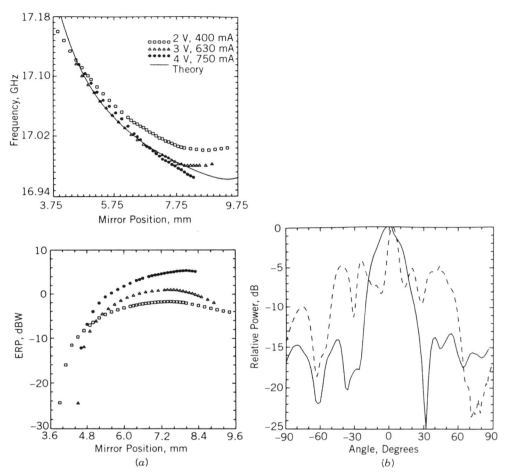

FIGURE 8.10. Ku-band grid oscillator performance: (*a*) Frequency and power tuning curves of the Ku-band MESFET grid for three different bias points. The measurements were taken without the front dielectric slab. (*b*) Measured far-field patterns in the *H*-plane (——) and *E*-plane (- - -) for the Ku-band grid oscillator. (From Ref. 31 with permission from IEEE.)

gain. To overcome gate coupling to the oscillation mode, Weikle et al. introduced a modified grid unit cell [31, 32]. The modification is possible through use of the transistor gate chip instead of a packaged device. The connections are such that the drain and source leads are oriented vertically to couple strongly to the incident field. The gate terminals are connected along

Twist reflector   Grid amplifier   Output polarizer   Tuning slab

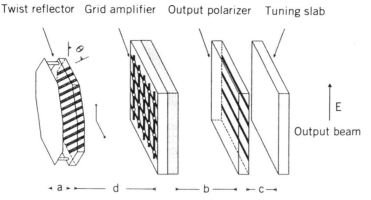

E

Output beam

◄ a ► ┣━━━ d ━━━┫   ┣━━ b ━━┫ ┣ c ┫

FIGURE 8.11. The external feedback oscillator. The twist reflector that provides the feedback replaces the input polarizer in a normal grid amplifier. The
distance between the amplifier surface and the twist reflector, d, was varied to
change the oscillation frequency. (From Ref. 33 with permission from IEEE.)

the horizontal traces and are thereby capacitively coupled to the vertically
polarized mode. A 16-element X-band grid produced 335 mW of power at
11.6 GHz with a 20% conversion efficiency.

Experimental results of a scaled grid operating at Ku-band are shown in
Figure 8.10. Figure 8.10a shows frequency and power versus mirror position
without the front dielectric slab, while the far-field patterns are shown in Figure
8.10b. The 36-element scaled grid produces 235 mW at 17 GHz with a 7%
conversion efficiency. The modified unit cell allows this scaling flexibility [31].
Good radiation performance is shown in the H-plane, but the E-plane pattern
is highly distorted.

A wideband source was created by Kim et al. in 1993, using a grid amplifier
and external feedback [33]. Feedback is accomplished by using a twist reflector
which sets the feedback phase and changes the output polarization to obtain
a good input match. The arrangement provides a very wide tuning range, more
so than previously reported with conventional grid oscillators. Adjustments of
the reflector and polarizer allow frequency tuning from 6.5 to 11.5 GHz with a
maximum EIRP of 6.3 W at 9.9 GHz. Smooth tuning can be accomplished
from 8.2 to 11.5 GHz by changing the grid to twist reflector spacing. Figure
8.11 shows the external feedback grid oscillator configuration, which uses a
twist reflector, output polarizer, and tuning slab. Power and frequency tuning
results are shown in Figure 8.12 for various twist reflector spacings.

A 10-W grid oscillator was demonstrated by Hacker et al. [34]. An EIRP
of 660 W at 9.8 GHz was demonstrated with an overall conversion efficiency of

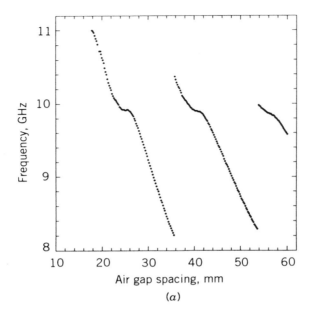

(a)

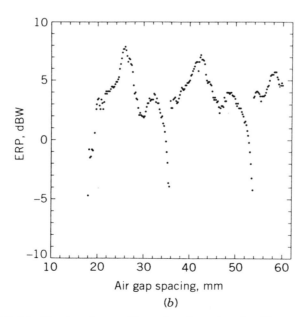

(b)

FIGURE 8.12. Tuning the oscillator by changing the distance between the grid amplifier and the twist reflector (dimension *d* in Fig. 8.11): (*a*) frequency; (*b*) ERP. (From Ref. 33 with permission from IEEE.)

23%. Figure 8.13a shows the grid oscillator unit cell and equivalent circuit. The calculated and measured radiation performance are shown in Figure 8.13b.

An interesting development in grid oscillators has been the concept of placing several grid oscillators back to back to form the "three-dimensional" power combiner [35]. A 3-dB increase in EIRP was shown for a dual grid over a single grid at 5 GHz. Figure 8.14 shows the dual-grid oscillator configuration. This idea was extended to a quadruple-grid oscillator combiner. The EIRP of the quadruple grid is 6.5 dB higher than a single grid, with less than 5% change in operating frequency [36]. The configuration is shown in Figure 8.15.

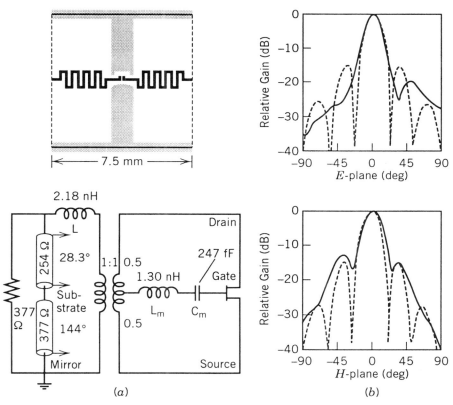

FIGURE 8.13. A 100-element grid oscillator: (a) Equivalent-waveguide unit-cell metal pattern for the grid oscillator. Boundary conditions are imposed by grid symmetry. The solid lines are electric walls and the dashed lines are magnetic walls. Equivalent-circuit model used to predict the theoretical performance of the grid at 10 GHz. (b) Measured E-plane and H-plane patterns (solid lines) and theoretical patterns (dashed lines) for the grid oscillator with 0.4-mm mirror spacing. Aperture blockage occurs in the E-plane for angles between −60° and −90°. (From Ref. 34 with permission from IEEE.)

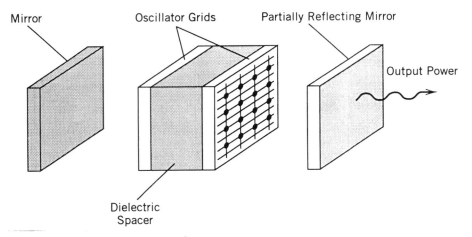

FIGURE 8.14. Three-dimensional grid oscillator. Two planar oscillator grids are placed back to back against a dielectric spacer. The two mirrors form the Fabry-Perot cavity. (From Ref. 35 with permission from IEEE.)

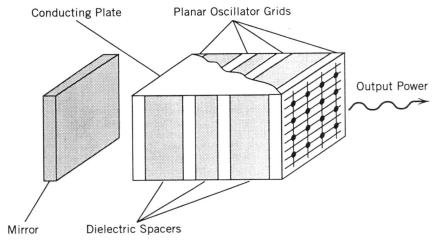

FIGURE 8.15. A quadruple grid oscillator, consisting of four planar oscillator grids stacked in parallel separated by dielectric spacers. An external mirror provides passive cavity feedback, while the grids mutually injection-lock each other through active feedback. (From Ref. 36 with permission from IEEE.)

8.4   GRID MULTIPLIERS

A novel quasi-optical multiplier design was shown by Archer in 1984 [37]. Grid multipliers were introduced at nearly the same time as the grid oscillator in 1988 by Hwu et al. [38, 39]. A grid of 1000 Schottky diodes used as a frequency doubler produced 0.5 W at 66 GHz with a doubling efficiency of 9.5%. The input source was pulsed at 33 GHz. Figure 8.16 shows the double configuration and the barrier-intrinsic-N^+ (BIN) diode used. The quasi-optical doubler uses two tuning slabs and filter to optimize the fundamental frequency input presented to the grid. Similar output filter and tuning slabs are used to match the output.

In 1989, a tripler array was demonstrated using the same BIN diode and concept [40]. Figures 8.17a and b show the tripler configuration and metal grid

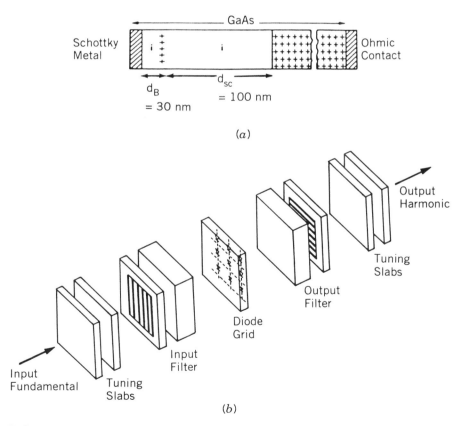

FIGURE 8.16. Quasi-optical diode grid doubler: (a) GaAs BIN diode. (b) Doubler configuration. (From Refs. 38 and 39 with permission from IEEE.)

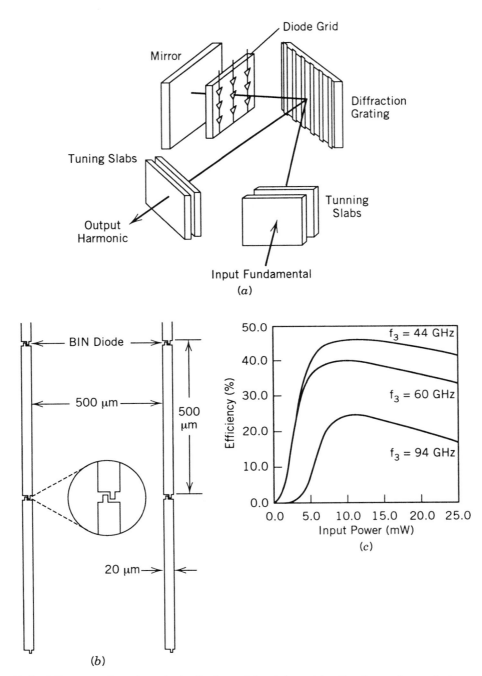

FIGURE 8.17. Quasi-optical diode grid tripler: (*a*) Configuration of the diode-grid tripler. (*b*) Design of metal grid for the BIN diode tripler array. (*c*) Tripling efficiency versus input power at various input frequencies. (From Ref. 40 with permission from IEEE.)

integrated with diodes. This configuration is a more compact structure, which takes advantage of a diffraction grating in conjunction with tuning slabs at specific angles to provide its tripler performance. Figure 8.17c shows various efficiency curves for the grid tripler.

8.5   GRID AMPLIFIERS

Kim et al. introduced another important quasi-optical grid component in 1991 with amplification function [41]. A 50-MESFET grid amplifier demonstrated 11-dB gain at 3.3 GHz. Figure 8.18 shows the grid amplifier configuration, unit-cell, and transmission line model. As shown, the amplifying grid accepts a vertically polarized input and produces a horizontally amplified output, thereby providing some input-to-output isolation. The unit cell is made up of a two-MESFET differential amplifier. Vertical leads attached to the gates of the transistors pick up incident radiation. The sources of the transistors are connected together to perform the differential amplification. Horizontal leads are attached to the drain of the transistors. The 5×5 array of differential amplifiers is tuned together with input and output polarizer layers. Figure 8.19 shows the amplifier gain as a function of frequency for the S-band grid amplifier. A narrow-band gain of 11 dB at 3.3 GHz was demonstrated.

A gain of 11 dB was reported at 9.9 GHz with a 16-element amplifier grid [27]. The unit cell used two HBT transistors to make the differential pair amplifiers. This design was improved to develop a 100-element HBT amplifying grid [42]. The grid uses an input and output polarizer and a tuning slab to optimize performance. Figure 8.20 shows a unit-cell and quasi-optical grid configuration. The radiation pattern of the grid amplifier was shown using a small grid oscillator as an input to the grid amplifier. Figure 8.21 shows the grid amplifier radiation patterns with the oscillator on and off. Figure 8.22 shows radiation patterns as a function of amplifier input incident angle. As shown, beam shifts of $\pm 20°$ are possible with little pattern degradation. Also shown in Figure 8.22 is a pattern as a function of grid rotation, which demonstrates the ability to receive a $\pm 30°$ polarized input and maintain within 3 dB of the maximum gain level.

8.6   GRID PHASE SHIFTERS, TRANSMITTANCE CONTROLLERS, AND SWITCHES

A natural extension of the use of frequency-selective surfaces would be to make a tunable FSS. A $70°$ phase-shift range with 6.5-dB loss was shown by Lam et al. in 1988 [43]. Sjogren et al. proposed a W-band $360°$ phase shifter [44]. Figure 8.23 shows the grid unit cell, array configuration, and simulated

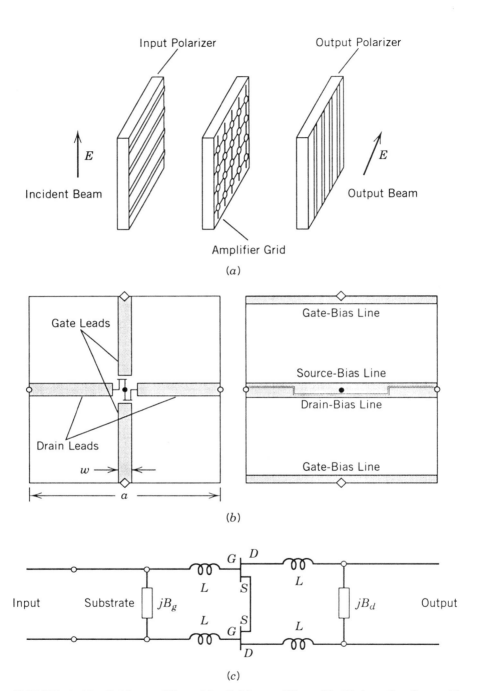

FIGURE 8.18. Grid amplifier: (*a*) Grid amplifier. (*b*) Unit cell of a grid amplifier: (left) front view, (right) back view. Symbols indicate different connections between the front and the back of the grid amplifier. ●: 120-Ω source-bias resistor, ◇: 1-kΩ gate-bias resistor, ○: drain-bias pin. (c) Transmission line model for the unit cell. (From Ref. 41 with permission from IEEE.)

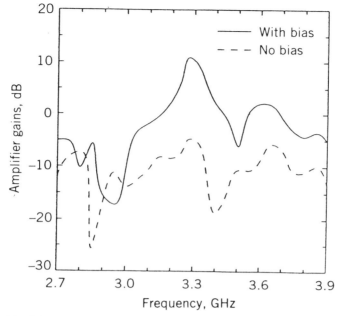

FIGURE 8.19. Measured grid amplifier gain with bias (solid line) and without bias (dashed line). The total incident power was about 300 μW. (From Ref. 41 with permission from IEEE.)

operation results. This investigation was pursued and later demonstrated in 1994 [45]. Measured results of reflectance magnitude and phase are shown in Figure 8.24 for frequencies of 118, 121, and 126 GHz. The arrays were used to control the power patterns through focusing.

Sjogren et al. showed results for a 4800-diode array in 1993 [46]. Figure 8.25 shows the array and the measurement setup. The measured data are given in Figure 8.26. Results are shown for frequencies from 99 to 165 GHz. Transmittance curves for the large array show the grid's crossover from capacitive to inductive at around 137 GHz. Also reported is power output versus bias modulation frequency. The array easily follows modulation control signals up to 100 MHz with less than 3-dB reduction in output power.

Bae et al. introduced a metal mesh coupler which uses an optical tunneling effect of evanescent waves present between a metal mesh and dielectric slab [47]. Figures 8.27a,b, and c show the optical-tunneling-type metal mesh coupler (OTM) configuration, experimental test setup, and transmittance results at 58 GHz, respectively. The metal mesh dimensions are optimized to change the transmittance by nearly 50%.

Transmittance control in a quasi-optical beam was also shown by Stephan and Goldsmith, using an array of 464 *pin* diodes [48]. The 464-*pin* diode quasi-optical reflection/transmission switch shows a reflection loss less than

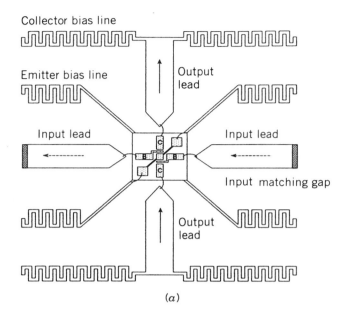

(a)

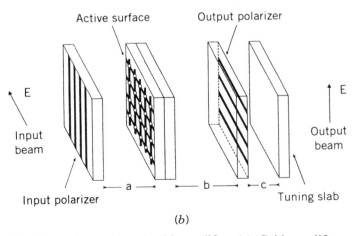

(b)

FIGURE 8.20. A 100-element HBT grid amplifier: (a) Grid-amplifier unit-cell design. The arrows indicate the direction of current flow. The unit cell size is 8 mm. The radiating leads are 0.8 mm wide, and the meandering bias lines are 0.1 mm wide. The span of the meander lines is 0.6 mm from top to bottom with 0.1-mm gaps. The input-matching gap is 0.1 mm wide. (b) View of the assembled grid amplifier. (From Ref. 42 with permission from IEEE.)

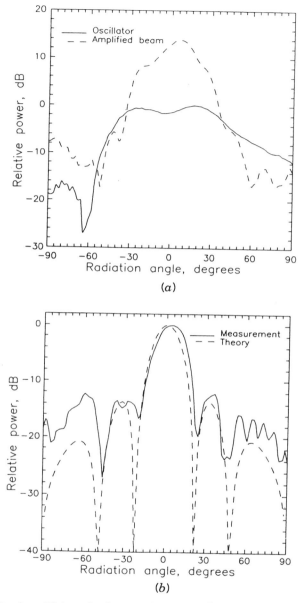

FIGURE 8.21. Amplifying the beam of a small 10-GHz grid oscillator with a grid amplifier. (*a*) H-plane pattern for the oscillator grid (solid line), and the *E*-plane pattern for the amplified beam (dashed line). The distance between the two grids was 1.6λ. Notice that both the grid-oscillator beam and the grid-amplifier beam are broad. (*b*) Spacing increased to 11.4λ. The solid line is the *E*-plane pattern of the amplified beam, now narrowed because the input phase and amplitude are uniform. The dashed line shows the theoretical pattern of a uniform array of 10 elementary dipoles 8 mm apart. (From Ref. 42 with permission from IEEE.)

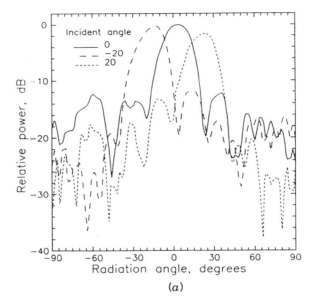

(a)

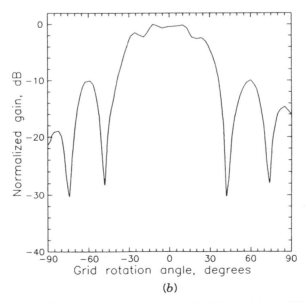

(b)

FIGURE 8.22. Amplifying beams at angles. (a) The grid-amplifier radiation patterns for three different incident beams. (b) Gain variation with incidence angle. (From Ref. 42 with permission from IEEE.)

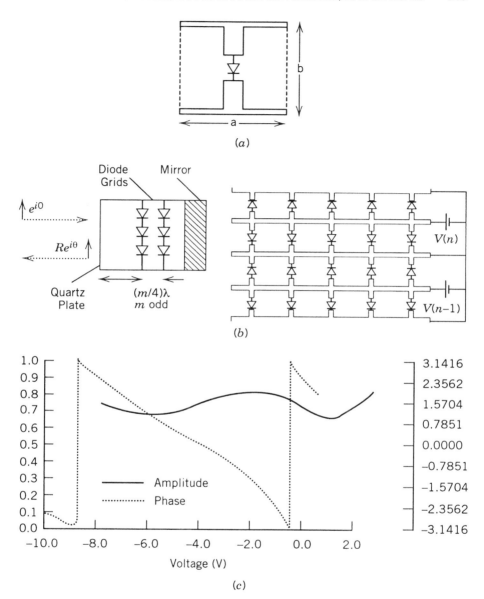

FIGURE 8.23. Diode grid phase shifter: (*a*) Unit cell. (*b*) Schematic illustration of the stacked 360° phase shifter. On the left, the cross-sectional structure of the phase shifter is shown. On the right, the arrangement of the cells in the array is shown. (*c*) Simulated performance of the stacked phase shifter. Diode resistance is assumed to be 10 Ω: parasitic capacitance is assumed to be 2 F. The left axis indicates reflection coefficient; the right axis indicates phase in radians.

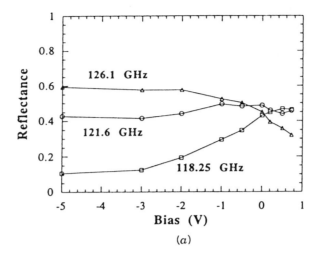

(a)

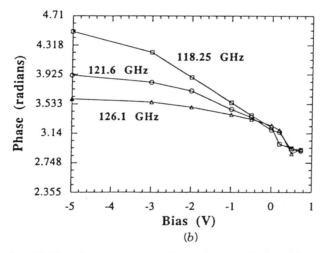

(b)

FIGURE 8.24. Grid reflectance controller: (a) magnitude of beam reflectance for the array. (b) Reflected beam phase curves for the array. Phase less than π represents net inductive reactance. Phase greater than π represents net capacitive reactance. (From Ref. 45 with permission from IEEE.)

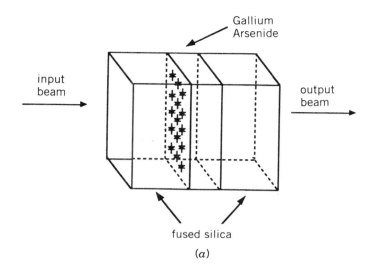

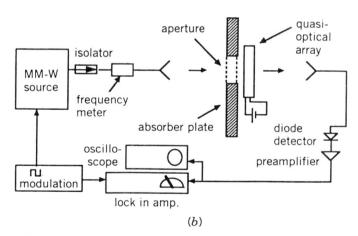

FIGURE 8.25. Array and test system: (*a*) Stacking configuration for transmission testing of the array. Array thickness is 0.635 mm. Fused-silica plate thickness is 1.16 mm. These thicknesses both correspond to $3\lambda/4$ at 99 GHz. (*b*) Quasi-optical system for measurement of transmitted beam amplitude. With the array stack removed, a reference beam of unity transmittance is measured. (From Ref. 46 with permission from IEEE.)

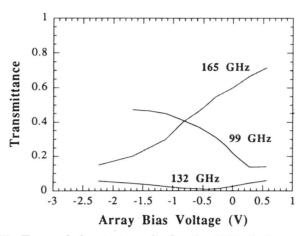

FIGURE 8.26. Transmission test results for the 4800-diode array. (From Ref. 46 with permission from IEEE.)

0.5 dB at 94 GHz when the diodes are off with a transmission loss of over 20 dB over a 12-GHz bandwidth. When the diodes are on, the transmission loss was 3.7 dB and the return loss was 9 dB at 94 GHz. Figure 8.28 shows the configuration and an equivalent transmission line model [49].

8.7 GRID MIXERS

The use of grid for mixers has truly shown the concept's versatility. Hacker and colleagues demonstrated a 100-element Schottky diode grid mixer in 1992 [50]. The quasi-optical grid mixer uses Schottky diodes integrated with a bow-tie grid. The local oscillator and rf signal strike the grid from the substrate side, while a metal mirror serves to tune the diodes reactance at the operating frequency. The if is easily extracted from the grid terminals.

Figure 8.29 shows the quasi-optical grid mixer configuration and equivalent circuit. The conversion loss is shown in Figure 8.30. The grid mixer's power handling and dynamic range increase with the number of devices in the grid. The 100-element grid shows an improvement of up to 19.8 dB over an equivalent single-diode mixer. The conversion loss and noise figure remain equal to a conventional mixer approach. The X-band grid mixer demonstrates one of the most important microwave components with reasonable results. At lower frequencies where tolerances are not as critical, grids offer higher power handling and redundancy but at a higher cost. At millimeter and submillimeter wavelengths where conventional approaches are not easily repeated due to tolerances and losses, the grid may be able to demonstrate its usefulness.

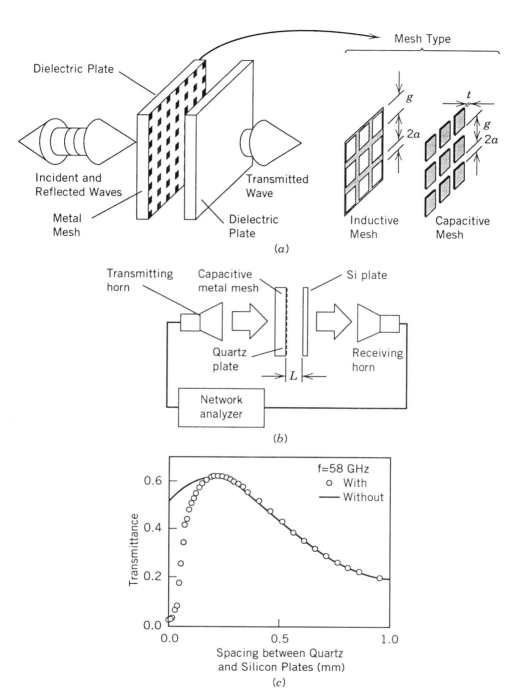

FIGURE 8.27. Metal mesh coupler for beam control: (*a*) Configuration of the OTM coupler. (*b*) Experimental setup. (*c*) Transmittance measured with and without a tunneling effect for the OTM coupler with a mesh plate of $g = 1.70$ mm.

223

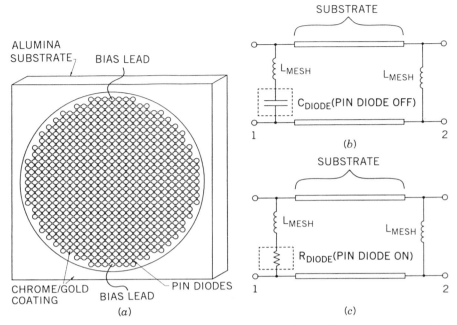

FIGURE 8.28. *pin* diode switch array: (*a*) View of 464-diode array with bias connections. (*b*) Plane-wave equivalent transmission line circuit of switch in reflection mode and (*c*) transmission mode. (From Ref. 49 with permission from IEEE.)

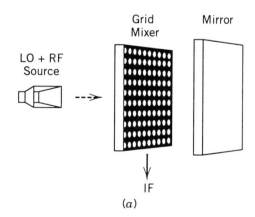

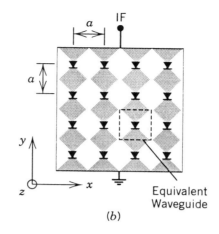

(b)

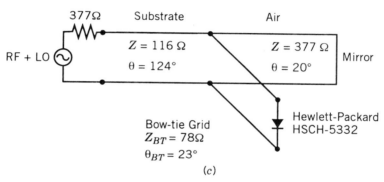

(c)

FIGURE 8.29. Diode grid mixer configuration and equivalent circuit: (a) Quasi-optical grid mixer configuration. The grid is mounted with the diodes facing the mirror. The incident rf and LO signals pass through the substrate which acts as an impedance transformer. The mirror is used to tune out the capacitive reactance of the Schottky diodes for a better match to free space. (b) Layout of the Schottky diode grid mixer. Boundary conditions are imposed by the grid symmetry. The solid lines are electric walls ($E_{\text{tangential}} = 0$), and the dashed lines are magnetic walls ($H_{\text{tangential}} = 0$). In our grid, $a = 3\,\text{mm}$. (c) Transmission line model for the grid mixer. The diode is modeled using the manufacturer's equivalent circuit. The reflection coefficient of the grid was matched to free space at the design frequency of 10 GHz. (From Ref. 50 with permission from IEEE.)

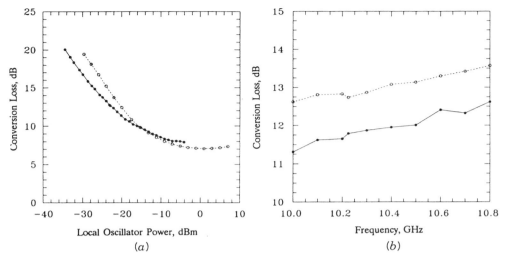

FIGURE 8.30. (*a*) Measured grid mixer conversion loss (——) and equivalent single-diode mixer conversion loss (- - -) as a function of LO power per diode for an LO frequency of 10.225 GHz and an if frequency of 214 MHz. The conversion loss of the grid mixer is comparable to the single-diode mixer. (*b*) Measured grid mixer conversion loss (——) and equivalent single-diode mixer conversion loss (- - -) as a function of frequency for a local oscillator power of −20 dBm per diode. The bandwidth of the grid mixer is primarily limited by the reactive tuning mirror. (From Ref. 50 with permission of IEEE.)

All of the grids described in this chapter provide a good starting point to the development of a useful quasi-optical system which can be constructed from the components described.

REFERENCES

1. J. D. Kraus, "A Backward Angle-Fire Array Antenna," *IEEE Transactions on Antennas and Propagation*, Vol. 12, No. 1, pp. 48–50, January 1964.

2. M. Tiuri, S. Tallgrist, and S. Urpo, "Chain Antennas," *IEEE Antennas and Propagation Symposium Digest*, pp. 274–277 (1974).

3. R. Conti, J. Toth, T. Dowling, and J. Weiss, "The Wire Grid Microstrip Antenna," *IEEE Transactions on Antennas and Propagation*, Vol. 29, No. 1, pp. 157–166, January 1981.

4. C. Chen, "Transmission of Microwaves through Perforated Flat Plates of Finite Thickness," *IEEE Transactions on Microwave Theory and Techniques*, Vol. 21, No. 1, pp. 1–6, January 1973.

5. J. A. Arnaud and F. A. Pelow, "Resonant-Grid Quasi-Optical Diplexers," *Bell System Technical Journal*, Vol. 54, No. 2, pp. 263–283, February 1975.

6. I. Anderson, "On The Theory of Self-Resonant Grids," *Bell System Technical Journal*, Vol. 54, No. 10, pp. 1725–1731, December 1975.

7. R. Watanabe, "A Novel Polarization Independent Beam Splitter," *IEEE Transactions on Microwave Theory and Techniques*, Vol. 8, No. 7, pp. 685–689, July 1980.

8. E. M. Sabbagh and R. George, "Microwave Energy Conversion," WADD Technical Report, Part I, April 1961.

9. E. M. Sabbagh and R. George, "Microwave Energy Conversion," WADD Technical Report, Part II, November 1961.

10. W. C. Brown, "Experiments in the Transportation of Energy by Microwave Beam," *IEEE MTT-S International Microwave Symposium*, pp. 8–17 (1964).

11. W. C. Brown, "Experimental Airborne Microwave Supported Platform," Final Report RADC-TR-65-188, December 1965.

12. W. C. Brown, "The History of Power Transmission by Radio Waves," *IEEE Transactions on Microwave Theory and Techniques*, Vol. 32, No. 9, pp. 1230–1242, September 1984.

13. S. W. Lee and T. T. Fong, "Electromagnetic Wave Scattering from an Active Corrugated Structure," *Journal of Applied Physics*, Vol. 43, No. 2, pp. 388–396, February 1972.

14. D. B. Rutledge and M. S. Muha, "Imaging Antenna Arrays," *IEEE Transactions on Antennas and Propagation*, Vol. 30, No. 4, pp. 535–540, July 1982.

15. S. W. Lee, "Scattering by Dielectric-Loaded Screen," *IEEE Transactions on Antennas and Propagation*, Vol. 19, No. 5, pp. 656–665, September 1971.

16. R. Kastner and R. Mittra, "Iterative Analysis of Finite-Sized Planar Frequency Selective Surfaces with Rectangular Patches or Perforations," *IEEE Transactions on Antennas and Propagation*, Vol. 35, No. 4, pp. 372–377, April 1987.

17. L. Brillouin, "Origin of Radiation Resistance," *Radioelectricite*, Vol. 3, pp. 147–152 (1922).

18. P. S. Carter, "Circuit Relations in Radiating Systems and Applications to Antenna Problems," *Proceedings IRE*, Vol. 20, No. 6, pp. 1004–1041, June 1932.

19. C. T. Tai, "A Study of the EMF Method," *Journal of Applied Physics*, Vol. 20, No. 7, pp. 717–723, July 1949.

20. R. S. Elliot, *Antenna Theory and Design*, Prentice-Hall, Englewood Cliffs, NJ, 1981, pp. 297–305.

21. Z. B. Popovic and D. B. Rutledge, "Diode Grid Oscillators," *IEEE Antennas and Propagation Symposium*, pp. 442–445, 1988.

22. Z. B. Popovic, M. Kim, and D. B. Rutledge, "Grid Oscillators," *International Journal of Infrared and Millimeter Waves*, Vol. 9, No. 7, pp. 647–654, July 1988.

23. L. B. Sjogren and N. C. Luhmann, Jr., "An Impedance Model for the Quasi-optical Diode Array," *IEEE Microwave and Guide Wave Letters*, Vol. 1, No. 10, pp. 297–299, October 1991.

24. A. Pance and M. J. Wengler, "Microwave Modeling of 2-D Active Grid Antenna Arrays," *IEEE Transactions on Microwave Theory and Techniques*, Vol. 41, No. 1, pp. 20–28, January 1993.

25. S. C. Bundy and Z. B. Popovic, "Analysis of Planar Grid Oscillators," *IEEE MTT-S International Microwave Symposium Digest*, pp. 827–830 (1994).

26. D. B. Rutledge, Z. B. Popovic, R. M. Weikle, M. Kim, K. A. Potter, R. C. Compton, and R. A. York, "Quasi-Optical Power-Combining Arrays," *IEEE MTT-S International Microwave Symposium Digest*, pp. 1201–1204 (1990).

27. D. B. Rutledge, J. B. Hacker, M. Kim, R. M. Weikle, R. P. Smith, and E. Sovero, "Oscillator and Amplifier Grids," *IEEE MTT-S International Microwave Symposium Digest*, pp. 815–818 (1992).

28. S. Bundy, T. B. Mader, and Z. B. Popovic, "Quasi-Optical Array VCOs," *IEEE MTT-S International Microwave Symposium Digest*, pp. 1539–1542 (1992).

29. T. Mader, S. Bundy, and Z. B. Popovic, "Quasi-Optical VCOs," *IEEE Transactions on Microwave Theory and Techniques*, Vol. 41, No. 10, pp. 1775–1781, October 1993.

30. Z. B. Popovic, R. M. Weikle, M. Kim, and D. B. Rutledge, "A 100-MESFET Planar Grid Oscillator," *IEEE Transactions on Microwave Theory and Techniques*, Vol. 39, No. 2, pp. 193–200, February 1991.

31. R. M. Weikle, M. Kim, J. B. Hacker, M. P. DeLisio, Z. B. Popovic, and D. B. Rutledge, "Planar MESFET Grid Oscillators Using Gate Feedback," *IEEE Transactions on Microwave Theory and Techniques*, Vol. 40, No. 11, pp. 1997–2003, November 1992.

32. R. M. Weikle, M. Kim, J. B. Hacker, M. P. DeLisio, Z. B. Popovic, and D. B. Rutledge, "Transistor Oscillator and Amplifier Grids," *Proceedings of the IEEE*, Vol. 80, No. 11, pp. 1800–1809, November 1992.

33. M. Kim, E. A. Sovero, J. B. Hacker, M. P. DeLisio, J. J. Rosenberg, and D. B. Rutledge, "A 6.5 GHz–11.5 GHz Source Using a Grid Amplifier with a Twist Reflector," *IEEE Transactions on Microwave Theory and Techniques*, Vol. 41, No. 10, pp. 1772–1774, October 1993.

34. J. B. Hacker, M. P. DeLisio, M. Kim, C. M. Liu, S. J. Li, S. W. Wedge, and D. B. Rutledge, "A 10 Watt X-Band Grid Oscillator," *IEEE MTT-S International Microwave Symposium Digest*, pp. 823–826 (1994).

35. W. A. Shiroma, B. L. Shaw, and Z. B. Popovic, "Three-Dimensional Power Combiners," *IEEE MTT-S International Microwave Symposium Digest*, pp. 831–834 (1994).

36. W. A. Shiroma, B. L. Shaw, and Z. B. Popovic, "A 100-Transistor Quadruple Grid Oscillator," *IEEE Microwave and Guided Wave Letters*, Vol. 4, No. 10, pp. 350–351, October 1994.

37. J. W. Archer, "A Novel Quasi-Optical Frequency Multiplier Design for Millimeter and Submillimeter Wavelengths," *IEEE Transactions on Microwave Theory and Techniques*, Vol. 32, No. 4, pp. 421–427, April 1984.

38. R. J. Hwu, C. F. Jou, W. W. Lam, U. Lieneweg, D. C. Streit, N. C. Luhmann, J. Maserjian, and D. B. Rutledge, "Watt-Level Millimeter-Wave Monolithic Diode-Grid Frequency Multipliers," *IEEE MTT-S International Microwave Symposium*, pp. 533–536 (1988).

39. C. F. Jou, W. W. Lam, H. Z. Chen, K. S. Stolt, N. C. Luhmann, and D. B. Rutledge, "Millimeter-Wave Diode-Grid Frequency Doubler," *IEEE Transactions on Micro-*

wave Theory and Techniques, Vol. 36, No. 11, pp. 1507–1514, November (1988).

40. R. J. Hwu, L. P. Sadwick, N. C. Luhmann, D. B. Rutledge, M. Sokolick, and B. Hancock, "Quasi-Optical Watt-Level Millimeter-Wave Monolithic Solid-State Diode-Grid Frequency Multipliers," *IEEE MTT-S International Microwave Symposium*, pp. 1069–1072 (1989).

41. M. Kim, J. J. Rosenberg, R. P. Smith, R. M. Weikle, J. B. Hacker, M. P. DeLisio, and D. B. Rutledge, "A Grid Amplifier," *IEEE Microwave and Guide Wave Letters*, Vol. 1, No. 11, pp. 322–324, November 1991.

42. M. Kim, E. A. Sovero, J. B. Hacker, M. P. DeLisio, J. C. Chiao, S. J. Li, D. R. Gagnon, J. J. Rosenberg, and D. B. Rutledge, "A 100-Element HBT Grid Amplifier," *IEEE Transactions on Microwave Theory and Techniques*, Vol. 41, No. 10, pp. 1762–1771, October 1993.

43. W. W. Lam, C. F. Jou, H. Z. Chen, K. S. Stolt, N. C. Luhmann, and D. B. Rutledge, "Millimeter-Wave Diode-Grid Phase Shifters," *IEEE Transactions on Microwave Theory and Techniques*, Vol. 36, No. 5, pp. 902–907, May 1988.

44. L. B. Sjogren, R. J. Hwu, H. X. King, W. Wu, X. H. Qin, N. C. Luhmann, M. Kim, and D. B. Rutledge, "Development of a Monolithic 94 GHz Quasi-Optical 360 Degree Phase Shifter," *International Infrared and Millimeter-Wave Conference Digest*, pp. 696–698 (1990).

45. L. B. Sjogren, H. X. Liu, X. H. Qin, C. W. Domier, and N. C. Luhmann, "Phased Array Operation of a Diode Grid Impedance Surface," *IEEE Transactions on Microwave Theory and Techniques*, Vol. 42, No. 4, pp. 565–572, April 1994.

46. L. B. Sjogren, H. X. Liu, F. Wang, T. Liu, X. H. Qin, W. Wu, E. Chung, C. W. Domier, and N. C. Luhmann, "A Monolithic Diode Array Millimeter-Wave Beam Transmittance Controller," *IEEE Transactions on Microwave Theory and Techniques*, Vol. 41, No. 10, pp. 1782–1790, October 1993.

47. J. Bae, J. C. Chiao, K. Mizuno, and D. B. Rutledge, "Metal Mesh Couplers Using Optical Tunneling Effect at Millimeter and Submillimeter Wavelengths," *IEEE MTT-S International Microwave Symposium Digest*, pp. 787–790 (1994).

48. K. D. Stephan and P. F. Goldsmith, "W-band Quasi-Optical Integrated PIN Diode Switch," *IEEE MTT-S International Microwave Symposium Digest*, pp. 591–594 (1992).

49. K. D. Stephan, F. H. Spooner, and P. F. Goldsmith, "Quasioptical Millimeter-Wave Hybrid and Monolithic PIN Diode Switches," *IEEE Transactions on Microwave Theory and Techniques*, Vol. 41, No. 10, pp. 1791–1798, October 1993.

50. J. B. Hacker, R. M. Weikle, M. Kim, M. P. DeLisio, and D. B. Rutledge, "A 100-Element Planar Schottky Diode Grid Mixer," *IEEE Transactions on Microwave Theory and Techniques*, Vol. 40, No. 3, pp. 557–562, March 1992.

CHAPTER NINE

Endfire Notches and Other Slotline Active Antennas

9.1 INTRODUCTION

As shown in Chapter 7, many investigators use the microstrip patch antenna for all types of passive and active device integrations. Microstrip patches make ideal planar, low-cost radiating elements. They provide resonant structures for devices to oscillate and a ground plane for efficient heat sinking. However, the majority of active microstrip patch antenna integrations have exhibited limited tuning ranges, high cross-polarization levels, and wide power output deviations. At millimeter-wave frequencies, small patch antenna dimensions cause difficulties during device integration. Dc bias lines also cause problems and degrade the performance. Furthermore, patches are not *easily* integrated with several devices for multiple functions due to their small size and the need for several distinct biasing pads. There are alternative configurations to the microstrip patch. Each has its pros and cons when applied to active and integrated antennas. The notch antenna is an attractive alternative which offers several advantages over the patch for solid-state device integration.

The notch antenna is the planar equivalent to a waveguide horn, and it is capable of very broad impedance bandwidths. If the length of the notch is short with respect to wavelength, then it behaves much like a resonant antenna with very broad half-power beamwidths. As the length is increased, it behaves more like a traveling-wave antenna. The slotline width taper provides a very broad impedance bandwidth, as shown by Gibson's exponentially tapered Vivaldi antenna [1]. Similarly, others have demonstrated nearly 5 to 1 impedance bandwidths [2]. These types of antennas, unlike microstrip patches, have been demonstrated up to 800 GHz.

When using very thin low-ε_r substrates with a reasonable impedance taper, notches can provide excellent radiation patterns over very wide bandwidths. Unlike the microstrip patch, the notch radiates primarily in the endfire

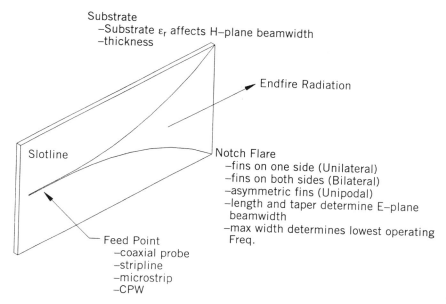

FIGURE 9.1. Notch antenna parameters.

direction. This fact is a mixed blessing since it provides a large area for device integration, biasing, and thermal dissipation, but its endfire radiation requires depth. When constructing arrays, this approach is often described as a brick style. In contrast, a patch antenna radiates in the broadside radiation, allowing the construction of planar, conformal arrays. This type of array, described as tile types, often makes a patch antenna more attractive than an endfire notch. However, for large-bandwidth applications where control over the radiation pattern is needed, notches are very easy to construct and provide very good performance.

Figure 9.1 shows the notch antenna configuration and the important parameters which determine its circuit and radiation properties. The length and flare of the taper can increase directivity as well as improve impedance-matching bandwidth [3]. Many variations of the flare of this antenna have been investigated. One such configuration used in several spatial power-combining schemes is referred to as the linear tapered slot antenna (LTSA) [4, 5, 6].

Since the notch antenna uses slotline, it is ideal for integration with two-terminal devices. The two fins form inherent biasing pads for integrating two-terminal devices. For three-terminal devices, the notch can be modified to accommodate another dc biasing terminal. The endfire orientation of the antenna gives greater volume for solid-state devices and heat sinking. The width of the slotline is used to develop matching sections and filters for integrated devices and is easily transformed into a radiating space wave [7]. Since it is a planar structure which can be photolithographically reproduced, integrated notch antennas have been scaled well into submillimeter wavelengths [8, 9].

9.2 ACTIVE NOTCH ANTENNAS AND POWER COMBINERS

For active antenna applications, the notch antenna, unlike the resonant microstrip patch, is seldom used by itself to induce oscillations. Active notch antennas typically use some sort of slotline or coplanar waveguide resonator to induce oscillations. These oscillations are consequently coupled to the notch antenna for radiation. It is interesting that one of the first Gunn diode integrated active antennas occurred in monolithic form, using a tapered slotline antenna [10]. The novel monolithic configuration used a planar CPW slotline resonator with a Gunn diode. A slotline and linear tapered slot antenna was loosely coupled to the Ka-band oscillator to allow easy testing of the 35-GHz oscillator frequency. Although it was not intentional, Figure 9.2 shows that the

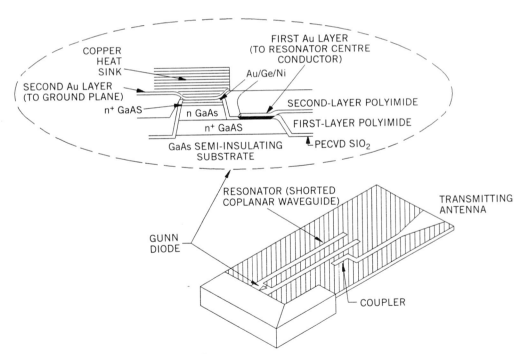

FIGURE 9.2. Monolithically Gunn integrated active notch antenna. (From Ref. 10 with permission from *Electronics Letters*.)

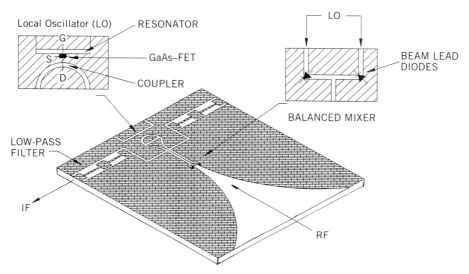

FIGURE 9.3. A 20-GHz notch-integrated receiver. (From Ref. 11 with permission from Wiley.)

planar monolithic integration can be considered an active integrated antenna. Since it was not meant to be a radiating oscillator or active antenna, its radiation properties were not mentioned.

One of the most complete hybrid integrations was demonstrated by Guttich in 1989 [11]. The notch antenna was integrated with coupled slotlines to create a complete rf front end. An FET transistor and a slotline resonator make the local oscillator (LO), which feeds a pair of diode 180° out of phase. The diodes are positioned at a slotline T-junction with one slotline which tapers out to a notch antenna. The rf comes in from the notch antenna and mixes with the diodes in phase to make a 20-GHz receiver front end, as shown in Figure 9.3. The local oscillator provides about 3 mW to the diode mixers, which provide 10- to 13-dB conversion loss for the 0- to 8-GHz intermediate frequencies. The radiation patterns are not discussed.

A simple bias-tuned Gunn oscillator was given by Navarro et al. in 1990. The active antenna uses a CPW resonator and a stepped-notch antenna [12]. Figures 9.4a and b show the configuration and equivalent circuit. The circuit consists of a stepped-notch antenna coupled to a coplanar waveguide (CPW) resonator via slotline. A Gunn diode is placed in a heat sink at the open terminals of the resonator. The notch incorporates many step transformers which match the slotline impedance to free space. The complete circuit is etched on 0.060-in. RT-Duroid 5870 substrate with ε_r of 2.3.

The configuration exhibits a clean and stable bias-tuned signal. The bias tuning creates a wide deviation in power output, and the resonator orientation

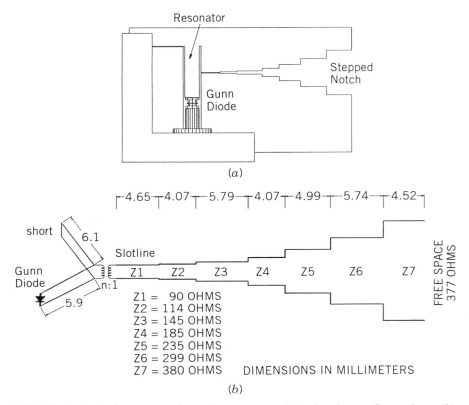

FIGURE 9.4. Active stepped-notch antenna: (*a*) circuit configuration; (*b*) equivalent circuit. (From Ref. 12 with permission of IEEE.)

introduces a strong cross-polarization component. Figure 9.5 shows the frequency and power output versus bias voltage. The 3-dB bias-tuning bandwidth is 275 MHz centered at 9.3 GHz with a maximum power output of 37.5 mW at 9.328 GHz. The radiation patterns exhibited good principal planes, but the cross-polarization levels were only 10 dB below the maximum. The heat sink introduces some asymmetry in the *E*-plane pattern, while the orientation of the CPW resonator increases the cross-polarization component.

A variation on the Gunn integrated notch, which included a varactor, was demonstrated by Navarro et al. the following year [13]. Figures 9.6*a* and *b* show the active varactor-tunable coupled slotline-CPW notch antenna configuration and its equivalent circuit. The circuit consists of a notch antenna integrated with a varactor-tuned coupled slotline-CPW resonator. A Gunn and a varactor diode are placed at either end of the coupled slotline-CPW resonator. The notch antenna couples to the resonator via slotline near the

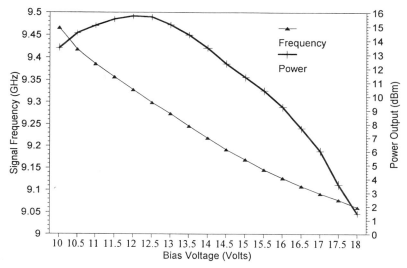

FIGURE 9.5. Frequency and power vs. bias voltage for the active stepped-notch antenna. (From Ref. 12 with permission from IEEE.)

center. The difference between the original CPW active antenna and the coupled slotline-CPW design is in the Gunn diode placement and resonator orientation. In the CPW design the Gunn is placed in the resonator to feed both slotlines symmetrically. In the coupled slotline-CPW design, the Gunn feeds only one slotline of the CPW which couples to the adjacent slotline where the varactor is mounted. The coupled slotline resonator orientation alleviates many of the cross-polarization problems encountered in the original design [14].

The coupled slotline-CPW resonator eases the integration and biasing of multiple devices. The Gunn diode is kept at a constant bias level, while the varactor voltage changes the oscillating frequency. Varactor tuning keeps the output power more level over frequency by maintaining the Gunn bias constant. The resonator configuration and antenna fins are used to provide bias to each element separately without filtering capacitors or filters. The slotline steps are modified to a smooth taper for improved radiation [15].

The wideband VCO active antenna uses a Gunn diode rated at 80 mW in an optimized waveguide circuit and an abrupt-junction varactor rated at 1.6 pF at 0 V. Figure 9.7 shows the theoretical and experimental frequency and power output results versus varactor tuning voltage for a Gunn bias of 13.5 V. A theoretical tuning curve derived from the equivalent circuit shown in Figure 9.6b along with Equations (9.1) and (9.2) predicted approximately 2% larger tuning range. Differences between the theory and experiment may be due to

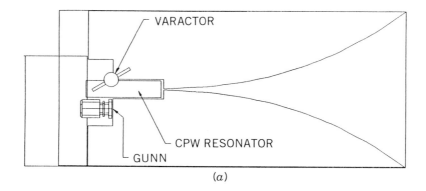

(a)

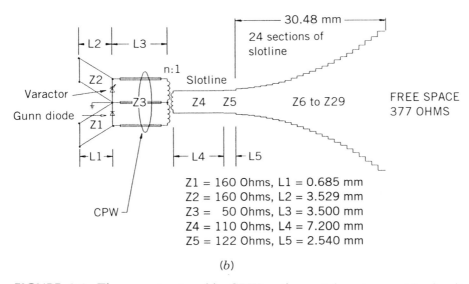

Z1 = 160 Ohms, L1 = 0.685 mm
Z2 = 160 Ohms, L2 = 3.529 mm
Z3 = 50 Ohms, L3 = 3.500 mm
Z4 = 110 Ohms, L4 = 7.200 mm
Z5 = 122 Ohms, L5 = 2.540 mm

(b)

FIGURE 9.6. The varactor-tunable CPW active notch antenna: (a) circuit configuration; (b) equivalent circuit. (From Refs. 13 and 14 with permission from IEEE.)

device integration discontinuities and placement errors within the resonator. The configuration exhibits a tuning bandwidth from 8.9 to 10.2 GHz with an output power of 14.5 ± 0.8 dBm. From Figure 9.6, the oscillation conditions are

$$|\mathrm{Re}(Z_{\mathrm{diode}})| \geqslant \mathrm{Re}(Z_{\mathrm{circuit}}) \tag{9.1}$$

$$\mathrm{Im}(Z_{\mathrm{diode}}) = -\mathrm{Im}(Z_{\mathrm{circuit}}) \tag{9.2}$$

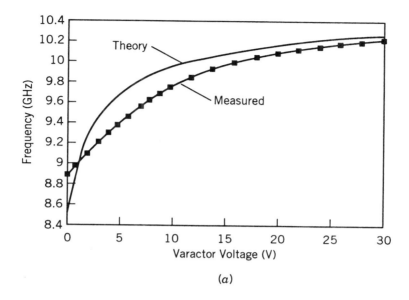

(a)

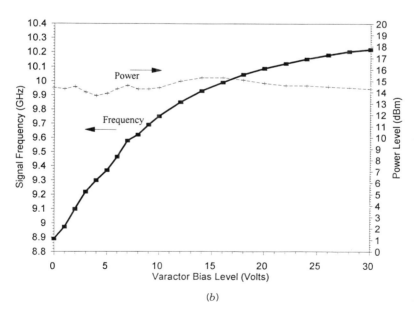

(b)

FIGURE 9.7. (a) Theoretical and experimental varactor timing results. (b) Experimental frequency and power vs. varactor bias voltage. (From Refs. 13 and 14 with permission from IEEE.)

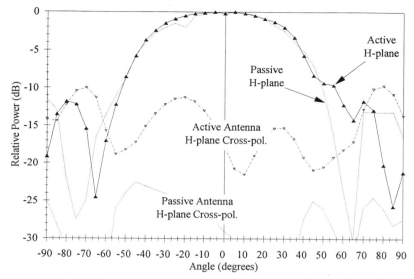

FIGURE 9.8. *H*-plane pattern comparison of a passive and active notch antenna at 10.2 GHz.

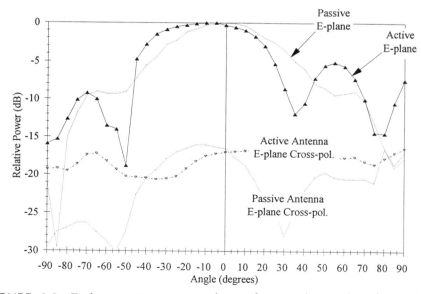

FIGURE 9.9. *E*-plane pattern comparison of a passive and active notch antenna at 10.2 GHz.

where $\text{Re}(Z_{\text{diode}})$ of the Gunn is assumed to be from -8 to $-10\,\Omega$ and the oscillations occur when condition 1 of Equation (9.1) is satisfied at the frequency specified by condition 2 of Equation (9.2).

Varactor voltages of 0 to 30 V achieve a frequency tuning range of 8.9 to 10.2 GHz. This is equivalent to over 14% electronic tuning bandwidth. There are no mode jumps and the signal spectrum remains clean and very stable. The output power varies only ± 0.8 dBm throughout the 1.3 GHz of frequency tuning range. The spectrum of the received signal from the varactor-tunable active notch antenna is comparable to other active antennas. Figure 9.8 shows a comparison between a passive and active notch antenna's H-plane radiation pattern at 10.2 GHz. The maximum cross-polarization level is higher at the lower frequencies but diminishes to less than -15 dB at 10.2 GHz. As shown, they are almost identical within $\pm 45°$ of the principal copolarization component. Differences outside this range as well as a large increase in the cross-polarization component are due to the active antenna device and resonator integration. Similarly, Figure 9.9 compares the passive and active notch antenna's E-plane pattern.

The wide tuning range for each individual active antenna provided the means to spatially power combine over a wide bandwidth. A two-element, H-plane power combiner showed efficient power combining over a 1.1-GHz tuning bandwidth from 9.0 to 10.1 GHz. Figure 9.10 shows the individual active antenna power levels, the measured power combiner power, and the calculated arithmetic sum of the individual oscillators. As shown, several points

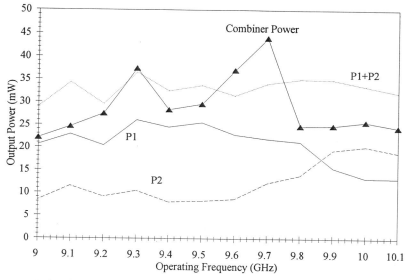

FIGURE 9.10. Two-element H-plane power-combiner results.

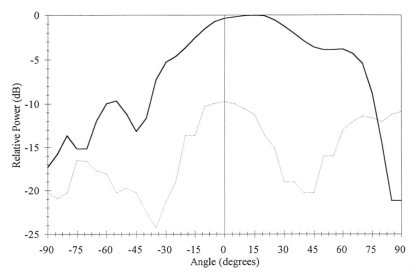

FIGURE 9.11. Measured H-plane pattern for two-element active notch spatial power combiner: (solid line — copolarization; dotted line — cross-polarization). (From Ref. 14 with permission from IEEE.)

exhibit over 100% efficiency. Specifically, over 129% is demonstrated near the middle operating frequency at 9.7 GHz. This point corresponds to approximately a quarter-wavelength separation and appears to provide the best mutual interaction for power combining. If each device had been individually optimized for highest output power, one would not expect to see higher than 100% efficiency. Due to the strong interaction between antennas, especially for an endfire design, optimization in an isolated environment will not provide an optimized spatial combiner array. The H-plane spatial power combiner pattern at 9.6 GHz is shown in Figure 9.11. Slight differences in the individual oscillating frequencies steer the spatial power combiner's beam off boresight.

To demonstrate the millimeter-wave application of this technology, a Ka-band active notch antenna was fabricated on a 0.762-mm-thick RT-Duroid 5880 substrate. A M/ACOM Gunn diode was integrated into the CPW resonator in the same fashion as shown in Figure 9.6. An output power of 15 mW was achieved at 37 GHz [15].

In order to provide the possibility of multiple functions and higher conversion efficiencies, the notch antenna can be integrated with a transistor or a complete monolithic chip. A notch antenna which uses a FET to form an oscillator at C-band was shown by Leverich et al. in 1992 [16, 17]. The active antenna configuration is shown in Figure 9.12a. The FET integrated notch antenna operates at 6.98 GHz with an output power of 8.9 mW and a 2% bias-tuning bandwidth. The dc-to-rf conversion efficiency is 7.4%. Figures 9.12b

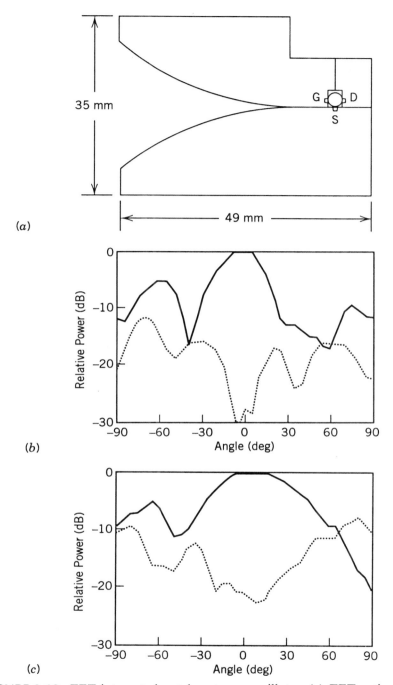

(a)

(b)

(c)

FIGURE 9.12. FET integrated notch antenna oscillator: (*a*) FET active notch antenna circuit layout; (*b*) *E*-plane radiation pattern (—— copolar, - - - cross-polar); (*c*) *H*-plane radiation pattern (—— copolar, - - - cross-polar). (From Ref. 16 with permission from *Electronics Letters*.)

show the radiation patterns of the active antenna. As shown, the E- and H-plane half-power beamwidths are 40° and 65°, respectively. The cross-polarization levels are -10 dB below the maximum.

These notch antennas were configured in 1×4 and 2×2 spatial power combiners [17]. The antenna spacing of a 1×4 H-plane linear array was nearly one wavelength. A 5-mm-thick Plexiglass ($\varepsilon_r = 2.6$) at one half-wavelength in front of the linear array was used to synchronize the spatial combiner. A 93% rf combining efficiency was measured for the four-element active antennas. The H-plane beam was narrowed from 65° to 15°. A 1×4 E-plane array demonstrated a 15° half-power beamwidth and an 83% rf combining efficiency.

These antennas were then set up in a planar 2×2 array with E- and H-plane spacing of one wavelength. The combining efficiency of the 2×2 array was 72%. The E- and H-plane half-power beamwidths are both 25°.

9.3 ACTIVE NOTCH ANTENNA AMPLIFIERS AND MULTIPLIERS

The integration of FETs provided a more efficient alternative to the Gunn integrated notch antenna. With a slight increase in biasing complexity, the transistor also provided the ability for amplification as shown by Wu and Chang in 1993 [18]. The CPW-fed FET integrated slot-CPW antenna amplifier configuration is shown in Figure 9.13a. The passive antenna has a gain of 7 dB at 9.2 GHz with a return loss of 20 dB over a bandwidth of 1.55 GHz. The active antenna amplifier has a gain of over 14 dB over a 1.75-GHz bandwidth. It added over 7 dB of gain over a passive antenna at 9.2 GHz. The active amplifying antenna gain versus frequency is shown in Figure 9.13b, while the radiation patterns are shown in Figures 9.13c and d. As shown, the principal planes are smooth and well behaved and cross-polarization levels are 13 dB below the maximum. The configuration makes an ideal single-antenna amplifier, but its use in a spatial array would be limited in its present form. A large array would require an rf feed distribution network in a brick-style array approach which increases complexity and losses.

The integration of a complete monolithic chip at the notch terminals for a variety of functions would then seem to be the next logical step in the evolution of this active antenna. Simons and Lee integrated a MMIC amplifier with an antipodal notch in 1993 [19]. Each module consists of a receive antenna, an amplifier, and a transmit antenna as shown in Figure 9.14a. The combination has a power added efficiency of 14% and a gain of 11 dB over the frequency band as shown in Figure 9.14b. Figure 9.14c shows the single-element radiation pattern with and without the amplifier. Figure 9.14d shows the radiation patterns for spatial power combiner with and without the use of the amplifiers. This integration is an ideal active antenna with potential for multiple functions. The monolithic chip can be miniaturized and optimized on gallium arsenide or

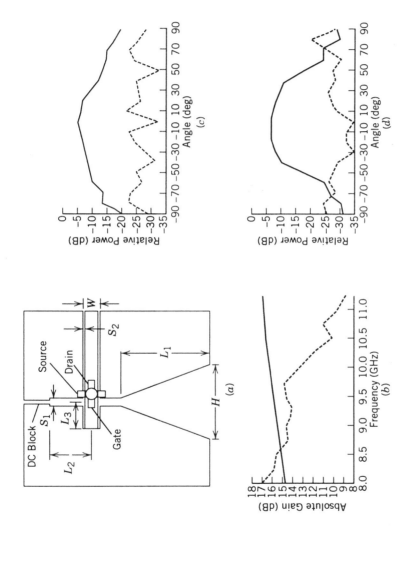

FIGURE 9.13. FET integrated notch amplifier: (*a*) Circuit configuration of **CPW** cross-feed linear tapered slot antenna amplifier. (*b*) Absolute gain of active antenna amplifier (——— NARDA model 640, standard horn, ——— active slot antenna amplifier). (*c*) Measured *E*-plane pattern of active antenna amplifier (——— copolarization, ——— cross-polarization). (*d*) Measured *H*-plane pattern of active antenna amplifier (——— copolarization, ——— cross-polarization). (From Ref. 18 with permission from *Electronics Letters*.)

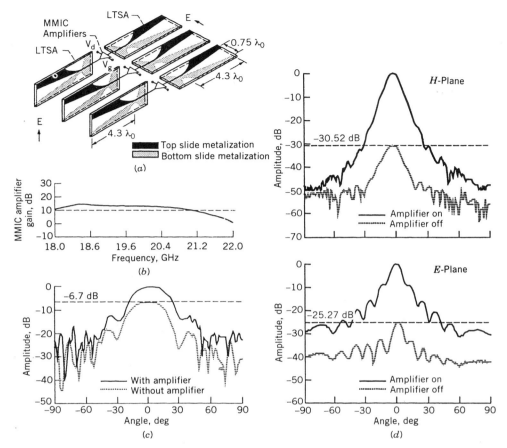

FIGURE 9.14. Notch antenna spatial amplifier. (*a*) Schematic illustrating the three-element array module (λ_0: free-space wavelength). (*b*) Typical measured gain of MMIC amplifier. (*c*) Measured *H*-plane radiation pattern of a single LTSA with and without the amplifier. (*d*) Measured radiation pattern of the horn antenna showing space power amplification. (From Ref. 19 with permission from IEEE.)

silicon and later integrated with an antenna whose characteristics are optimized on a low-ε_r substrate.

Simons and Lee later showed a similar setup to demonstrate a spatial frequency multiplier [20]. The multiplier uses a GaAs MMIC amplifier chip with a 5 ± 1 dB gain from 2 to 22 GHz. The power added efficiency at a power output of 0.5 W is 14%. The antenna configuration as shown in Figure 9.15*a* uses a linear tapered slot antenna (LTSA). The spatial multiplier maintains

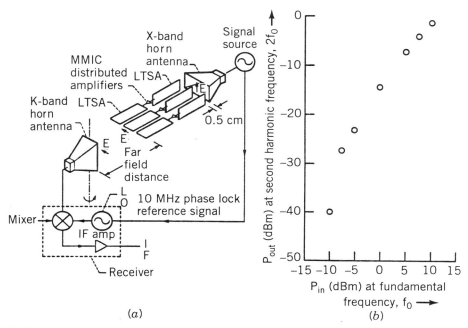

(a)   *(b)*

FIGURE 9.15. Spatial frequency multiplier using notch antennas: (*a*) Schematic illustrating an experimental setup for frequency multiplication and space power combining. (*b*) Measured second harmonic output power as a function of input power at a fixed fundamental frequency of 9.3 GHz. (From Ref. 20 with permission from IEEE.)

relatively constant second harmonic power as the input is varied from 2 to 10 GHz. The maximum output power occurs at 9.3 GHz. Figure 9.15*b* shows the second harmonic power output varied from -40 to 0 dBm for a -10 to $+10$ dBm variation of the fundamental input power at a frequency of 9.3 GHz. The maximum conversion efficiency of 8.1% occurs at an input power of 10 dBm. The hybrid integration of MMICs and antennas greatly increases the multifunction capabilities of integrated antennas systems.

9.4   ACTIVE SLOT OR SLOTLINE ANTENNAS

The notch antenna with all of its flexibility and advantages requires the use of the brick-style approach in the development of an array. This approach has shown increased surface area for devices and heat sinking, but it requires depth. For many applications where depth is not available, the alternative tile approach is required. In these instances, planar, broadside radiators such as

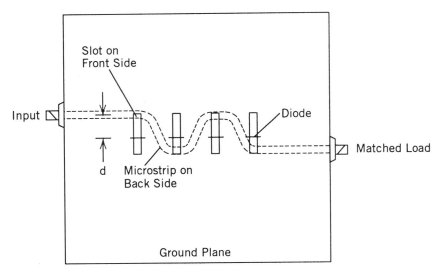

FIGURE 9.16. Structure of multiplying slot array fed by microstrip. (From Ref. 22 with permission from IEEE.)

microstrip patches or dipoles are used. A planar broadside radiator can also be constructed from a resonant slotline. Several investigators have used the active slotline antenna for spatial power combination. There are several possible configurations such as slotline dipoles, loops, or rings. Unlike a microstrip patch, this antenna is a bidirectional radiator which makes it ideal for spatial amplifier applications. However, it presents some problems for distributed oscillators. The use of a reflector plane or cavity has demonstrated some success for using this type of antenna in power combining.

Slot antennas for active antenna quasi-optical multipliers have been proposed since the mid-1980s [21]. Another spatial multiplying diode array was demonstrated in 1987 by Nam et al. [22]. The configuration shown in Figure 9.16 uses a microstrip to feed an array of diode-loaded slotline antennas. The diodes are positioned to maximize the radiation performance of the multipler. A 26-dBm, 5.4-GHz input is multiplied to 10.8 GHz. In the 1990s, Kawasaki and Itoh have used a similar configurations for microstrip oscillators [23, 24, 25]. A microstrip oscillator using slot antennas has advantages of simplicity and compactness. Figure 9.17 shows the configurations of a single element and six-element quasi-optical oscillators [25].

Active slotline antennas integrated with an IMPATT diode was investigated by Luy et al. in 1993 [26]. The authors model the real and imaginary device impedance by two closed-form equations. The equations can be used with (9.1) and (9.2) to obtain an oscillation frequency. The use of these equations to model the diode impedance in a transmission line program is ideally suited for other types of configuration.

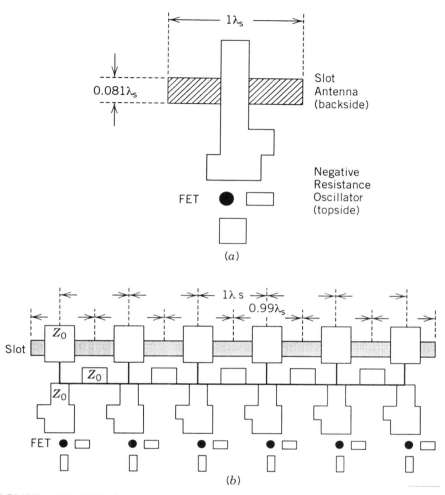

FIGURE 9.17. FETs integrated with slot antennas. (*a*) Configuration of active antenna with single FET and single slot. (*b*) Configuration of six-element periodic linear power-combining array. (From Ref. 25 with permission from IEEE.)

An active slot dipole antenna amplifier was shown by Wu and Chang in 1993 [27]. The CPW-fed FET integrated slot-CPW antenna amplifier configuration is shown in Figure 9.18*a*. The passive antenna operates at 7.1 GHz with a return loss of 18 dB over a bandwidth of 800 MHz. The FET integrated slot dipole amplifier added 7 dB of gain over a passive antenna. The active amplifying antenna gain versus frequency is shown in Figure 9.18*b*, while the radiation patterns are shown in Figures 9.18*c* and *d*. As shown, the principal

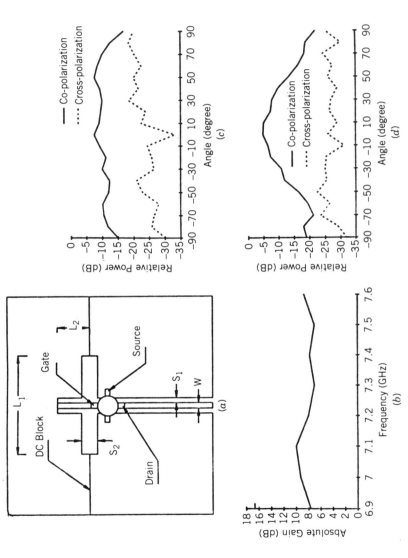

FIGURE 9.18. FET integrated slot antenna amplifier: (*a*) Circuit configuration of a CPW cross-fed slot dipole antenna amplifier. (*b*) Absolute gain of the active antenna amplifier. (*c*) Measured *E*-plane pattern of the active antenna amplifier. (*d*) Measured *H*-plane pattern of the active antenna amplifier. (From Ref. 27 with permission from Wiley.)

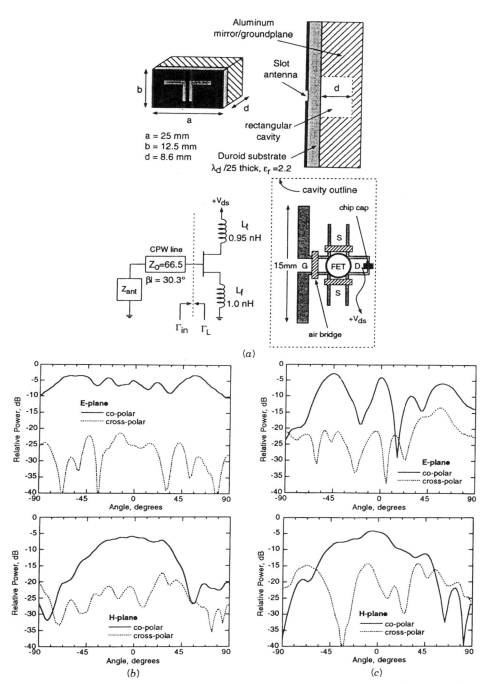

FIGURE 9.19. Active cavity-backed slot antenna using FETs. (*a*) Circuit and antenna configuration. (*b*) Measured radiation patterns for an X-band proto-type with cavity backing. Ripples in the *E*-plane are due to ground plane edge diffraction. (*c*) Measured radiation patterns with cavity backing replaced by a ground plane. Note the change in the *E*-plane copolar pattern as compared with the cavity-backed slot, and the increased cross-polarization in both pattern planes. (From Ref. 28 with permission from IEEE.)

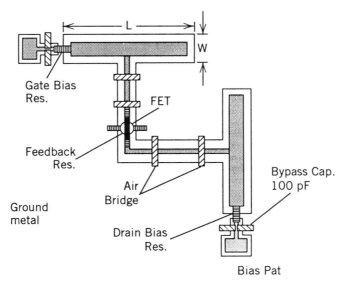

FIGURE 9.20. Plan view of the folded-slot amplifier cell. The drain-gate feedback resistor is soldered piggy-back to the FET. (From Ref. 29 with permission from IEEE.)

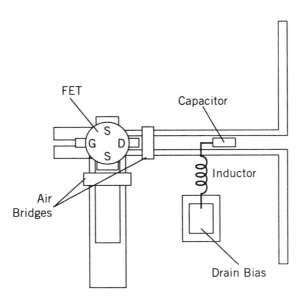

FIGURE 9.21. CPW oscillator designed for 10 GHz constructed using a packaged MESFET. (From Ref. 30 with permission from *Electronics Letters.*)

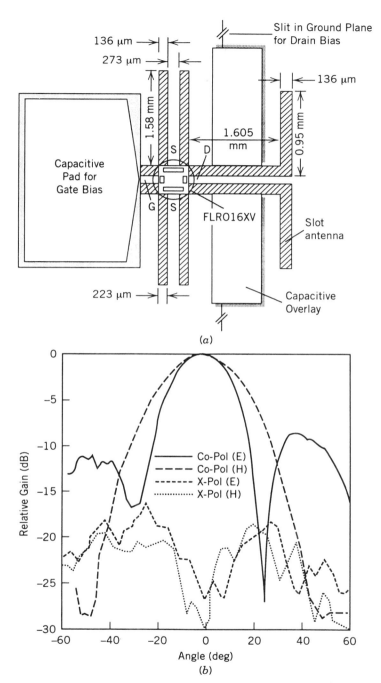

FIGURE 9.22. FET integrated slot antenna oscillator: (*a*) CPW circuit layouts for the 20 GHz oscillators showing the capacitive bypassing to allow application of dc bias. (*b*) Measured radiation patterns of a 20-GHz slot oscillator on a 2.6-cm silicon substrate lens. (From Ref. 32 with permission from IEEE.)

planes are smooth and well behaved, and cross-polarization levels are 11 dB below the maximum.

A FET integrated slotline antenna oscillator was demonstrated by Moyer and York in 1993 [28]. The cavity-backed slot antennas used MESFETs and a combination of CPW and slotline to develop the active antenna. The configuration demonstrates improvements of a cavity over a ground plane. Figure 9.19 shows the FET integrated CPW/slotline active antenna configuration. It also shows the differences of using a ground plane or a cavity to back the slotline antenna. As shown, the cavity removes problems encountered in the elevation plane as well as reducing H-plane cross-polarization levels.

A folded slot antenna configuration was shown by Tsai et al. in 1994 [29]. Unlike the previously mentioned amplifier antenna, the input and output ports occur through folded slot antennas. The input and output are polarized orthogonally to each other for isolation. Figure 9.20 shows the configuration of the folded slot amplifier. The peak effective isotropic power gain in the transmission mode is 11 dB at 4.3 GHz with 10% bandwidth for a single cell. A 4×4 array was also built with a gain of 32 dB at 4.24 GHz and 8% bandwidth.

Vaughan and Compton showed a 10-GHz active slot dipole antenna fed by a CPW oscillator in 1993 [30]. Figure 9.21 shows the configuration. A similar antenna was used to obtain the highest-frequency quasi-optical slot oscillator in 1994 [31]. InP HFET devices were used. Active slot antennas operated at 155 GHz with 10 μW of output power. Similarly, 215-GHz active antennas with 1 μW of output power were demonstrated. The same configuration was

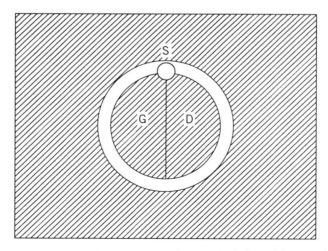

FIGURE 9.23. Active slotring antenna. (From Ref. 33 with permission from *Electronics Letters.*)

investigated at microwave frequencies by Kormanyos et al. [32]. Figures 9.22*a* and *b* show the K-band configuration and radiation performance of the structure.

Finally, a slotline ring resonator antenna was integrated with an FET by Ho and Chang in 1993 [33]. Figure 9.23 shows the circuit configuration. Very good radiation patterns and performance were demonstrated at 7.73 GHz. A simple transmission line method was used to predict the resonant frequency. The active antenna radiated 21.6 mW with 18% efficiency.

REFERENCES

1. P. J. Gibson, "The Vivaldi Aerial," *Proceedings of 9th European Microwave Conference*, Brighton, UK, pp. 101–105, 1979.

2. S. N. Prasad and S. Mahapatra, "A New MIC Slot-Line Aerial," *IEEE Transactions on Antennas and Propagation*, Vol. 31, pp. 525–527, May 1983.

3. H. Jingxi and F. Zhibo, "Analysis of Vivaldi Antennas," *IEEE 6th International Conference on Antennas and Propagation*, Part 1, pp. 206–208, 1989.

4. K. S. Yngvesson, T. L. Korzeniowski, Y. S. Kim, E. L. Kollberg, and J. F. Johansson, "The Tapered Slot Antenna—A New Integrated Element for Millimeter-Wave Applications," *IEEE Transactions on Microwave Theory and Techniques*, Vol. MTT-37, No. 2, pp. 365–374, February 1989.

5. R. Janaswamy, D. H. Schaubert, and D. M. Pozar, "Analysis of the Transverse Electromagnetic Mode Linearly Tapered Slot Antenna," *Radio Science*, Vol. 21, No. 5, pp. 797–804, September-October 1986.

6. K. S. Yngvesson, D. H. Schaubert, T. L. Korzeniowski, E. L. Kollberg, T. Thungren and J. F. Johansson, "Endfire Tapered Slot Antennas on Dielectric Structures," *IEEE Transactions on Antennas and Propagation*, Vol. 33, No. 12, pp. 1392–1400, December 1985.

7. R. Janaswamy and D. H. Schaubert, "Characteristic Impedance of a Wide Slotline on Low Permittivity Substrates," *IEEE Transactions on Microwave Theory and Techniques*, Vol. 34, No. 8, pp. 900–902, August 1986.

8. H. Ekstrom, S. Gearhart, P. R. Acharya, G. M. Rebeiz, E. L. Kollberg, ancd S. Jacobsson, "348 GHz Endfire Slotline Antennas on Thin Dielectric Membranes," *IEEE Microwaves and Guided Wave Letters*, Vol. 2, No. 9, pp. 357–358, September 1992.

9. P. R. Acharya, H. Ekstrom, S. S. Gearhart, S. Jacobson, J. F. Johansson, E. L. Kollberg, and G. M. Rebeiz, "Tapered Slotline Antennas at 802 GHz," *IEEE Transactions on Microwave Theory and Techniques*, Vol. 41, No. 10, pp. 1715–1719, October 1993.

10. N. Wang, S. E. Schwarz, and T. Hierl, "Monolithically Integrated Gunn Oscillator at 35 GHz," *Electronics Letters*, Vol. 20, No. 14, pp. 603–604, July 1984.

11. U. Guttich, "Planar Integrated 20 GHz Receiver in Slotline and Coplanar Waveguide Technique," *Microwave and Optical Technology Letters*, Vol. 2, No. 11, pp. 404–406, November 1989.

12. J. A. Navarro, Y. H. Shu and K. Chang, "Active Endfire Antenna Elements and Power Combiners Using Notch Antennas," *IEEE MTT-S International Microwave Symposium Digest*, pp. 793–796, 1990.

13. J. A. Navarro, Y. H. Shu, and K. Chang, "Wideband Integrated Varactor-Tunable Active Notch Antennas and Power Combiners," *IEEE MTT-S International Microwave Symposium Digest*, pp. 1257–1260, June 1991.

14. J. A. Navarro, Y. Shu, and K. Chang, "Broadband Electronically Tunable Planar Active Radiating Elements and Spatial Power Combiners Using Notch Antennas," *IEEE Transactions on Microwave Theory and Techniques*, Vol. MTT-40, No. 2, pp. 323–328, February 1992.

15. J. A. Navarro and K. Chang, "Broadband Electronically Tunable Planar Active Radiating Elements and Spatial Power Combiners Using Notch Antennas," *Microwave Journal*, Vol. 35, No. 10, pp. 87–101, October 1992.

16. K. Leverich, X. D. Wu, and K. Chang, "New FET Active Notch Antenna," *Electronics Letters*, Vol. 28, No. 24, pp. 2239–2240, November 1992.

17. K. Leverich, X. D. Wu, and K. Chang, "FET Active Slotline Notch Antennas for Quasi-Optical Power Combining," *IEEE Transaction on Microwave Theory and Techniques*, Vol. 41, No. 9, pp. 1515–1518, September 1993.

18. X. D. Wu and K. Chang, "Compact Wideband Integrated Active Slot Antenna Amplifier," *Electronics Letters*, Vol. 29, No. 5, pp. 496–497, March 1993.

19. R. N. Simons and R. Q. Lee, "Space Power Amplification with Active Linearly Tapered Slot Antenna Array," *IEEE MTT-S International Microwave Symposium Digest*, pp. 623–626, 1993.

20. R. N. Simons and R. Q. Lee, "Spatial Frequency Multiplier with Active Linearly Tapered Slot Antenna Array," *IEEE MTT-S International Microwave Symposium Digest*, pp. 1557–1560, 1994.

21. N. Camilleri and T. Itoh, "A Quasi-Optical Multiplying Slot Array," *IEEE Transactions on Microwave Theory and Techniques*, Vol. 33, No. 11, pp. 1189–1195, November 1985.

22. S. Nam, T. Uwano, and T. Itoh, "Microstrip-Fed Planar Frequency Multiplying Space Combiner," *IEEE Transactions on Microwave Theory and Techniques*, Vol. 35, No. 12, pp. 1271–1276, December 1987.

23. S. Kawasaki and T. Itoh, "A Layered Negative Resistance Amplifier and Oscillator Using FETs and a Slot Antenna," *IEEE MTT-S International Microwave Symposium Digest*, pp. 1261–1264, 1991.

24. S. Kawasaki and T. Itoh, "40 GHz Quasi Optical Second Harmonic Spatial Power Combiner Using FETs and Slots," *IEEE MTT-S International Microwave Symposium Digest*, pp. 1543–1546, 1992.

25. S. Kawasaki and T. Itoh, "Quasi-Optical Planar Arrays with FETs and Slots," *IEEE Transactions on Microwave Theory and Techniques*, Vol. MTT-41, No. 10, pp. 1838–1844, October 1993.

26. J. F. Luy, J. Bueschler, M. Thieme, and E. Biebl, "Matching of Active Millimetre-Wave Slot-Line Antennas," *Electronic Letters*, Vol. 29, No. 20, pp. 1772–1774, September 1993.

27. X. D. Wu and K. Chang, "Compact Wideband Integrated Active Slot Dipole Antenna Amplifier," *Microwave and Optical Technology Letters*, Vol. 6, No. 15, pp. 856–857, December 1993.

28. H. P. Moyer and R. A. York, "Active Cavity-Backed Slot Antenna Using MESFETs," *IEEE Microwave and Guided Wave Letters*, Vol. 3, No. 4, pp. 95–97, April 1993.

29. H. S. Tsai, M. J. W. Rodwell, and R. A. York, "Planar Amplifier Array with Improved Bandwidth Using Folded-Slots," *IEEE Microwave and Guided Wave Letters*, Vol. 4, No. 4, pp. 112–114, April 1994.

30. M. J. Vaughan and R. C. Compton, "Resonant-Tee CPW Oscillator and the Application of the Design to a Monolithic Array of MESFETs," *Electronics Letters*, Vol. 29, No. 16, pp. 1477–1479, August 1993.

31. B. K. Kormanyos, S. E. Rosenbaum, L. P. B. Katehi, and G. M. Rebeiz, "Monolithic 155 GHz and 215 GHz Quasi-Optical Slot Oscillators," *IEEE MTT-S International Symposium Digest*, pp. 835–838, 1994.

32. B. K. Kormanyos, W. Harokopus, L. P. B. Katehi and G. M. Rebeiz, "CPW-Fed Active Slot Antennas," *IEEE Transactions on Microwave Theory and Techniques*, Vol. MTT-42, No. 4, pp. 541–545, April 1994.

33. C. H. Ho, L. Fan and K. Chang, "New FET Active Slotline Ring Antenna," *Electronics Letters*, Vol. 29, No. 6, pp. 521–522, March 1993.

CHAPTER TEN

Integrated and Active Inverted Stripline Antennas and Other Active Antenna Configurations

10.1 INTRODUCTION

Besides microstrip patches, grids, tapered notches, and slotlines, there are many other types of integrated and active integrated antennas. In this chapter, the inverted stripline antenna and other designs are described. These configurations are viable alternatives to microstrip for integration. The inverted stripline antenna has the advantages of easy integration with solid state devices, low effective dielectric constant, low harmonic radiation and low cross-polarization radiation for an active antenna, good heat dissipation, and easy hermetic sealing capability.

The active antennas can be used to form a spatial power combiner. Other applications where integrated and active antennas are well suited include miniaturized transceivers, Doppler sensors, decoys, RFID, wireless communications, beacons, and repeaters.

10.2 INTEGRATED INVERTED STRIPLINE ANTENNAS

An important milestone in integrated antennas occurred in the early 1980s when Bhartia and Bahl introduced solid-state varactor diodes on a rectangular microstrip patch antenna to provide resonance tunability (see Ref. 15 of Chap. 6). The integrated configuration is shown in Figure 6.8 along with a transmission line equivalent circuit. This frequency agility allows a narrow-band antenna to operate over a wide range of frequencies. More importantly, it brought solid-state devices (which were previously confined to guided-wave circuits) together with antennas. The approach makes an antenna be another circuit component whose properties can also be electronically controlled.

Although varactors were used in this integration for frequency tuning, many other devices and functions can also be realized. Also, the addition of an electronically tuned reactance to the rectangular patch antenna can be analyzed by using a simple transmission line approach. The varactors are variable reactive components which load the transmission line [1, 2]. Fringing fields are accounted for with an equivalent length of line, and radiation is modeled by a resistor. Although this method ignores package parasitics, integration discontinuities, and cavity perturbations, it goes a long way toward predicting the antenna frequency response. A more intensive full-wave approach such as finite difference time domain (FDTD) method can provide an improved model with more radiation effects due to the interaction of the solid-state package and the antenna. However, this would also incur more time and computational costs.

The following equations describe models used to determine characteristics of microstrip patch antennas. As far as these circuit and cavity models are concerned, the differences between a microstrip patch and an inverted stripline antenna (ISA) can be consolidated by a simple scaling factor. For the ISA, the relative dielectric constant (ε_r) is replaced by an effective dielectric constant (ε_{eff}), which is determined from the inverted configuration.

For a rectangular microstrip patch antenna loaded with solid-state devices shown in Figure 10.1, the transmission line model can be used to determine the input admittance at an edge feed:

$$Y_{\text{in}} = G + j(B + B_v) + Y_0 \frac{G + j(B + B_v + Y_0 \tan \beta L)}{Y_0 + j(G + j(B + B_v)) \tan \beta L} \tag{10.1}$$

where G is the radiation conductance given by [3]

$$G = \begin{cases} \dfrac{W^2}{90\lambda_0^2}, & W \ll \lambda_0 \\[3mm] \dfrac{W}{120\lambda_0^2}, & W \gg \lambda_0 \end{cases} \tag{10.2}$$

and the susceptance B of the open end is approximated by

$$B = \frac{k_0 \Delta l \sqrt{\varepsilon_{\text{eff}}}}{Z_0} \tag{10.3}$$

where Y_0 is the characteristic admittance of the transmission line, β is the transmission line propagation constant, Δl is the extension which accounts for fringing fields, L is the physical resonant length, and B_v is an electronically variable susceptance provided by the solid-state diode. Figure 10.1 shows the antenna configuration dimensions and approximate transmission line model.

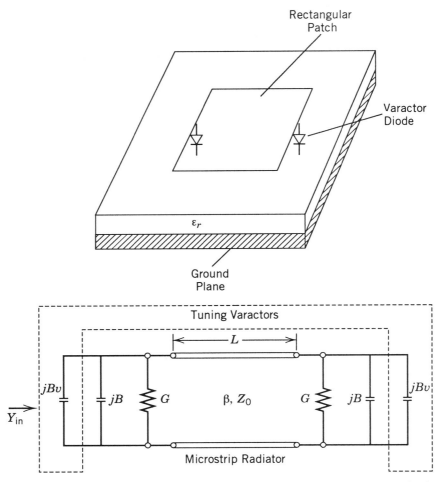

FIGURE 10.1. Varactor-tuned rectangular patch antenna and its equivalent circuit (same as Figure 6.8).

Circular patch antennas can also be used for integration and analyzed in a similar manner. Figure 10.2 shows the configuration dimensions, coordinate system used, and the electrical parameters of the substrate. The cavity model assumes that the circular patch antenna has perfect electric conductors and a perfect magnetic conductor which encloses the patch conductor and ground plane. The fields within the cavity must satisfy these boundary conditions. If the cavity is very thin, the z-directed electric field is described by

$$E_z = E_0 J_1(k\rho) \cos(\phi), \qquad H_\phi = -\frac{jk}{\omega\mu_0} J_1'(k\rho) \cos(\phi) \qquad (10.4)$$

where $J_n(k\rho)$ are Bessel functions of the first kind of order n.

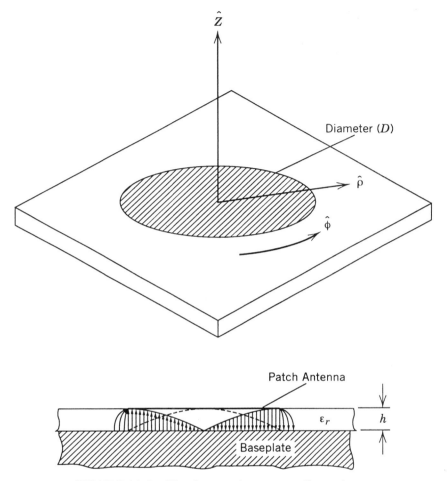

FIGURE 10.2. Circular patch antenna dimensions.

The field distribution for a given mode indicates places where a solid-state device may provide the most effect on the frequency or radiating characteristics the patch. First, the resonant frequency of a given circular patch can be calculated from the equation

$$F_0 = \frac{\alpha_{mn}}{\pi D \sqrt{\mu \varepsilon}} \tag{10.5}$$

where α_{mn} is the mth zero of the first derivative of the Bessel function of order n (i.e., 1.8411 for the dominant mode), μ and ε are the permeability and permittivity of the substrate, and D is the physical patch diameter.

It can be seen that the higher-order modes are not integer multiples of the fundamental mode. This makes the structure an attractive choice for the incorporation of active devices, since any harmonics produced by the active devices will not occur at one of the resonant frequencies of the antenna.

Since fields fringe over the antenna edges as shown in Figure 10.2, the patch diameter, D, in Equation (10.5) is replaced by an effective diameter D_{eff}. The effective diameter compensates for the fringing fields:

$$D_{eff} = \frac{D}{2} \sqrt{1 + \left(\frac{4h}{\pi D \varepsilon_r}\right) \left(\ln\left(\frac{D}{4h}\right) + 1.77 + 1.41\varepsilon_r + \frac{2h}{D}(0.268\varepsilon_r + 1.65)\right)}$$

(10.6)

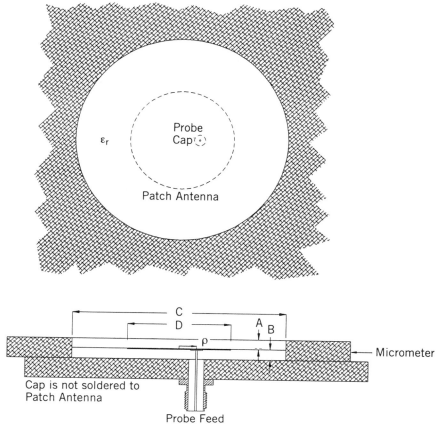

FIGURE 10.3. A passive probe-fed ISA.

where h is the substrate height and ε_r is the relative dielectric constant of the substrate. Since the system operating frequency (F_0) is known, Equation (10.5) is often rearranged to obtain the patch diameter required to operate at that frequency.

Once the patch dimensions have been finalized to ensure dominant mode operation at F_0, the impedance characteristics within the antenna cavity can be found. Although several assumptions have been made, these equations give insight into methods and places to best integrate passive and active devices. During integration, matching becomes critical, and, to this end, the real part of the input impedance at ρ from the center can be approximated with

$$R_{\text{in}} = R_0 \left(\frac{J_n^2(k\rho)}{J_n^2(ka)} \right) \tag{10.7}$$

R_0 is the resistance at resonance for $\rho = a = D/2$. The cavity model of Equations (10.4) to (10.7) have been used to calculate the resonant frequencies of passive probe-fed ISAs. Empirically, a novel test fixture was devised to nondestructively determine input impedance as a function of probe position. The fixture also allows testing of different antenna dimensions, cavity diameters, and substrate characteristics (i.e., dielectric constants, thickness, loss tangent, etc.). The fixture was used to verify the resonant frequencies calculated using the model from the equations. Figure 10.3 shows a picture of the ISA fixture and its flexibility.

Figure 10.4 shows the critical dimensions of the cavity-enclosed ISA structure. As an example, the $A = h = 0.060$-in. thick substrate has an $\varepsilon_2 = \varepsilon_r \approx 2.2$ and $C = 62$ mm, $B = 1.5$ mm. In the inverted configuration, the majority of the fields are concentrated within the region defined by the patch and the ground plane. Since air fills the region below the patch, the effective dielectric constant (ε_{eff}) of the structure is very close to 1. An empirically determined

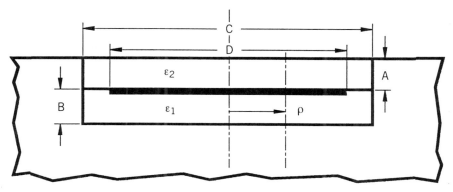

FIGURE 10.4. Inverted stripline antenna (ISA) configuration dimensions.

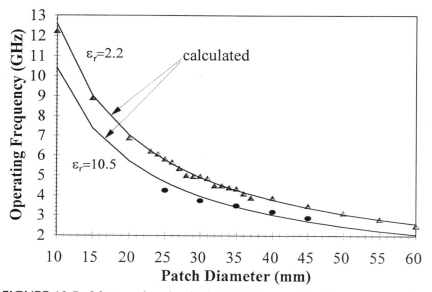

FIGURE 10.5. Measured and calculated ISA operating frequency results.

value of 1.1 gives resonant frequency values within 3% of the measured results over a wide range of frequencies from 3 to 12 GHz. Figure 10.5 shows measured versus calculated operating frequencies of the ISA structure as a function of patch diameter. Patch diameters (D) ranged from 10 to 60 mm with a patch to ground spacing (B) of 1.5 mm.

The operating frequencies are primarily determined by the patch size and patch-to-ground separation. These operating frequencies are typically far away from the nearest cavity mode. Errors in correlation may be due to variations in patch-to-ground separation, probe discontinuities, and metallic wall effects on the antenna radiating edges. Wall proximity effects on the patch resonance were not modeled or accounted for.

Figure 10.6 shows the ISA input impedance of the fundamental mode versus probe position from the center for a 30-mm-diameter patch. The best 50-Ω input match for this diameter antenna occurs at 5.21 mm off-center where the VSWR is 1.0003 and the 2:1 bandwidth is 3.77%. The dominant mode appears as a short-circuit at the center of the patch and increases in impedance at the edge of the antenna. Similar results are obtained for other patch diameters. This impedance data can be deembedded and used for device integration in the design of oscillators or amplifiers.

Although effects of the wall proximity are not directly noticeable on the patch input impedance, they can be readily seen on the radiation patterns. The radiation patterns for the 30-mm-diameter antenna are shown in Figure 10.7.

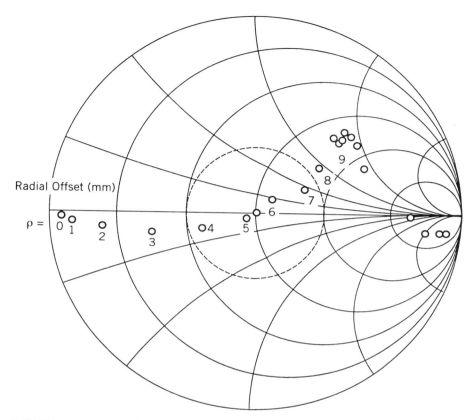

FIGURE 10.6. Input impedance versus probe position. The dotted line circle denotes the 2:1 VSWR.

The E- and H-plane half-power beamwidths (HPBWs) are 57.5° and 61.5°, respectively. The cross-polarization levels are 19.3 dB below the measured gain of 10.5 dBi. These HPBWs are clearly narrower than in a conventional microstrip patch antenna. Tabulating the HPBWs as a function of patch diameter shows that the HPBW varies with the patch-to-cavity diameter ratio. Figure 10.8 shows the HPBWs of the E- and H-plane patterns of several antennas tested in the 62-mm-diameter cavity. The inverted substrate and enclosure combination increase the antenna directivity. Noticeable beam sharpening occurs near patch-to-cavity diameter ratios of ≈ 0.5.

The inverted configuration offers good circuit and antenna performance, and it has flexibility in operating bandwidth and radiating beamwidth. Just as the coaxial probe was moved under the antenna for impedance measurement and impedance matching, solid-state device positions can also be optimized for best impedance match.

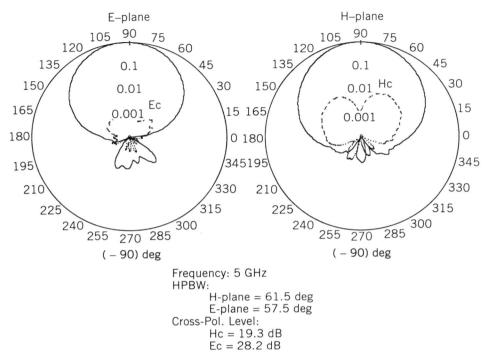

Frequency: 5 GHz
HPBW:
    H-plane = 61.5 deg
    E-plane = 57.5 deg
Cross-Pol. Level:
    Hc = 19.3 dB
    Ec = 28.2 dB

FIGURE 10.7. Co-pol and cross-polarization components of a passive probe-fed inverted stripline antenna. (Solid line) copolarization; (dotted line) cross-polarization.

Equation (10.7) is used to match probe feeds or devices to the antenna. Integrated devices can be modeled as lumped loads which affect the resonant frequencies of the circular patch antenna. Since the microstrip patch is a resonant cavity, there are a large number of modes which can be excited within the antenna due to discontinuities or perturbations. Excitation of these modes often degrades the circuit and radiation performance of the patch antenna. Shorting pins can be used to suppress unwanted modes [4, 5, 6]. External elements such as capacitors and inductors can also be strategically placed to enhance or suppress different modes. In the same manner, *pin* diodes can provide electronic control over the radiation efficiency of a microstrip patch. The zero and forward bias turns the antenna on and off, respectively. The integration can serve as a microwave switch and/or modulator [7].

Diodes are easily connected across the patch to ground with a low-pass filter to provide the necessary biasing voltage. Varying the diode position under the patch allows the impedance presented to the diode terminals to be changed. Alternatively, the diode position determines the diode's effect on the antenna

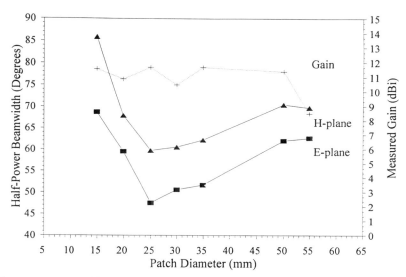

FIGURE 10.8. Half-power beamwidth and gain as a function of patch diameters.

cavity. Since the electric field is a maximum at the radiating edges of the antenna, *pin* and varactor diodes will have the most effect at these positions. Diodes placed near the radiating edges can easily tune the resonance over 30%. This is a significant range, since a typical microstrip patch has less than 3% instantaneous impedance bandwidth.

An inverted microstrip antenna has been used extensively to integrate devices. The structure has demonstrated good radiation performance and design flexibility. Figure 10.9 shows a circular ISA integrated with two packaged diodes. The diodes can either be *pin* or varactors. In the case of two *pin* diodes, the integrated antenna will behave according to the voltage used to bias the devices. No voltage represents an open circuit state, while 1 V turns the *pin*s into a short-circuit.

The 30-mm-diameter patch normally operates around 4.97 GHz. At zero-voltage bias (ON state), the open-circuited *pin*s have parasitic reactances which detune the ISA F_0 about 20%. The ON state allows the antenna to operate at 4.125 GHz, with a VSWR of 1.013 and a 2:1 impedance bandwidth of 130 MHz (3.15%). When the *pin* diodes are biased on, they provide very good short at the operating frequency. For a 1.2-V bias level (100 mA), the OFF state shows a VSWR of 22.87. Figure 10.10 shows the performance of the *pin* integrated ISA. General-purpose *pin* diodes used in this investigation are housed in the case style shown in Figure 10.9. The probe-fed antenna can also be amplitude-modulated by varying the bias continuously from 0 to 100 mA (0 to 1.2 V).

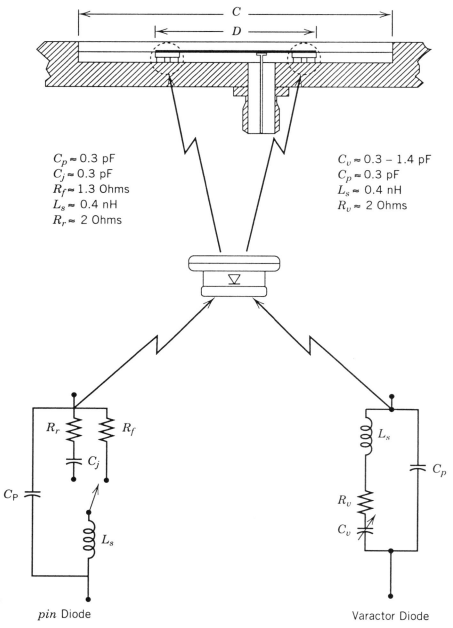

$C_p \approx 0.3$ pF
$C_j \approx 0.3$ pF
$R_f \approx 1.3$ Ohms
$L_s \approx 0.4$ nH
$R_r \approx 2$ Ohms

$C_v \approx 0.3 - 1.4$ pF
$C_p \approx 0.3$ pF
$L_s \approx 0.4$ nH
$R_v \approx 2$ Ohms

pin Diode

Varactor Diode

FIGURE 10.9. Solid-state device packages and equivalent circuits.

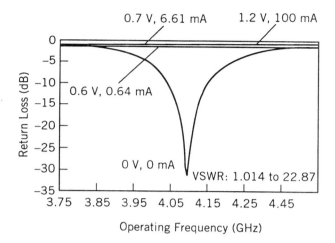

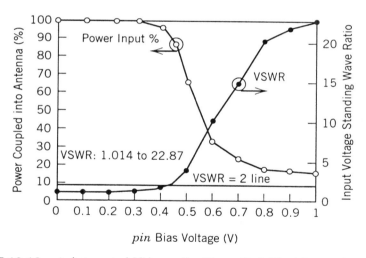

FIGURE 10.10. *pin* integrated ISA results. (From Ref. 17 with permission from IEEE.)

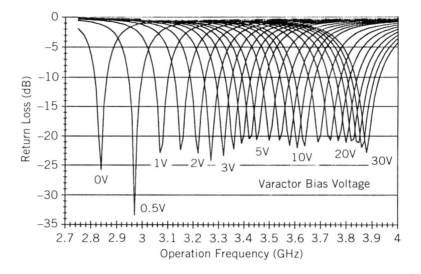

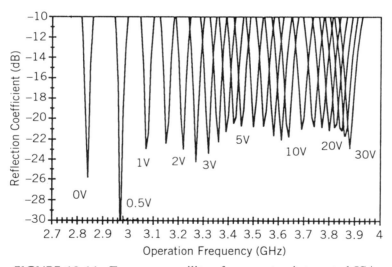

FIGURE 10.11. Frequency agility of a varactor integrated ISA.

In a similar maner, *pin* diodes can be replaced with varactors to provide the frequency agility shown in Figure 10.11. As shown, the operating frequency is shifted or tuned with respect to the varactor bias voltage. This change can be accomplished electronically within a couple of nanoseconds with very little current drain. The concept can be used for channelized or tunable receiver, frequency modulation, and so on.

Radiation for the *pin* integrated antenna can be operated over a relatively narrow band of frequencies. However, the varactor integrated antenna operates over a very broad frequency range. The voltage-variable reactance which loads the integrated antenna can be modeled as a lumped load on the resonant circuit. Device parasitics which depend on case styles of the packages will affect the antenna cavity. Depending on the frequency of operation and package style, the integration may introduce higher cross-polarization and a change in the principal planes. However, from a simple lumped model, the variable junction capacitance of the varactor can provide a good description of the frequency-tuning characteristics of the integrated antenna.

The varactor junction capacitance varies as a function of bias voltage. The bias voltage (V) is described by

$$C(V) = \frac{C(0)}{(1 + V/V_{bi})^{\gamma}} \tag{10.8}$$

where $C(0)$ is the capacitance at 0 V, V_{bi} is the built-in potential (GaAs = 1.3 V), and γ is 0.5 for abrupt junctions.

Varactor diodes will have maximum effect on the antenna if they are placed at electric field maxima. At the radiating edges the varactors couple very strongly with the radiating electric field loading the antenna and lowering the operating frequency. When biased, the varactors can quickly tune over a range of frequencies. Varactor integrated ISAs have exhibited a wide operating tuning range of 31% centered at 3.4 GHz. As shown in Figure 10.12, at 30 V the operating frequency is at 3.88 GHz with a VSWR of 1.0105. At 0 V the operating frequency is at 2.84 GHz with a VSWR of 1.0053. Figure 10.13 shows the ISA instantaneous 2:1 impedance bandwidth versus operating frequency. As shown, the VSWR remains below 1.02, while the impedance bandwidth varies from 1.8% to 2.5%. This integrated antenna can be used to rapidly scan over its tuning bandwidth for wideband receiver or transmitter applications. The small instantaneous bandwidth ($\sim 2\%$) can reduce the noise pickup in a wideband channelized receiver system.

As in most integrated antennas, there may be a degradation in radiation performance due to antenna modifications and device perturbation of the antenna fields. For example, the cross-polarization for the varactor integrated patch is -10 dB, compared to -19 dB for the nonintegrated patch. Smaller devices, improved biasing schemes, and optimized diode positions reduce these adverse effects on the radiation pattern. In identical fashion, detectors and mixers can be integrated within the confines of the antenna to provide various circuit functions.

A different integrated antenna which used *pin* diodes to electronically affect the frequency characteristics of a patch antenna was demonstrated by Daryoush et al. in 1986 [8]. In this investigation a stub was loosely coupled to the radiating edge of a patch antenna. A *pin* diode is placed from the antenna to the probe, as shown in Figure 10.14. Forward and reverse biasing of the *pin*

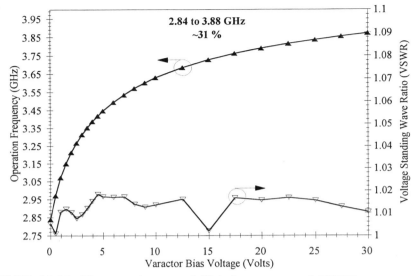

FIGURE 10.12. Varactor integrated ISA frequency and VSWR vs. varactor voltage. (From Ref. 7 with permission from IEEE.)

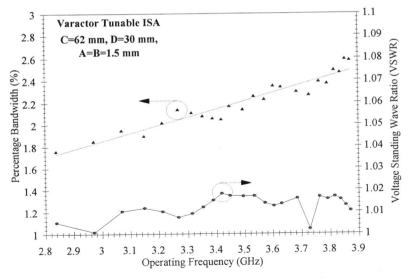

FIGURE 10.13. Varactor integrated ISA %BW vs. frequency.

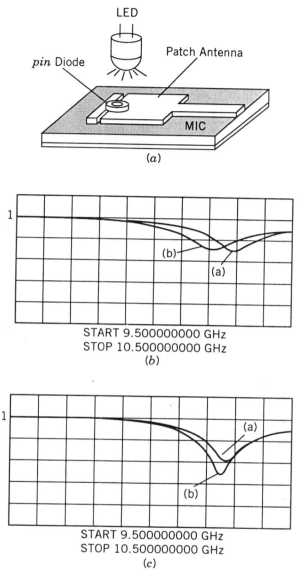

FIGURE 10.14. (*a*) Conceptual presentation of optically tuned patch antenna. (*b*) Return loss of patch antenna vs. frequency; a) zero-biased *pin* diode, b) forward-biased *pin* diode (vertical scale 10 dB/div, horizontal scale 100 MHz/div). (*c*) Return loss of patch antenna vs. frequency for 0.5-V forward-biased diode; a) dark, b) 1 W/cm² white light illumination (vertical scale 10 dB/div, horizontal scale 100 MHz/div). (From Ref. 8 with permission from IEEE.)

diode provide two different operating frequencies for the antenna at the input port. The zero-bias operating frequency is 10.285 GHz, while the forward-bias operating frequency is 10.207 GHz. The radiation patterns show 70° H-plane half-power beamwidth and 7-dB gain. In this case, the *pin* can also be optically controlled. White light illumination varies the diode's reactance enough to tune the operating frequency 15 MHz at 10.270 GHz. Optical control of the antenna can provide a very useful method of modulation.

A drawback in the use of integrated antennas is that they still require some type of feed network to distribute the rf signal to the terminals of each integrated antenna. Consequently, a large array of antennas integrated with varactors or *pin* diodes would still need a lossy rf network. The array aperture, however, would become agile either in frequency, phase, or amplitude according to the biasing of each device at each antenna.

Integrated and active integrated antennas require the ability to combine different types of devices easily. Specifically, microstrip requires that shunt connections be drilled in hybrid MICs while in monolithics via holes must be processed. Some alternatives to microstrip which do not require drilling include coplanar waveguide, slotline, and coplanar strips. Inverted microstrip also alleviates hybrid integration discontinuities encountered in microstrip. Inverted microstrip is not a true uniplanar line, but it has several advantages. Since the majority of the fields propagate through air, it has potential for lower transmission loss. It also offers some integration advantages over a conventional microstrip line. Inverted microstrip does not require drilling for shunt connections, which allows nondestructive experimental testing as well as position optimization of diodes and coaxial probe inputs. This trait makes inverted microstrip attractive for many hybrid applications. When used for integrated antennas, inverted microstrip provides a built-in radome for protection. As previously shown by *pin* and varactor integrated patches, these integrations provide very good circuit performance and moderate radiation performance. It has also been used to demonstrate Gunn and FET integrated antennas.

The ISA structure has been extensively integrated with passive, active, and combinations of devices to make transmitters, receivers, modulators, and transceivers. The following section reviews the integration with the active devices.

10.3 ACTIVE INTEGRATED INVERTED STRIPLINE ANTENNAS

For active integrated devices, excess heat must be dissipated efficiently. By carefully matching the thermal expansion and conductivity coefficients of substrate materials and housing alloys, the integrated antenna can be hermetically sealed for improved system durability and reliability [9]. Dc biasing of devices can be achieved on the substrate or through the ground plane underneath. To increase metal volume, the antenna can be enclosed by metal walls. The resulting structure can be classified as trapped inverted microstrip; however, it is a special case of general stripline-type transmission lines.

Gunn integrated active inverted stripline antennas have been demonstrated for beam steering and spatial power combiners [10, 11]. These active antennas exhibit good radiation patterns, low cross-polarization levels, easy device integration, and good heat-sinking capacity.

Unlike *pin*- and varactor integrated antennas, Gunn integrated antennas generate rf power output. The antenna serves as the resonator as well as the radiator for the oscillator. Oscillations occur at the frequency where the diode and the antenna reactances cancel out. Oscillation start-up occurs if the placement of the diode is such that the magnitude of the diode negative resistance is greater than the circuit resistance presented to its terminals. Figure 10.15*a* shows the Gunn integrated ISA using the adjustable test fixture. The package connection to the heat sink and the added metal volume of the metallic walls provide efficient heat sinking for the active device.

Typical screw-type packaged Gunn diodes were integrated with antennas ranging from 8 to 11 mm in diameter. The smaller antenna diameters are used for X-band operation. Gunn integrated ISAs differ from the *pin* and varactor integrated ISAs in that the active antennas are mainly intended for spatial and quasi-optical power combining. In these applications, packing density is increased by placing metallic walls very near the edges of the antenna. The enclosure proximity to the antenna radiating edges affects the radiation performance. The diode package also presents a significant perturbation on the X-band antenna volume, which adversely affects the radiation characteristics and disturbs the oscillating frequency. The Gunn oscillator frequency and deembedded power outputs are listed in Table 10.1.

Radiation patterns for the Gunn integrated ISA are shown in Figure 10.15*b*. The cross-polarization level is at least $-10\,dB$ for a Gunn diode 2 mm off-center. The HPBWs are 100° in the *E*-plane and 70° in the *H*-plane. A similar passive antenna using a probe feed exhibited HPBWs of 105° and 80° in the *E*- and *H*-plane, respectively, with a cross-polarization level of $-16\,dB$. The differences in radiating performance are attributed to the active antennas' diode package and bias lines. Improved performance can be achieved by reducing the size of the device packages.

The power-combining concept uses many low-power sources to provide a single coherent, higher-power output. To demonstrate spatial power combining via mutual coupling, two four-element arrays were constructed. Eight active inverted stripline antennas were built and tested for the square and diamond array. Since mutual coupling is used for injection locking, the locking gain level between two of these active antennas would be high with a corresponding narrow locking bandwidth. Mechanical tuning of the diodes ensured that several antennas would operate within the narrow locking bandwidth. This allows the use of a single power supply at a fixed bias voltage for all diodes and successful power combining. Antennas 1–4 were used in a four-element square array, while antennas 5–8 were used in a four-element diamond array.

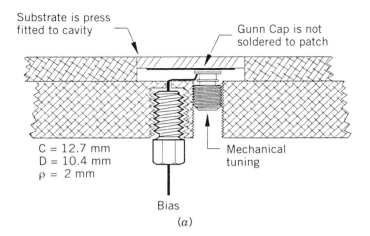

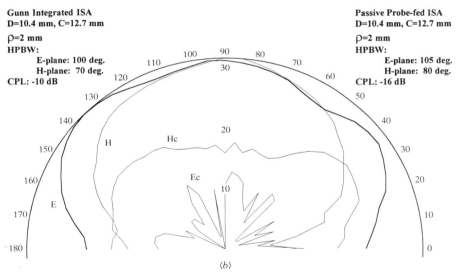

FIGURE 10.15. Gunn integrated active ISA: (*a*) configuration; (*b*) radiation patterns. (From Ref. 11 with permission of IEEE.)

Table 10.2 lists the operating frequency, oscillator power, and efficiency of each antenna and array as well as the EIRP. The square array uses 17-mm spacing between elements. The diamond array spacing between elements is 17 mm along the diagonal. Figure 10.16*a* shows the array configurations, spacing, and diode position.

The square array was used to test two-element injection locking and power combining via mutual coupling for *E*- and *H*-plane elements. The two-element

TABLE 10.1. Gunn Integrated ISA Operating Frequencies, Power versus Diameter [11]

Patch Diameter (mm)	Measured F_0 (GHz)	Oscillator Power* (mW)
8	9.997	63.99
9	9.637	56.76
10	9.438	57.04
10.4	9.240	63.98
11	8.908	66.98

*Calculated using a passive antenna gain of 6.65 dBi. Cavity diameter is 12.7 mm.

H-plane array shows a power-combining efficiency of 86.4%, while the E-plane array's efficiency was 99.8%. The four-element square array exhibited a power-combining efficiency of 89%. The overall dc-to-rf efficiency of the four-element array is 1.63%. The two-element patterns for E- and H-plane power combining show nearly 2:1 beamwidth sharpening of over a single element. The antenna patterns for the square power-combiner array are shown in Figure 10.16b. The array can be bias-tuned from 10.7 to 14 V without losing injection lock over a 60-MHz bandwidth from 9.467 to 9.527 GHz. The output power level varied less than 0.8 dB over the bias-tuned range.

The four-element diamond array was also tested for injection-locked power combining. The diamond array power-combining efficiency is 86.57% with an overall dc-to-rf conversion efficiency of 1.47%. The array can be bias-tuned from 9.5 to 12.2 V without losing injection lock over a 50-MHz bandwidth.

The ISA structure has also been integrated with an FET transistor [12]. The use of the FET transistor improves conversion efficiency while it reduces thermal requirements. However, the antenna must be modified to accommodate the FET as shown in Figure 10.17. FET integration requires three dc blocks for the drain, gate, and source terminals. Dc biasing can be achieved from behind the ground plane or etched to the nonradiating edges of the antenna. A chip resistor connected across the source to gate simplifies biasing to the device. The operating frequency is determined by the loads at each transistor port. The ISA test fixture can be used with a coaxial probe feed to deembed approximate impedance levels at each transistor port. The data are then used in a lumped model to determine oscillating frequencies for the active antenna. Another advantage of the transistor is that it can be induced to oscillate at lower frequencies to reduce effects from the device package and bias lines. Lower frequencies of operation can be accomplished with larger patch diameters.

A 30-mm diameter circular patch was modified to insert the FET. At the center of the patch, 0.4-mm gaps isolate the source from the gate and drain terminals. A 0.1-mm gap is etched from the center to the nonradiating edges

TABLE 10.2. Operating Frequency, Oscillator Power, and Efficiency of Each Antenna and Array

Antenna	1	2	3	4	Square Array*	5	6	7	8	Diamond Array†
Oscillator frequency (GHz)	9.498	9.499	9.498	9.497	9.511	9.330	9.340	9.330	9.340	9.380
Oscillator power (mW)	52.63	56.80	59.32	60.14	204	50.55	49.51	56.72	55.55	184
EIRP (mW) [dBm]	243.4 [23.9]	262.6 [24.2]	274.3 [24.4]	278.1 [24.4]	3773 [35.8]	233.7 [23.7]	228.9 [23.6]	262.3 [24.2]	256.8 [24.1]	3403 [35.3]
Combining efficiency (%)	—	—	—	—	89.03	—	—	—	—	86.57
Dc to rf efficiency (%)	1.75	1.89	1.98	2.00	1.63	1.69	1.65	1.89	1.85	1.47

*Antennas 1–4 are used in the square array in 17-mm spacing.
†Antennas 5–8 are used in the diamond array with 17-mm spacing along the diagonal.
Source: Ref. 11.

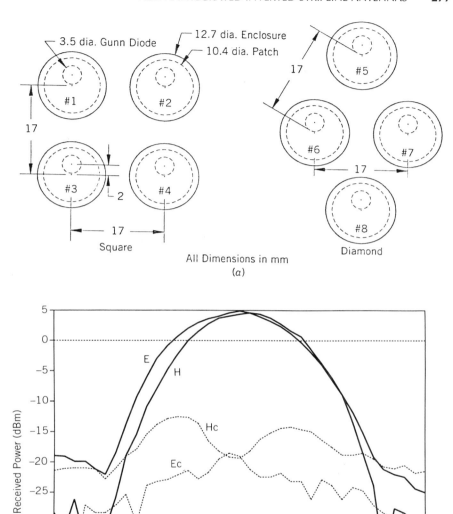

FIGURE 10.16. Four-element power combiners: (*a*) configurations; (*b*) radiation patterns of the square array. (From Ref. 11 with permission from IEEE.)

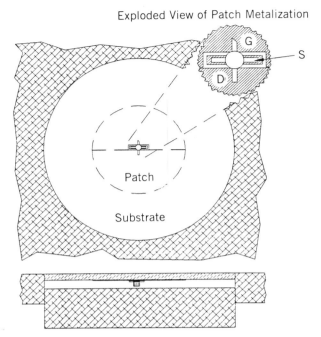

FIGURE 10.17. FET integrated ISA. (From Ref. 12 with permission from *Electronics Letters.*)

of the patch to provide dc isolation between the gate and drain terminals. The FET drain lead is soldered at the center of the patch, and the gate lead is approximately 2 mm off-center. Three dc lines bias the source, gate, and drain of the transistor. Alternatively, the source bias line can be replaced by a resistor. The frequency of oscillation depends on the impedance loads at the FET ports. The loads are a function of the position along the inverted patch antenna. Heat generated is dissipated by the patch. The lack of a low-thermal-impedance path from the device to the metal heat sink may cause thermal dissipation problems in higher-power devices. This could be alleviated with a shorting pin at the center of the patch to provide a low thermal impedance path to the housing.

Since the antenna was modified and the device terminals are soldered, it is not possible to experimentally optimize the FET position as in the Gunn integrated design. However, since the transistor is not directly attached to the base like the Gunn diode, the base can be used as a variable ground plane. The ground plane is essentially a sliding short which changes the antenna cavity depth and alters the oscillation frequency of the oscillator. The cavity depth is also instrumental in improving spectral and radiation characteristics of this type of active antenna.

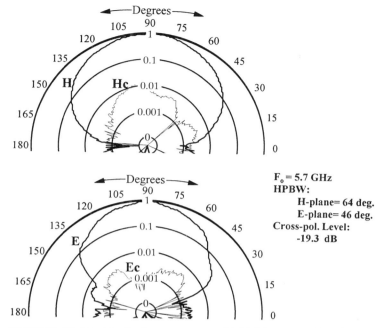

FIGURE 10.18. Radiation patterns of FET integrated ISA.

The FET provides 57 mW at 5.69 GHz when biased at 3.8 V and 26 mA. The 3-dB bias-tuning range is approximately 1% for a 1-V change in V_{ds}. The sliding ground plane allows a mechanical tuning range of nearly 6%. At a cavity depth of 4.15 mm, the measured oscillation frequency remains stable at 5.695 ± 0.002 GHz over the antenna test sweep. The HPBWs in the E- and H-plane patterns are 46° and 64°, respectively. The cross-polarization levels are -19.3 dB below the maximum. Figure 10.18 shows principal plane and cross-polarization patterns of the active antenna. The smooth radiation patterns and low cross-polarization levels compare favorably with previously reported active antennas. Probe-fed passive antennas with cavity depths of 3 mm exhibited HPBWs of 51° and 61° in the E- and H-plane patterns, respectively. The cross-polarization levels of the passive antenna are also lower than -19.3 dB with a gain of 10.2 dBi. Biasing modifications and cavity depth differences in the active antenna may account for the changes. The effective isotropic radiated power of the active antenna is 594 mW. Approximating the active antenna gain with the passive antenna gain of 10.2 dB results in an oscillator power of 57 mW and a dc-to-rf conversion efficiency of 57%.

Further integrations have since been carried out by Flynt et al. [13]. A completely integrated ISA transceiver has been demonstrated. It uses the FET ISA as the transmitter and local oscillator. The LO pumps a pill-type mixer

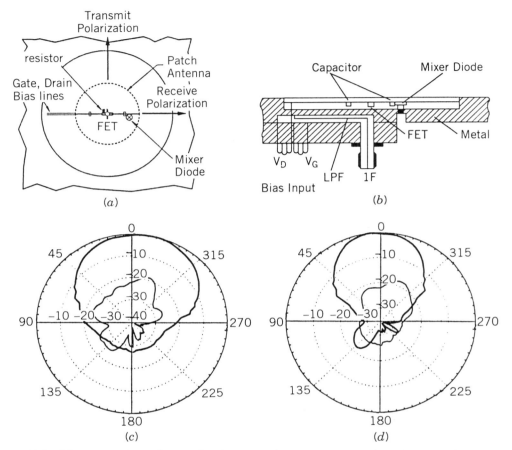

FIGURE 10.19. Complete active integrated antenna transceiver: (a) top view, (b) side view. (c) *H*-plane pattern with the mixer diode in place. The cross-polarization level = − 18.84 dB, and the HPBW = 67.0°. (d) *E*-plane pattern with the mixer diode in place. The cross-polarization level = − 17.99 dB, and the HPBW = 49.3°. (From Ref. 13 with permission from IEEE.)

diode. The transceiver is unique in that it integrates oscillator and mixing functions within the antenna volume. Isolation is provided through polarization. Preliminary results at 5.8 GHz exhibit a 5.5-dB isotropic mixer conversion loss for a 200-MHz if frequency. Figure 10.19 shows the complete ISA transceiver configuration and the transceiver radiation performance.

The addition of another device has affected the radiation performance of the antenna. The *E*- and *H*-plane HPBWs are 49° and 67°, which differ from the 46° and 64° measured without the mixer diode. Differences may be due to

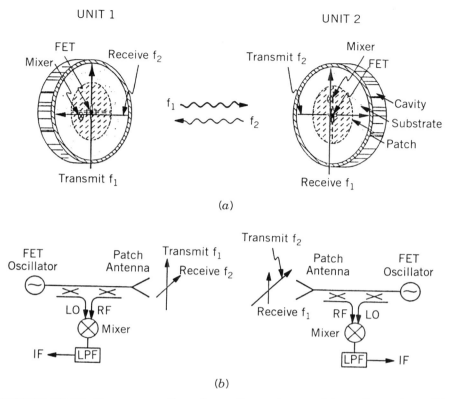

FIGURE 10.20. Two-way radio using active antennas: (*a*) configurations, (*b*) block diagram. (From Ref. 13 with permission from IEEE.)

differences in the antenna dimensions, feedlines, cavity depth, and diode package. Figure 10.20 shows how to use these transceivers for typical communication link.

10.4 ACTIVE INTEGRATED RAMPART LINE ANTENNAS

As shown earlier, active devices can be integrated directly at the antenna to create amplifiers and oscillators. In the case of a large array of amplifier antennas, an rf feed distribution is still required, but the power levels distributed to amplifying antennas are relatively low. Losses in the feed network occur at relatively low power levels, which improve the overall aperture efficiency. Each amplifier is combined with the others to provide a high spatial power output.

To completely avoid rf distribution networks, sources must be distributed across the aperture. Each low-power source radiates into space where spatial power combining takes place. The problem for such an array is the synchronization of each relatively low power source. Schemes have been devised to synchronize these sources through circuit or space-wave coupling or through an external source.

The use of an active integrated antenna as a source has been shown in several chapters. In the late 1980s, Birkeland and Itoh used active antennas for transceivers as well as radiating oscillators and multipliers. Instead of Gunn diodes, FETs were used in single- and dual-device transceivers. Instead of a single patch antenna, a linear series-fed microstrip patch antenna array was used to control the radiation pattern. This active antenna was previously described in Chapter 7 and shown in Figure 7.21. A more unique integration

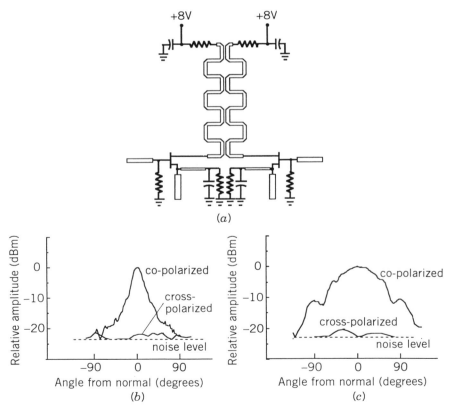

FIGURE 10.21. Active integrated rampart line antennas: (*a*) schematic view of oscillator circuit; (*b*) *H*-plane antenna pattern; (*c*) *E*-plane antenna pattern. (From Ref. 14 with permission from SPIE.)

was shown using a rampart line for the active antenna. This active rampart antenna used a pair of FETs as shown in Figure 10.21*a* [14]. The dual FETs can be used to oscillate in a push-pull or push-push mode. In the push-pull mode, the active rampart antenna radiates as shown in Figures 10.21*b* and *c*.

The integrations make ideal transceivers and oscillators, which can be used in quasioptical power-combining configurations. These transistor-based radiating oscillators or active antennas provide increased dc-to-rf efficiency and lower operating voltage and current levels. The use of rampart lines provides design flexibility to control beamwidth and polarization.

10.5   ACTIVE RING-DISK ANTENNAS AND DIPOLE ANTENNAS

A very interesting and useful configuration was demonstrated by Haese et al. in 1992 [15]. It involved the use of a ring-disk active antenna, as shown in Figure 10.22. A disk antenna is used in the TM_{11} mode, while a ring antenna is used in the TM_{12} modes. These two coupled antennas were designed to resonate at nearly the same frequency. The ring-disk is then an ideal feed for a Fresnel lens. The ring-disk is used as a pulsed active antenna operating at 60 GHz. This integration demonstrates the ability of an active antenna to serve as a critical component in a very difficult application. Very good isotropic conversion loss was also demonstrated for the system with a minimum of 5 dB for a 400-MHz offset at 60 GHz.

Another antenna, which is a common dipole, presents one of the most clever active antenna designs to date [16]. Using a typical packaged HEMT, the device is oriented and biased such that the structure becomes an active dipole antenna polarized along the drain-to-gate direction.

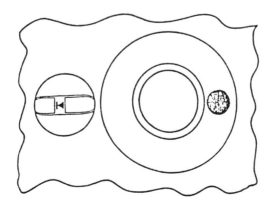

FIGURE 10.22. 60-GHz ring-disk active antenna. (From Ref. 15 with permission from IEEE.)

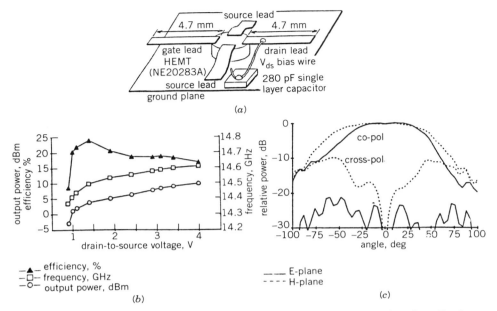

FIGURE 10.23. HEMT integrated active dipole: (*a*) structure of active dipole antenna; (*b*) measured output power, efficiency, and oscillation frequency; (*c*) radiation pattern of active dipole antennas. (From Ref. 16 with permission from *Electronics Letters.*)

Figure 10.23 shows the active antenna configuration, dimensions, circuit performance, and radiation characteristics. As shown, the structure comprises inverted packaged HEMT devices with source leads grounded to the ground plane, equal-length gate and drain leads form the dipole antenna, and a slender bond wire provides bias voltage to the drain terminal. An output power of 2.43 mW was measured at 14.5 GHz with a 24.1% dc-to-rf conversion efficiency. The configuration shows a very good principal plane pattern with an *H*-plane cross-polarization level of −10 dB. Given consistency in the device characteristics as well as good tolerance limits on the critical package dimensions, the design (except for the bondwire) is as simple as an active antenna can get. It seems ideal for an array in a quasi-optical resonator or other spatial power-combining applications.

10.6 ACTIVE DIELECTRIC ANTENNAS

Another unique active antenna integration uses a dielectric resonator and a transistor. A dielectric resonator and an active bipolar junction transistor were integrated to operate as a stable radiating DRO [17]. The active antenna

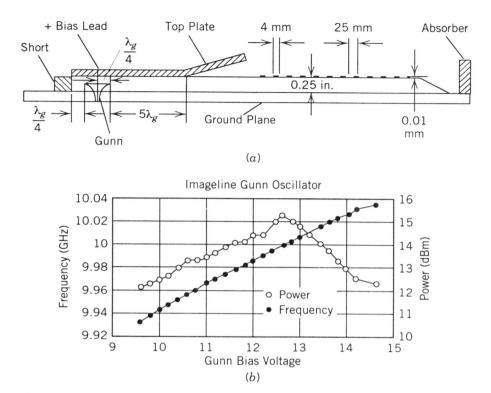

FIGURE 10.24. Gunn integrated imageline active antenna: (*a*) configuration; (*b*) frequency and power output vs. Gunn bias voltage. (From Ref. 20.)

operates at 500 MHz and exhibits low phase noise (-90 dBc/Hz) and high-frequency stability over temperature (0.0126 MHz/°C; -15°C to 40°C). The configuration makes an ideal Doppler sensor with low-cost potential. Another Gunn integrated dielectric resonator antenna operated at 5.55 GHz, but little is said of radiation properties of the device [18]. The pulsed diode did not generate as much power in the dielectric resonator as in a conventional waveguide due to field and impedance mismatching problems as well as losses in the circuit. However, the integration has potential for much higher operating frequencies with improved performance over conventional antennas. It is not unlike dielectric lenses which enclose many integrated antennas at millimeter and submillimeter wavelengths.

Along the same lines, if you replace microstrip with a dielectric transmission line, a relatively efficient active radiator can be realized. The active dielectric antenna uses a Gunn diode and an imageline with a set of periodic disturbances for controlled radiation [19, 20]. The disturbances are designed to radiate a pencil beam in a given direction at a given frequency. The periodic separation

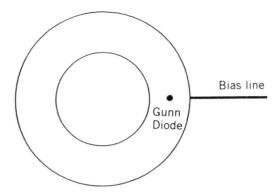

FIGURE 10.25. Active annular ring antenna integrated with Gunn diode. (From Ref. 21 with permission by Wiley.)

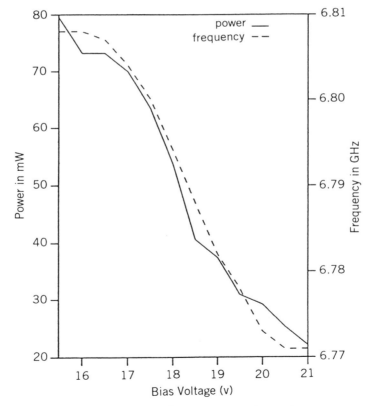

FIGURE 10.26. Power output and frequency vs. bias voltage. (From Ref. 21 with permission by Wiley.)

changes slightly as a function of wavelength, which shifts the output beam with respect to frequency. This property allows beam scanning of a pencil beam over a range of frequencies. A Gunn diode provides the structure with a negative resistance for rf power generation. The use of imageline for traveling-wave active antennas is interesting due to its low loss and its potential use well into the submillimeter wavelengths. Figure 10.24 shows the X-band Gunn integrated imageline array antenna along with power and frequency response of the oscillator.

10.7   ACTIVE MICROSTRIP RING ANTENNAS

An active antenna was developed by the direct integration of a Gunn device with a ring antenna as shown in Figure 10.25 [21]. The radiated output power level and frequency response of the active antenna are shown in Figure 10.26 [21]. The Friis transmission equation was used to calculate the power radiated from the active ring (see Chap. 5). It can be seen that over 70 mW output power was achieved with a bias of 16 V at 6.805 GHz. The Gunn diode used produced a maximum of 100 mW in an optimized waveguide circuit.

REFERENCES

1. A. G. Derneryd, "A Theoretical Investigation of the Rectangular Microstrip Antenna Element," *IEEE Transactions on Antennas and Propagation*, Vol. 26, No. 4, pp. 532–535, July 1978.

2. R. B. Waterhouse, "Modelling of Schottky-Barrier Diode Loaded Microstrip Array Elements," *Electronics Letters*, Vol. 28, No. 19, pp. 1799–1801, September 1992.

3. R. E. Munson, "Conformal Microstrip Antennas and Microstrip Phased Arrays," *IEEE Transactions on Antennas and Propagation*, Vol. 22, No. 1, pp. 74–78, January 1974.

4. D. L. Sengupta, "Resonant Frequency of a Tunable Rectangular Patch Antenna," *Electronics Letters*, Vol. 20, No. 15, pp. 614–615, July 1984.

5. G. L. Lan and D. L. Sengupta, "Tunable Circular Patch Antennas," *Electronics Letters*, Vol. 21, No. 22, pp. 1022–1023, October 1985.

6. D. J. Roscoe, L. Shafai, A. Ittipiboon, M. Cuhaci and R. Douville, "Tunable Dipole Antennas," *IEEE Antennas and Propagation Symposium*, Ann Arbor, Michigan, pp. 672–675, 1993.

7. J. A. Navarro and K. Chang, "Low-Cost Inverted Stripline Antennas with Solid-State Devices for Commercial Applications," *IEEE MTT-S International Microwave Symposium Digest*, pp. 1771–1774, 1994.

8. A. S. Daryoush, K. Bontzos, and P. R. Herczfeld, "Optically Tuned Patch Antenna for Phased Array Applications," *IEEE Antennas and Propagation Symposium*, pp. 361–364, 1986.

9. J. A. Navarro, K. Chang, J. Tolleson, S. Sanzgiri, and R. Q. Lee, "A 29.3 GHz Cavity Enclosed Aperture-Coupled Circular Patch Antenna for Microwave Circuit Integration," *IEEE Microwave and Guided Wave Letters*, Vol. 1, No. 7, pp. 170–171, July 1991.

10. J. A. Navarro and K. Chang, "Electronic Beam Steering of Active Antenna Arrays," *Electronics Letters*, Vol. 29, No. 3, pp. 302–304, February 1993.

11. J. A. Navarro, Lu Fan, and K. Chang, "Active Inverted Stripline Circular Patch Antennas for Spatial Power Combining," *IEEE Transactions on Microwave Theory and Techniques*, Vol. MTT-41, No. 10, pp. 1856–1863, October (1993).

12. J. Navarro, L. Fan, and K. Chang, "Novel FET Integrated Inverted Stripline Patch," *Electronics Letters*, Vol. 30, No. 8, pp. 655–657, April 1994.

13. R. Flynt, L. Fan, J. Navarro, and K. Chang, "Low Cost and Compact Active Integrated Antenna Transceiver for System Applications," *IEEE MTT-S International Microwave Symposium Digest*, pp. 953–956, 1995.

14. J. Birkeland and T. Itoh, "Power Combining FET Oscillators using Coupled Rampart-Line Antennas," *13th International Conference on Infrared and Millimeter-Waves*, pp. 7–8, December 1988.

15. N. Haese, M. Benlamlih, D. Cailleu, and P. A. Rolland, "Low-Cost Design of a Quasi-Optical Front-End for on Board mm-Wave Pulsed Radar," *IEEE MTT-S International Microwave Symposium Digest*, pp. 621–623, 1992.

16. O. P. Lunden, "Ku-band Active Dipole Antenna," *Electronics Letters*, Vol. 30, No. 19, pp. 1560–1561, September 1994.

17. Y. Yu-On, L. Man-Lung, and Ng. Tsz-Bun, "Active Integrated Radiator for Microwave Sensing System," China International Conference on Circuits and Systems, Shenzhen, China, pp. 952–955, June 1991.

18. D. Jaisson and F. Bachelot, "An Oscillator with a Gunn Diode Integrated in a Dielectric Resonator," *Microwave Journal*, Vol. 37, No. 2, pp. 117–120, February 1994.

19. R. Fralich and J. Litva, "Beam-Steerable Active Array Antenna," *Electronics Letters*, Vol. 28, No. 2, pp. 184–185, January 1992.

20. A. M. Kirk and K. Chang, "Integrated Image-Line Steerable Active Antennas," *International Journal of Infrared and Millimeter Waves*, Vol. 13, No. 6, pp. 841–851, 1992.

21. R. E. Miller and K. Chang, "Integrated Active Antenna Using Annular Ring Microstrip Antenna and Gunn Diode," *Microwave and Optical Technology Letters*, Vol. 4, No. 2, pp. 72–75, January 1991.

CHAPTER ELEVEN

Integrated Antennas with Passive Solid-State Devices

11.1 INTRODUCTION

The integration of devices and antennas at millimeter and submillimeter wavelengths has been accomplished primarily through passive two-terminal devices. Large integrated antenna detector or mixer arrays can be used in radio astronomy [1], remote sensing, communications, and radar. With the increase of computation power, large imaging arrays [2] can develop an image faster than can conventional scanning systems. With the current FCC ruling [3] to open up millimeter wavelengths, many commercial applications will spring up for collision avoidance, perimeter surveillance, and high-data-rate wireless local area networks. This chapter reviews antennas which have been integrated with passive solid-state devices. These integrations face many of the same problems associated with active integrated antennas, but they have been used well into the submillimeter wavelengths. A comprehensive review of millimeter-wave integrated antennas is given by Rebeiz [4].

Similar to active antennas, the combination is defined by two key technologies: antennas and solid-state devices. The many kinds of antennas (i.e., wires, loops, dipoles, microstrip patches, horns, dielectric rods, etc.) provide control over polarization and directivity. A variety of passive solid-state devices (i.e., Schottky-barrier detectors, *pin* switches, varactors, etc.) are configured within the antenna structure to create detectors, modulators, tuners, and switches. Integrated antenna components require more stringent design considerations because they simultaneously include both guided- and space-wave propagation. Therefore, component design, circuit layout, and device packaging play a more critical role in determining the overall performance.

The combination requires knowledge about state-of-the-art antennas as well as solid-state devices. The state of the art of these technologies varies widely over different frequency ranges. For instance, at low frequencies, solid-state technology can reliably mass-produce complete monolithic transceivers at a relatively low cost. On the other hand, due to the long wavelengths, efficient low-frequency antennas may be large and bulky. While low-frequency systems can compensate for smaller, less-efficient antennas with better integrated circuits, higher operating frequencies pose more stringent requirements on the antennas performance. At millimeter and submillimeter wavelengths, more efficient antennas are required to provide more directionality and higher gain. It becomes important to optimize both the circuit and radiating structure.

This chapter reviews a wide range of millimeter- and submillimeter-wave antennas. These include waveguide horns, microstrip arrays, and, inevitably, integrated antennas. At very high millimeter and submillimeter frequencies, the use of integrated antennas becomes a matter of necessity due to tolerances, size, and losses encountered using conventional approaches. These approaches include quasi-optical imaging arrays and even grid-type antennas. The latter is included in Chapter 8 and will not be discussed here.

11.2   ANTENNA TECHNOLOGIES

Waveguides and horns are of reasonable size at and above microwave frequencies and they provide high-power handling, excellent radiation efficiency, and performance, but tolerance constraints and machining costs limit many applications above 100 GHz. Also, a large array of horns becomes quite bulky and requires a complex, lossy rf feed network.

In order to reduce the overall size, weight, and cost of low- to medium-power systems, microstrip antennas are used to replace waveguides and horns. Microwave hybrid and monolithic integrated circuits provide very good performance and can be easily integrated with microstrip antennas. Since microstrip antennas use photolithographic techniques, corporate, series, or any other feed network combinations are easily manufactured and reproduced, but these suffer from losses which reduce aperture efficiencies. Integrated antennas attempt to combine the advantages of printed antennas with those of solid-state devices and circuits (i.e., MICs, MMICs) to overcome high transmission losses and low aperture efficiencies. At higher W-band, for a large corporate fed array, this loss may reduce the efficiency of the aperture to 25%.

Low-loss dielectrics and conductors in different MIC line configurations can provide lower-loss transmission lines and more efficient radiators. Even with these improvements, a conventional approach requires long interconnection lengths in the rf feed distribution network. Figure 11.1 shows several antenna technologies used throughout the frequency spectrum. Also included are problems which often limit the use of a particular type of antenna. Typically, size, conductor losses, and tolerance constraints are major obstacles at very high frequencies of operation.

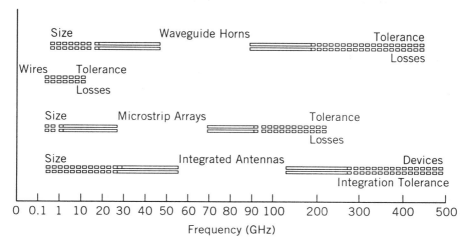

FIGURE 11.1. Antenna technology used over the microwave and millimeter-wave regions.

Alternatively, integrated antennas can be used to avoid rf transmission line losses altogether. In conjunction with quasi-optical power-combining techniques, integrated antennas can increase power output or signal-to-noise ratio. The technique is not limited to medium-power systems because power combining occurs in space, away from the individual sources. This is clearly shown for active antennas used in large spatial combining arrays. Individually, the active antennas can provide a relatively small power output due to the limitations of the solid-state device, substrate, and thermal dissipation of the structure. However, a large number of these sources can be synchronized together to work as a single coherent unit providing large amounts of power. Radio-frequency power combining occurs in space, which reduces losses associated with distribution networks, and it is not limited by the power limitations of the MICs.

In a similar manner, a large array of detectors can improve the signal-to-noise ratio over a conventional single-detector system. The combination of the antenna and the connecting line to the device is the first loss encountered in the receiver which contributes to the overall receiver noise figure. By detecting the rf at the antenna terminals, feed network losses are avoided. Furthermore, integrated antennas provide low-profile, compact systems for detection, mixing, radiating oscillators, etc. An effective integrated antenna can be effectively repeated over a large array to provide a planar integrated antenna array.

A major consideration of a large integrated antenna array, however, is the large number of devices required. Since solid-state devices are generally the cost drivers for a system, this may not always be a cost-effective solution for commercial applications. At very high frequencies ($>100\,$GHz) where current rf feed distribution networks are prohibitively lossy, the use of many devices can be justified. In certain applications, such as rf power transmission links, the use of integrated antennas for rectification has been shown to be very effective. However, solid-state devices drive the system cost up and the use of the least number of devices that meets system performance will lead to the most cost-effective solution.

11.3   ANTENNAS ABOVE 100 GHz

Conductor losses increase and dimensions become critical above $100\,$GHz. Bends, corners, steps, and other discontinuities cause erratic radiation patterns and higher cross-polarization levels. Discontinuities and losses reduce the antenna gain and lower the overall system efficiency. Radiation problems, such as high cross-polarization and erratic patterns, can be improved through feed networks placed on the backside. However, discontinuities and conductor/ dielectric losses remain to lower the overall efficiency.

Waveguide horns at millimeter and submillimeter wavelengths offer high efficiencies and excellent radiation patterns and functions as efficient feeds for reflectors and lenses. Some typical feedhorns include rectangular, conical, dual

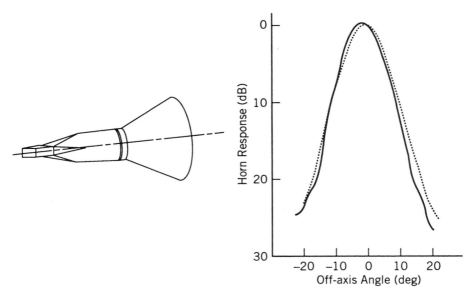

FIGURE 11.2. Dual-mode horn with a rectangular to circular waveguide transition and the measured patterns at $486\,$GHz. This horn had a 6% bandwidth. Even more symmetrical patterns are achieved between 200 and $300\,$GHz. [8].

mode, and scalar. Corrugated conical horns [5,6] exhibit high gain, low cross-polarization, rotationally symmetric patterns, and 40% impedance bandwidths. Similarly, dual-mode horns [7,8] exhibit high gain, symmetric patterns, and efficient radiation over an 8% impedance bandwidth. Figure 11.2 shows a dual-mode horns antenna and its radiation pattern [8].

The need for complex, high-tolerance machining limits many of these waveguide designs. Tolerances are such that these designs are not readily reproduced above 500 GHz. Simpler pyramidal or conical horns can be used up to 800 GHz, where, again due to small tolerances, they are not easily reproduced. For ease in construction, traveling-wave corner-cube antennas [9] can be fabricated through the 1000-GHz range. However, high side lobes and cross-polarization levels result in lower efficiencies ($\sim 50\%$). Figures 11.3a and b show the performance of several types of millimeter-wave antennas, including parabolic, Cassegrain, and lens [10], as well as microstrip arrays [11,12].

F_0 (GHz)	Aperture Size (inch) E-plane	H-plane	Half-Power Beamwidths (deg) E-plane	H-plane	Directivity (dBi) D_0	Gain (dBi) G_0
24.125	7	6	3.8	4	34.11	29.5
35.5	2	2	12	12	24.57	21
59.5	2.8	2.8	4	4	34.11	29.5
76.5	1.3	2.4	7	4	31.68	25.5

(a)

FIGURE 11.3. Millimeter-wave antennas: (a) microstrip arrays [11]; (b) beamwidth and gain for parabolic, Cassegrain, and lens antennas [12]. (*Continues on next page.*)

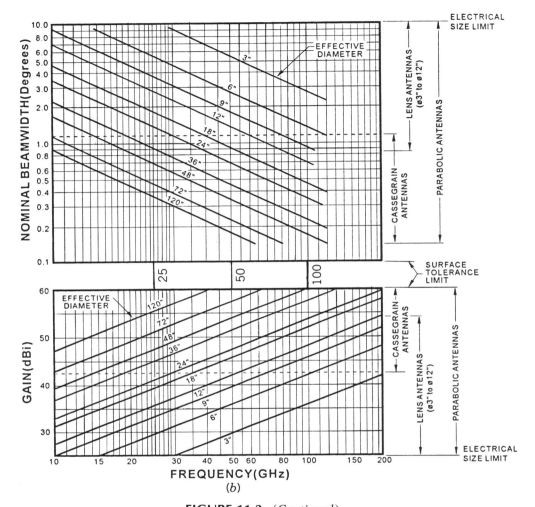

FIGURE 11.3. (*Continued*).

11.4   ANTENNAS INTEGRATED WITH MIXERS AND DETECTORS

At millimeter and submillimeter wavelengths, the overriding concern is the efficient transformation of the space wave to a guided wave very near the antenna. The use of a conventional rf feed distribution network becomes prohibitively lossy as well as dependent on very small dimensions and tight tolerances. The mechanical complexity and feed losses can be avoided by placing the detector or mixer directly to the antennas. The distribution network can be incorporated at a much lower intermediate frequency where losses are lower and circuit dimensions are not as critical. This avoids the losses

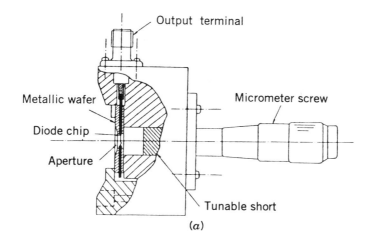

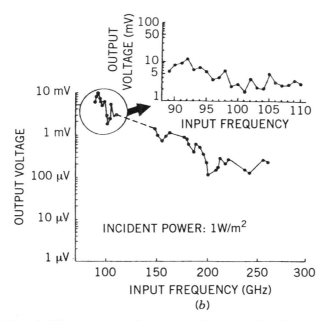

FIGURE 11.4. Millimeter-wave integrated detector. (*a*) Structure of the detector. A whisker diode is placed on the aperture of the metallic wafer. The metallic wafer is screwed to the oversized waveguide tuning section which has a short plane and a micrometer screw. The metallic post which supports the diode chip is insulated from the metallic wafer and is connected to the center conductor of the coaxial output terminal. (*b*) Output voltage versus input frequency. Wideband data over the short millimeter-wavelengths region and detailed data around 100 GHz measured with a distance of 1 GHz are shown. The data points are obtained by adjusting the tunable short for optimal sensitivity at each frequency. (From Ref. 14 with permission from IEEE.)

and tight tolerances and places the burden on the integrated antenna element design [13].

A transition or circuit interface to transfer the space wave to an integrated circuit for detection was shown by Kawasaki and Yamamoto [14]. The wideband quasi-optical detector uses a waveguide and Schottky-barrier diode to operate up to 300 GHz. The diode detector is integrated on a wafer placed over the mouth of an oversized waveguide. The waveguide has a mechanically adjustable sliding short which allows operation from 100 to 300 GHz. For an incident power level of 1 W/m^2, the detector provides 2.7, 0.6, 0.2, and 0.1 mV at 100, 180, 260, and 300 GHz, respectively. Other components can be integrated in a similar fashion. Figure 11.4 shows the integrated detector configuration and circuit performance. The waveguide and millimeter-wave detector integrated with a circuit card provide the means to easily achieve detection, mixing, and other circuit functions. Earlier, a twin-slotline mixer at 100 to 120 GHz was shown by Kerr et al. [15].

Many quasi-optical antenna designs rely on a lens to concentrate the rf energy over a planar antenna. A quasi-optical crossbar mixer was shown by Yuan, Paul, and Yen in 1982 [16]. The design uses beam lead diodes in a crossbar circuit layout and operates at 60 and 140 GHz. The rf input is inserted through a spherical lens which couples to slotline antennas which feed a pair of diodes in phase. The LO is fed through a waveguide input to the diodes 180° out of phase. For an LO frequency of 141 GHz, 7 dB of conversion loss was measured at 127 GHz. Figure 11.5 shows the crossbar mixer layout and its conversion loss performance. A detector integration using slot antennas through a large ground plane was shown in 1990 by Wentworth et al. [17] and by Heston et al. in 1991 [18]. The quasi-optical slot antenna detector is used for 94-GHz imaging. Multiple quarter-wave length layers are used to improve the slot antenna performance, as shown in Figure 11.6a. Figure 11.6b shows the detector performance for a quartz or GaAs cover layer. Another twin-slotline antenna configuration was demonstrated by Gearhart and Rebeiz in 1994 [19]. The monolithic receiver is integrated using CPW and slotline antennas fed by a high-resistivity silicon lens. At 258 GHz, the double-sideband (DSB) conversion loss is about 9 dB for a given LO power level at the lens aperture. Figures 11.7a and b show the integrated antenna configuration.

In 1982, Parrish et al. integrated a Schottky diode with a printed dipole for use at millimeter wavelengths [20]. Later, the planar twin-dipole antenna was integrated with diodes and FETs for detection from X-band through W-band by Chew and Fetterman [21, 22]. Figure 11.8 shows a four-element array integrated with diodes. In 1990, Filipovic et al. showed very good radiation performance for the detector integrated double-dipole antenna at 2 GHz [23]. Figure 11.9 shows the configuration.

Skalare et al. demonstrated two half-wave dipoles backed by a reflector and used with a lens to create a compact receiver from 7 to 14 GHz [24]. The configuration offers cross-polarization levels 20 dB below the maximum and side-lobe levels of −13 dB over an octave bandwidth. Figure 11.10 shows the configuration of the twin dipole integrated with a lens. Very good patterns are

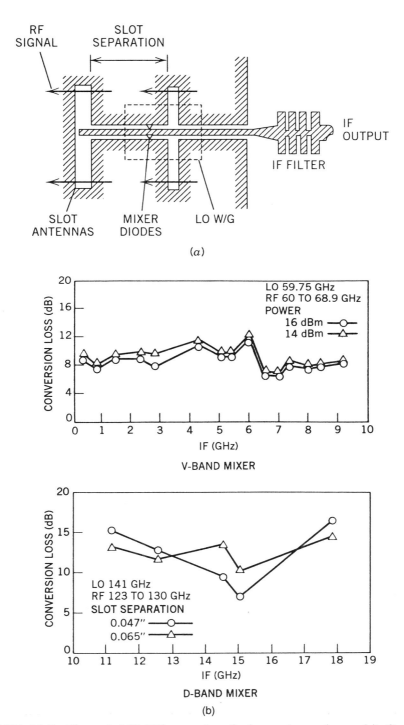

FIGURE 11.5. 60- and 140-GHz quasi-optical crossbar mixers. (*a*) Quasi-optical crossbar mixer circuit layout. (*b*) Conversion loss of quasi-optical crossbar mixers. (From Ref. 16 with permission from IEEE.)

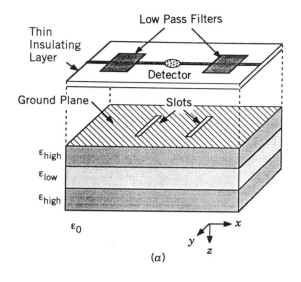

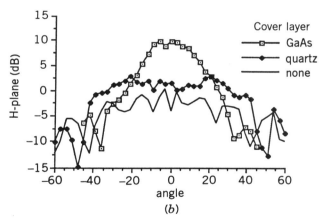

FIGURE 11.6. Integrated twin-slot antenna detector [18]. (*a*) Twin-slot antenna on a dielectric stack, where each layer is one-quarter wavelength thick. The slots couple power into a microstrip line supported over the slots by a thin insulator. This power is dissipated in a detector and is isolated from the rest of the feed network by low-pass filters. (*b*) 94-GHz beam patterns for the twin-slot structure with three different cover layers.

demonstrated for 7, 10, and 14 GHz. An integrated Fresnel zone plate antenna was demonstrated by Gouker and Smith in 1991 [25]. The integrated dipole and a Fresnel zone plate make a useful and simple planar antenna with very short focal lengths. The bolometer integrated dipole is centered on a set of rings

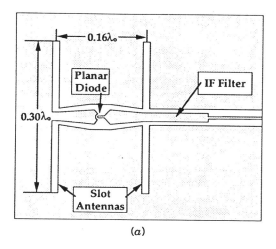

(a)

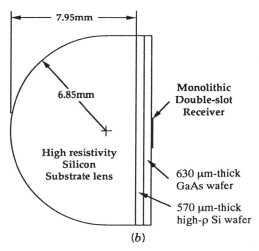

(b)

FIGURE 11.7. The 250-GHz monolithic double-slot antenna receiver: (a) top view, (b) side view illustrating the high resistivity silicon lens and GaAs wafer. (From Ref. 19 with permission from IEEE.)

which form the Fresnel zone plate as shown in Figure 11.11. Very good patterns and performance are obtained at 230 GHz. Similar dipoles for the 230-GHz range are shown in Figure 11.12 [26].

An integrated bow tie antenna was demonstrated for a subharmonically pumped mixer by Stephen and Itoh in 1984 [27]. Figure 11.13 shows the integrated mixer antenna configuration. A conversion loss of 8.6 dB was

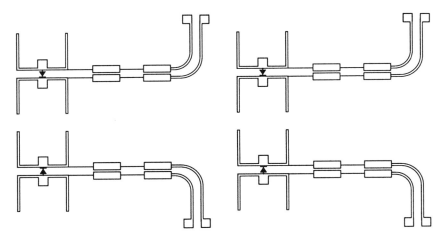

FIGURE 11.8. Four-element building block for large imaging array. Beam-lead Schottky diodes are placed between the twin dipoles [21, 22].

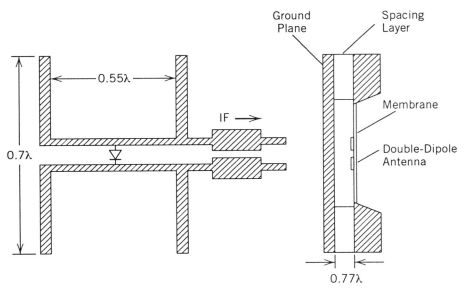

FIGURE 11.9. The twin-dipole antenna on a thin dielectric membrane and backed by a ground plane [23].

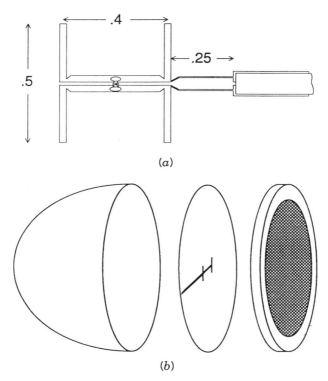

FIGURE 11.10. Diode integrated dual-dipole and lens antennas [24]. (*a*) The antenna with two dipoles. A conducting reflector plane at a quarter-wave behind the dipoles creates an image of two more elements. The detector was mounted midway between the two dipoles, and the dc/if leads are on the right. The drawing is correct to scale, and the dimensions are in units of λ. (*b*) An exploded view of the lens, the Kapton film with dipoles and a quarter-wavelength slab with the reflector plane. The diameter of the elliptical lens was 110 mm, and the eccentricity 0.43.

demonstrated at 14 GHz. In 1985, Taylor et al. demonstrated a planar dipole-fed mixer for imaging applications [28]. Jackson and Sun [29] showed a subharmonically pumped slotline mixer in 1986. Scattering from an array of diode-loaded dipoles has been considered by Janaswamy and Lee in 1988 [30]. Results show the ability to use the *pin* integrated dipole for an electrically controlled reflectance or transmittance array as in the grid arrays of Chapter 8. (See Sect. 8.6.)

In 1982, Stephen et al. used the advantages of a slotline ring antenna to develop a quasi-optical mixer [31, 32, 33]. The slotline ring was integrated with two mixer diodes. The LO and rf strike the plane of the mixer from either side,

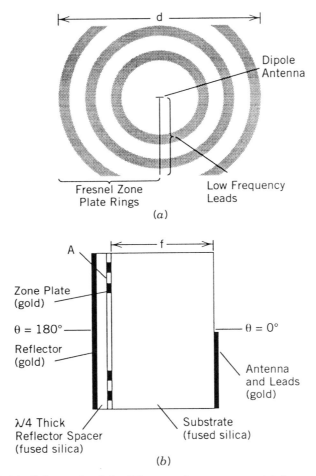

FIGURE 11.11. Schematic of the IC zone-plate antenna: (*a*) front view, (*b*) side view. (From Ref. 25 with permission from IEEE.)

and the if is extracted through a low-pass filter. Figure 11.14 shows the slotline ring mixer layout, and Figure 11.15 shows the radiation patterns at X-band. Patterns deteriorated for V- and W-band mixers using the same approach. However, a quasi-optical monopulse receiver front end has been shown by Gingras et al., using this approach [34]. Figure 11.16*a* shows the 35-GHz monopulse integrated receiver system configuration, and Figure 11.16*b* shows the diplexer configuration.

In 1988, Jackson introduced two integrated antenna configurations for quasi-optical mixing [35]. Figure 11.17*a* shows the fundamental mixer configuration which uses a patch antenna for the rf port, a waveguide LO, and

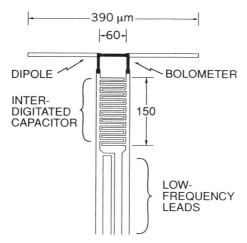

FIGURE 11.12. Schematic of the dipole structure. All dimensions are in micrometers. (From Ref. 26 with permission from IEEE.)

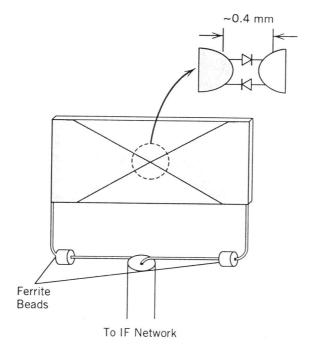

FIGURE 11.13. Subharmonic mixer using antiparallel diodes and bow tie antenna. (From Ref. 27 with permission from IEEE.)

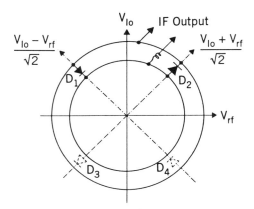

FIGURE 11.14. Slotline ring antenna integrated with mixers. (From Ref. 32 with permission from IEEE.)

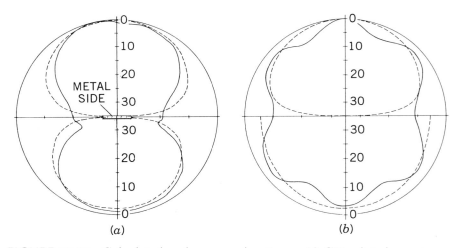

FIGURE 11.15. Calculated and measured patterns, 10-GHz slot-ring antenna. Inner ring radius $= 0.39$ cm, outer ring radius $= 0.54$ cm, dielectric $\varepsilon_r = 2.23$, thickness $d = 0.3175$ cm. All patterns are decibles down from maximum: (a) H-plane, (b) E-plane (---) calculated, (—) measured. (From Ref. 32 with permission from IEEE.)

microstrip if out. The patch antenna, rf matching network, and if matching section are integrated on a single 0.010-in. Duroid substrate, while the local oscillator signal is fed from a waveguide input perpendicular to the substrate. Figure 11.7b shows the conversion loss versus frequency. Isotropic conversion losses of 7 dB were measured over the 28- to 31-GHz range. Jackson also

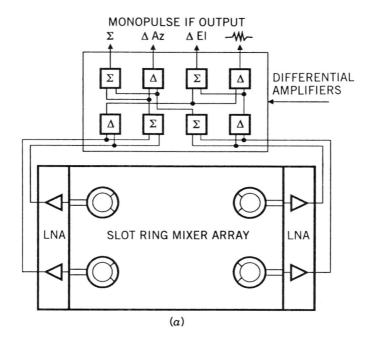

(a)

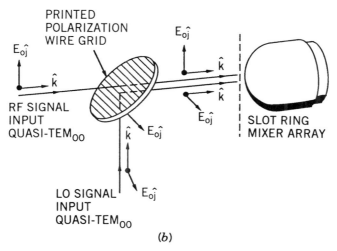

(b)

FIGURE 11.16. 35-GHz monopulse receiver [34]. (*a*) Circuit diagram of the four monolithic chips that comprise the slot ring mixer array receiver. (*b*) RF and LO signals diplexed by polarization grid.

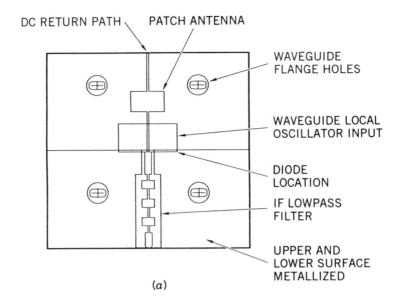

(a)

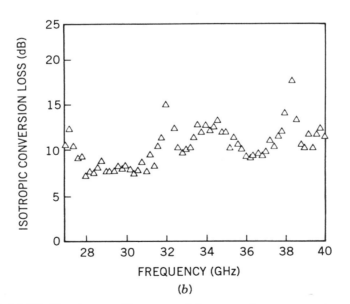

FREQUENCY (GHz)

(b)

FIGURE 11.17. Fundamentally pumped integrated antenna mixer [35]. (*a*) Circuit layout for a fundamentally pumped mixer configuration. (*b*) Fundamentally pumped mixer conversion loss vs. frequency.

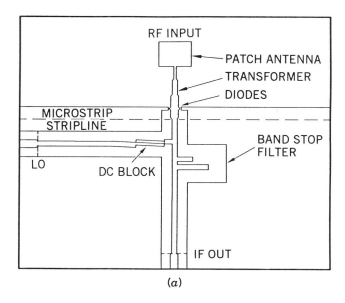

(a)

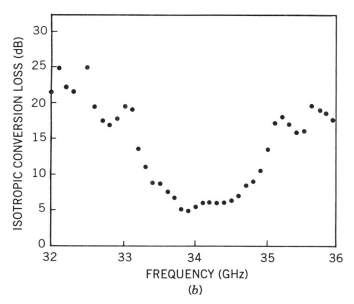

(b)

FIGURE 11.18. Subharmonically pumped integrated antenna mixer. (a) Circuit layout for a subharmonically pumped mixer with compact planar LO feed. (b) Subharmonically pumped mixer conversion loss vs. frequency. (From Ref. 35 with permission from Wiley.)

introduced a subharmonically pumped mixer whose configuration is shown in Figure 11.18*a*. The LO provides 17 GHz for the 33- to 36-GHz rf, which provides 0- to 2-GHz if signal. Figure 11.18*b* shows the measured isotropic conversion loss. Conversion losses of 5 dB were measured at 34 GHz.

The large span of the millimeter-wave region makes the use of a detector and narrow-band antenna somewhat impractical. The use of a wideband antenna such as a horn, notch, or spiral antenna appears more suited for detector applications above 100 GHz. In 1987, Rebeiz et al. used a reflector-

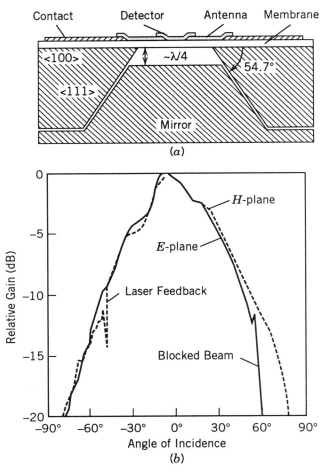

FIGURE 11.19. 700-GHz log-periodic integrated antenna [36]. (*a*) Cross section of an antenna on a thin membrane with a reflector. (*b*) The measured *E*- and *H*-plane patterns for a log-periodic antenna at 700 GHz. These patterns were taken from the top side of the wafer, without a reflecting mirror.

backed log-periodic antenna and bismuth-microbolometer detector to build a wideband submillimeter-wave integrated antenna [36]. Figure 11.19 shows the configuration cross section along with the radiation patterns at 700 GHz. Also in 1987, another log-periodic antenna was integrated with a Schottky detector at millimeter-wavelengths by Lee and Frerking [37]. A two-lobe log-periodic antenna was tested at 67 and 205 GHz with identical HPBW in the *E*- and *H*-plane of 30° and 42°, respectively. Similarly, a four-lobe log-periodic showed similar radiation patterns except for a higher

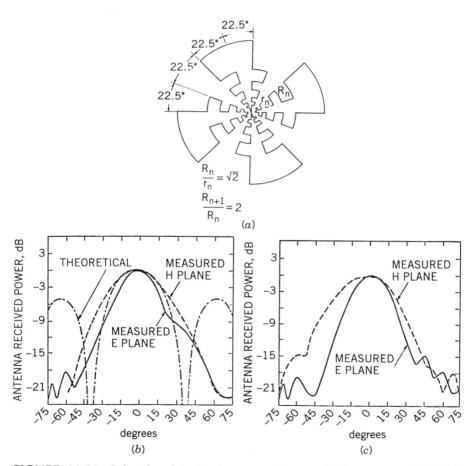

FIGURE 11.20. Schottky detector integrated log-periodic antenna [37]. (*a*) Planar four-lobed log-periodic structure. (*b*) Measured radiation patterns of the bilobed log-periodic antenna compared with that calculated for the equivalent double-dipole. (*c*) Measured radiation patterns of the four-lobed log-periodic antenna.

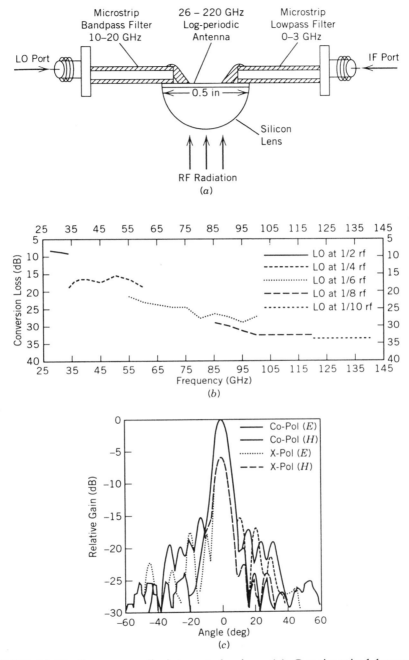

FIGURE 11.21. Log-permodic integrated mixer. (*a*) Quasi-optical harmonic mixer-receiver using a planar log-periodic antenna on a substrate lens. (*b*) Measured SSB conversion loss of the wideband harmonic mixer-receiver. (*c*) 180-GHz antenna pattern for the planar log-periodic antenna at the optimal position on a substrate lens. (From Ref. 38 with permission from IEEE.)

310

cross-polarization level. The log-periodic configuration and radiation patterns are shown in Figure 11.20. A similar log-periodic antenna using a lens and microstrip LO port is shown in Figure 11.21. It was demonstrated in 1992 by Kormanyos and Rebeiz [38]. The 30–180 GHz harmonic mixer-receiver's SSB conversion loss performance is plotted from 25 to 145 GHz. It shows less than

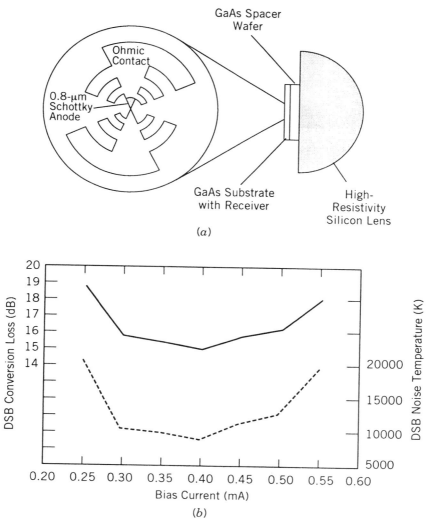

FIGURE 11.22. Log-periodic integrated antenna mixer. (*a*) Planar receiver with spacer wafers and silicon lens. (*b*) 761-GHz DSB conversion loss and noise temperature versus bias current. (From Ref. 39 with permission from IEEE.)

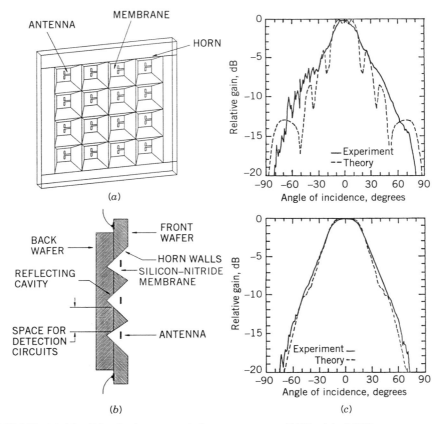

FIGURE 11.23. Dipole integrated horn antenna [43]. (*a*) Millimeter-wave imaging with a focal-plane imaging array. (*b*) Side view. (*c*) Typical *E*-plane pattern (top) and *H*-plane pattern (bottom) of a single element in a 9 × 9 array.

8 dB conversion loss from 28 to 35 GHz and greater than 15 dB for the rest of the range. Well-behaved patterns are shown for 180 GHz. In 1993, a wideband receiver at 760 GHz using an integrated spiral and lens was shown by Gearhart et al. [39]. Figure 11.22 shows the circuit layout and lens with the radiation patterns and DSB conversion loss. Another log-periodic integration for an 80- to 200-GHz planar subharmonic receiver was later shown by Kormanyos et al [40]. DSB conversion loss of 6.7 and 8.5 dB was achieved at 90 and 182 GHz, respectively. Lee et al. used a separately biased Schottky diode pair with the same log-periodic and lens antenna configuration [41,42]. An 8-dB SSB conversion loss was achieved at 90 GHz, and 9.7-dB conversion loss was demonatrated at 180 GHz.

Rebeiz et al. used thin membranes [43] to combine the integration advantages of microstrip with the radiation performance of a wavelength horn.

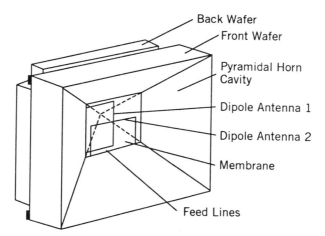

FIGURE 11.24. Monolithic dual-polarized horn antenna element with a novel bias and feeding structure. (From Ref. 45 with permission from IEEE.)

Figure 11.23 shows the millimeter-wave imaging array configuration and performance used at 242 GHz. The bismuth bolometer is integrated with dipoles on thin membranes and incorporated with pyramidal horn antennas. E- and H-plane half-power bandwidths are 35° and 46°, respectively. Similar horn antennas exhibit 10 to 12 dB of directive gain with nearly 72% efficiency [44]. A dual-polarized integrated horn was shown by Ali-Ahmad and Rebeiz in 1991 [45]. Polarization isolation of 20 dB was shown for two orthogonal dipoles. Figure 11.24 shows the configuration.

In 1992, a quasi-integrated horn was shown with 20 dB of gain, symmetrical patterns, low cross-polarization, and high aperture efficiency [46, 47], Figure 11.25 shows the configuration and radiation patterns at 91 GHz. Design information on these integrated horn designs is given by Rebeiz et al. [48]. Some improvements of these integrated horns were demonstrated by Guo et al. by optimizing dipole lengths and coating the horn walls with gold [49]. Figure 11.26 outlines antenna efficiencies for various parameters of the integrated design.

A different approach, which uses a corner reflector and a traveling-wave antenna instead of a resonant dipole and pyramidal horn, is shown in Figure 11.27. The figure shows the configuration and radiation patterns at 3.2 GHz [50]. Later, similar results for this design were reported with 93% efficiency at 180 GHz and 83% at 222 GHz [51, 52]. A directivity of 18 ± 0.5 dB is calculated from E- and H-plane measurements for the 222-GHz array.

A wideband alternative to the broadside log-periodic is the endfire notch antenna (see Chap. 9.) The endfire notch antenna has been integrated with diodes and transistors for active and passive antenna applications. In 1992, Ekstrom et al. demonstrated a detector integrated endfire notch antenna at

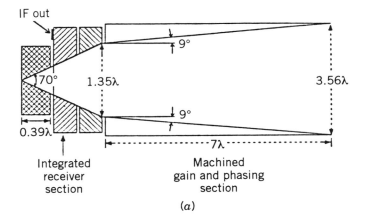

(a)

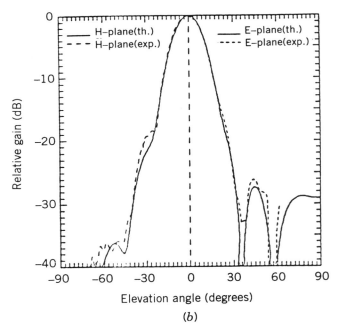

(b)

FIGURE 11.25. Quasi-integrated horn. (a) 20-dB quasi-integrated horn antenna. (b) Predicted and measured patterns at 91 GHz. (From Ref. 46 with permission from IEEE.)

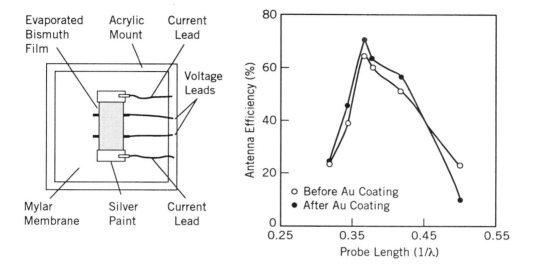

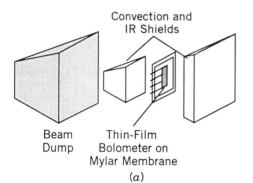

Loss component	loss (dB)
Intrinsic pattern loss	0.2
Mismatch loss	0.4
Cross-polarization loss	0.2
Horn-to-horn coupling loss	0.1
Total calculated loss	0.9
Measured loss	1.4

(a)   (b)

FIGURE 11.26. Improved integrated horn. (*a*) Thin-film power-density meter and assembly. (*b*) Measured aperture efficiencies at 93 GHz versus antenna length. The efficiencies were measured before and after coating the membrane wafer with evaporated gold. Summary of measured and calculated losses is also shown. (From Ref. 49 with permission from Wiley.)

348 GHz [53]. Figure 11.28 shows the bismuth-bolometer integrated antenna and the corresponding radiation patterns. Similarly, in 1993 Acharya et al. showed the detector integrated endfire notch antenna at 802 GHz [54]. Figures 11.29*a*, *b* and *c* show the configuration dimensions and submillimeter-wave radiation patterns. The microbolometer integrated notch antenna demonstrates very smooth patterns at 802 GHz, which agrees well with calculated results.

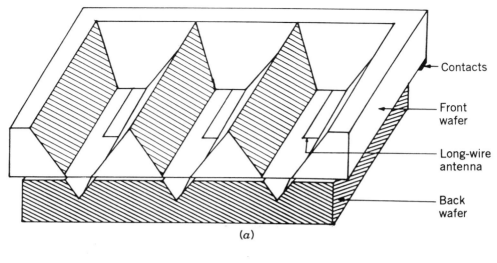

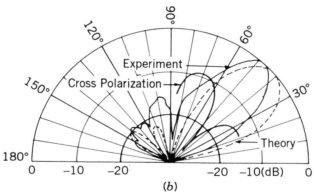

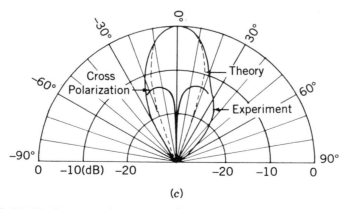

FIGURE 11.27. Integrated corner reflector antennas. (*a*) A monolithic corner-reflector imaging array. (*b*) The measured and predicted *E*-plane and (*c*) quasi-*H* plane patterns at 3.2 GHz. (From Ref. 50 with permission from Wiley.)

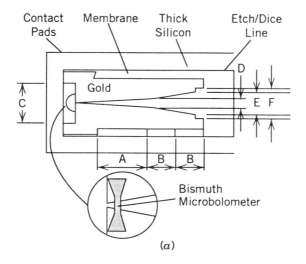

(a)

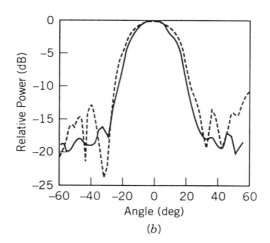

(b)

FIGURE 11.28. Bolometer integrated notch. (a) Antenna with bolometer on SiO_2–Si_3N^4 membrane. Dimensions of the 348-GHz antenna are $A =$ 3.23 mm, $B = 1.29$ mm, $C = 2.46$ mm, $D = 0.32$ mm, $E = 0.78$ mm, $F =$ 0.97 mm. Slot is approximately 10 μm wide at the bolometer, and the width of the bolometer is approximately 5 μm. (b) Antenna patterns from scale measurements at 45 GHz. Solid line is E-plane, and dotted line is H-plane. (c) Antenna patterns of the 348-GHz design on SiO_2–Si_3N_4 membrane, measured at 348 GHz. D is for diagonal plane and Dx represents cross-polarization in D-plane. (From Ref. 53 with permission from IEEE.) (*Continues on next page.*)

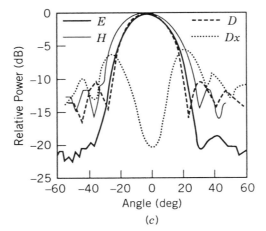

FIGURE 11.28. (*Continued*).

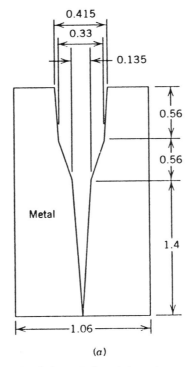

(*a*)

FIGURE 11.29. Bolometer integrated notch antenna. (*a*) Dimensions (in millimeters) of the broken linearly tapered slotline antenna (BLTSA) at 802 GHz. (*b*) BLTSA on a SiO2/Si$_3$N$_4$ membrane supported on three sides by a silicon substrate. (*c*) Radiation patterns of a BLTSA at 802 GHz. Calculated (solid line) and measured (dashed line). (From Ref. 54 with permission from IEEE.)

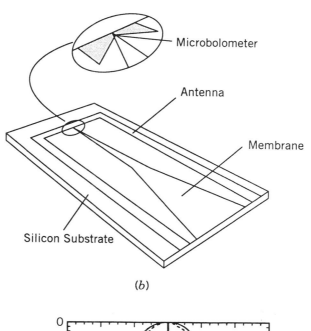

Microbolometer

Antenna

Membrane

Silicon Substrate

(b)

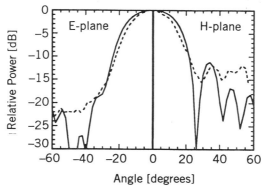

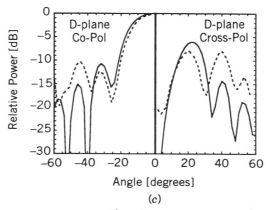

(c)

FIGURE 11.29. (Continued).

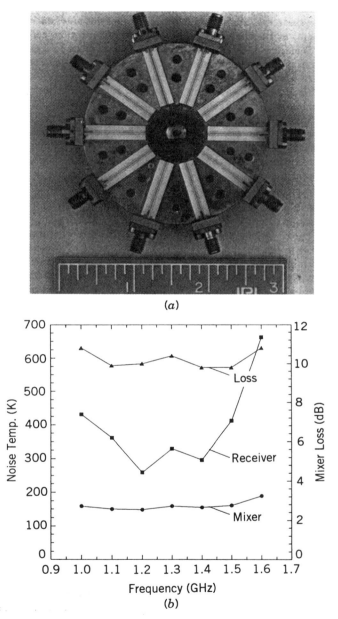

(a)

(b)

FIGURE 11.30. Quasi-optical SIS receiver. (*a*) The mixer block with the upper half removed. The central dielectric-filled parabola (dark), containing 10 antenna and mixer elements, is surrounded by 10 if baluns (light) and SSMA connectors at the edge of the block. (*b*) Mixer and receiver noise temperature and mixer loss as a function of if. The best results are obtained at 1.2 GHz, where a T_M of 148 K DSB, a T_R of 259 K DSB and conversion losses of 10 dB were measured. (From Ref. 55 with permission from IEEE.)

Planar quasioptical superconductor-insulator-superconductor (SIS) receivers were demonstrated by Stimson et al. in 1992 [55]. Ten niobium-aluminum oxide-niobium SIS junctions are integrated with a 2 × 5 dipole array, which are backed by a quartz-filled parabola. Coplanar strip transformers are used to extract the 1.2-GHz if from the 230-GHz rf signal. A 10-dB conversion loss has been demonstrated at 230 GHz. Figure 11.30 shows the configuration and performance. Similarly, in 1994, other results were presented on the SIS array receiver [56].

Rectennas were initially developed by W. C. Brown for the efficient conversion of rf to dc in microwave links [57, 58, 59]. The power converted by the rectenna was used to power a helicopter-type platform at the Raytheon company in the 1960s. A 680-diode grid antenna developed by Sabbagh and George to rectify a 2.45-GHz microwave signal is shown in Figure 8.1a [60]. Similarly, Bharj et al. have developed a 1000-element C-band rectenna array [61]. Unlike previous integrations which use integrated bolometers for thermal detectors [62], rectennas operate at very high field intensities in the diode's nonlinear region. Brown has made many contributions to rectenna development and microwave power transmission [59]. Figure 11.31 shows the typical 2.45-GHz rectenna layout. An rf-to-dc conversion efficiency of 85% to 90% has been achieved.

Yoo, McSpadden, and Chang introduced two 35-GHz rectenna designs in 1992 [63]. One of the designs uses a dipole not unlike that shown by Brown, and the other is a microstrip patch. Higher conversion efficiencies were demonstrated by the dipole over the patch design. Figure 11.32 shows the dipole and microstrip patch rectenna configurations and results. A complete analysis of this structure has been reported by Yoo and Chang [64].

Similar to thermal detectors, photodetectors have been used extensively in radiometry and imaging. Imaging arrays are an ideal application for an array

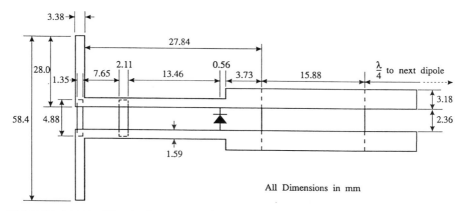

FIGURE 11.31. Physical rectenna layout and dimensions. The rectenna was built on a 1-mil-thick Kapton substrate.

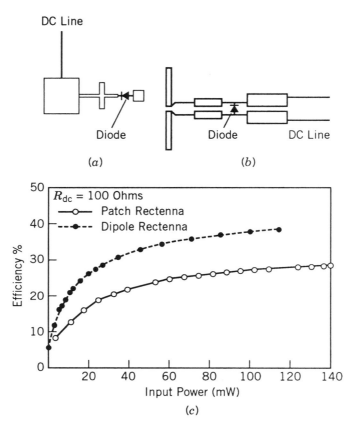

FIGURE 11.32. 35-GHz rectennas and their performance: (*a*) a patch rectenna. (*b*) A microstrip dipole rectenna. (*c*) Power conversion efficiency of the 35-GHz rectenna measured with waveguide array simulator. (From Refs. 63 and 64 with permission from IEEE.)

of photodetector integrated antennas. *Pin* photodetectors provide relatively low noise, high efficiency, and high operating frequencies and allow conversion of optical frequencies to microwave frequencies [65]. Many systems could then use the low-loss, wide-bandwidth fiber-optic interconnections up to the radiating element. At the antenna, the fiber-optic signal is translated to microwave frequencies for transmission. A receiving antenna can then demodulate the microwave signal back to a fiber-optic signal. Transistors can be used as detectors to replace diodes. Advancements in transistors such as HEMTs and HBTs allow the replacement of *pin*s with improvements in power and higher signal-to-noise ratios [66]. HEMTs have demonstrated superior noise performance over MESFETs and are very useful high-speed photodetectors. HBTs provide large photocurrent gains without sacrificing noise performance or using high bias voltages. These integrations provide another area for integrated antennas and quasi-optical systems.

REFERENCES

1. J. M. Payne, "Millimeter and Submillimeter Wavelength Radio Astronomy," *Proceedings of the IEEE*, Vol. 77, No. 7, pp. 993–1017, July 1989.

2. W. J. Wilson, R. J. Howard, A. C. Ibbott, G. S. Parks, and W. B. Ricketts, "Millimeter-Wave Imaging Sensor," *IEEE Transactions on Microwave Theory and Techniques*, Vol. 34, No. 10, pp. 1026–1035, October 1986.

3. FCC Ruling 1995.

4. G. M. Rebeiz, "Millimeter-Wave and Terahertz Integrated Circuit Antennas," *Proceedings of the IEEE*, Vol. 80, No. 11, pp. 1748–1770, November 1992.

5. P. J. B. Clarricoats and A. D. Olver, *Corrugated Horns for Microwave Antennas*, IEE Electromagnetic Waves, Series 18, Peter Perigrinus, London, 1984.

6. R. Wylde, "Millimeter-Wave Gaussian Beam-Mode Optics and Corrugated Feedhorns," *Proceedings of the IEE*, Part H, Vol. 131, pp. 258–262, August 1984.

7. P. D. Potter, "A New Horn with Antenna Suppressed Sidelobes and Equal Beamwidths," *Microwave Journal*, Vol. 6, No. 6, pp. 71–78, June 1963.

8. G. A. Ediss, "Dual-Mode Horns at Millimeter and Submillimeter Wavelengths," *Proceedings of the IEEE*, Part H., Vol. 132, No. 3, pp. 215–218, June 1985.

9. E. N. Grossman, "The Coupling of Submillimeter-Wave Corner-Cube Antennas to Gaussian Beams," *Infrared Physics*, Vol. 29, No. 5, pp. 875–885, July 1989.

10. F. Lalezari and C. D. Massey, "mm-Wave Microstrip Antennas," *Microwave Journal*, Vol. 30, No. 4, pp. 87–96, April 1987.

11. Epsilon-Lambda Electronics, 1995 Product Catalog, p. 10.

12. Millitech Corporation, 1995 Product Catalog, p. 14.

13. D. B. Rutledge, D. P. Neikirk, and D. P. Kasilingam, "Integrated Circuit Antennas," *Infrared and Millimeter-Waves*, (K. J. Button, ed.); Chap. 1, vol. 10, pp. 1–90, Academic Press, 1983.

14. R. Kawasaki and K. Yamamoto, "A Wide-Band Mechanically Stable Quasi-Optical Detector for 100–300 GHz," *IEEE Transactions on Microwave Theory and Techniques*, Vol. 27, No. 5, pp. 530–533, May 1979.

15. A. R. Kerr, P. H. Siegel, and R. J. Mattauch, "A Simple Quasi-Optical Mixer for 100–120 GHz," *IEEE MTT-S International Microwave Symposium Digest*, pp. 96–98 (1977).

16. L. Yuan, J. Paul, and P. Yen, "140 GHz Quasi-Optical Planar Mixers," *IEEE MTT-S International Microwave Symposium Digest*, pp. 374–375 (1982).

17. S. M. Wentworth, R. L. Rogers, J. G. Heston, D. P. Neikirk, and T. Itoh, "Millimeter Wave Twin Slot Antennas on Layered Substrates," *International Journal of Infrared and Millimeter Waves*, Vol. 11, No. 2, pp. 111–131, February 1990.

18. J. G. Heston, J. M. Lewis, S. M. Wentworth, and D. P. Neikirk, "Twin Slot Antenna Structures Integrated with Micro-bolometer Detectors for 94-GHz Imaging," *Microwave and Optical Technology Letters,* Vol. 4, No. 1, pp. 15–19, January 1991.

19. S. S. Gearhart and G. M. Rebeiz, "A Monolithic 250 GHz Schottky-Diode Receiver," *IEEE MTT-S International Microwave Symposium Digest,* pp. 1333–1336 (1994).

20. P. T. Parrish, T. C. L. G. Sollner, R. H. Mathews, H. R. Fetterman, C. D. Parker, P. E. Tannenwald, and A. G. Cardiasmenos, "Printed Dipole-Schottky Diode Millimeter Wave Antenna Array," *SPIE Millimeter Wave Technology,* Vol. 337, pp. 49–52 (1982).

21. H. R. Fetterman, T. C. L. G. Sollner, P. T. Parish, C. D. Parker, R. H. Matthews, and P. E. Tannenwald, "Printed Dipole Millimeter Wave Antenna for Imaging Array Applications," *Electromagnetics,* pp. 209–215 (1983).

22. W. Chew and H. R. Fetterman, "Printed Circuit Antennas and FET Detectors for Millimeter Wave Imaging," *12th International Conference on Infrared and Millimeter-Waves,* pp. 222–223 (1987).

23. D. F. Filipovic, W. Y. Ali-Ahmad, and G. M. Rebeiz, "Millimeter Wave Double Dipole Antennas for High Efficiency Reflector Illumination," *15th International Conference on Infrared and Millimeter Waves,* pp. 617–619 (1990).

24. A. Skalare, T. de Graauw, and H. Van de Stadt, "A Planar Dipole Array Antenna with an Elliptical Lens," *Microwave and Optical Technology Letters,* Vol. 4, No. 1, pp. 9–12, January 1991.

25. M. A. Gouker and G. S. Smith, "A Millimeter-Wave Integrated-Circuit Antenna Based on the Fresnel Zone Plate," *IEEE MTT-S International Microwave Symposium Digest,* pp. 157–160 (1991).

26. M. A. Gouker and G. S. Smith, "Measurements of Strip Dipole Antennas on Finite-Thickness Substrates at 230 GHz," IEEE *Microwave and Guided Wave Letters,* Vol. 2, No. 2, pp. 79–81, February 1992.

27. K. D. Stephan and T. Itoh, "A Planar Quasi-Optical Subharmonically Pumped Mixer Characterized by Isotropic Conversion Loss," *IEEE Transactions on Microwave Theory and Techniques,* Vol. 32, No. 1, pp. 97–102, January 1984.

28. J. A. Taylor, T. C. L. G. Sollner, C. D. Parker, and J. A. Calviello, "Planar Dipole-Fed Mixer Arrays for Imaging at Millimeter and Sub-Millimeter Wavelengths," *10th International Conference on Infrared and Millimeter Waves,* pp. 187–188, 1985.

29. C. M. Jackson and C. Sun, "Subharmonically and Fundamentally Pumped Slotline Quasioptical Mixer," *IEEE MTT-S International Microwave Symposium Digest,* pp. 293–295 (1986).

30. R. Janaswamy and S. W. Lee, "Scattering from Dipoles Loaded with Diodes," *IEEE Transactions on Antennas and Propagation,* Vol. 36, No. 11, pp. 1649–1651, November 1988.

31. K. D. Stephan, N. Camilleri, and T. Itoh, "Quasi-Optical Polarization-Duplexed Balanced Mixer," *IEEE MTT-S International Microwave Symposium Digest,* pp. 376–378 (1982).

32. K. D. Stephan, N. Camilleri, and T. Itoh, "A Quasi-Optical Polarization-Duplexed Balanced Mixer for Millimeter-Wave Applications," *IEEE Transactions on Microwave Theory and Techniques*, Vol. 31, No. 2, pp. 164–170, February 1983.

33. K. D. Stephan and G. Perks, "Quasioptical Slot Ring Mixer Noise Measurements," *IEEE MTT-S International Microwave Symposium Digest*, pp. 643–644 (1985).

34. R. L. Gingras, C. Drubin, B. Cole, W. Stacey, R. Pavio, J. Wolverton, K. S. Yngvesson, and A. Cardiamenos, "Millimeter-Wave Slot Ring Mixer Array Receiver Technology," *IEEE Microwave and Millimeter-Wave Monolithic Circuits Symposium*, pp. 105–107, 1992.

35. C. M. Jackson, "Patch Antenna Quasi-Optical Mixers," *Microwave and Optical Technology Letters*, Vol. 1, No. 1, pp. 1–4, March 1988.

36. G. M. Rebeiz, W. G. Regehr, R. C. Compton and D. B. Rutledge, "Integrated Circuit Antennas on Thin Membranes," SPIE Vol. 791, *MM-Wave Technology IV and Radio Frequency Power Sources*, p. 68, May 1987.

37. K. A. Lee and M. A. Frerking, "Planar Antennas on Thick Dielectric Substrates," *12th International Conference on Infrared and Millimeter Waves*, pp. 216–217, 1987.

38. B. K. Kormanyos and G. M. Rebeiz, "A 30–180 GHz Harmonic Mixer-Receiver," *IEEE MTT-S International Microwave Symposium Digest*, pp. 341–344 (1992).

39. S. S. Gearhart, J. Hesler, W. L. Bishop, T. W. Crowe, and G. M. Rebeiz, "A Wideband 760 GHz Planar Integrated Schottky Receiver," *IEEE Microwave and Guided Wave Letters*, Vol. 3, No. 7, pp. 205–207, July 1993.

40. B. K. Kormanyos, P. H. Ostdiek, W. L. Bishop, T. W. Crowe, and G. M. Rebeiz, "A Planar Wideband 80–200 GHz Subharmonic Receiver," *IEEE Transactions on Microwave Theory and Techniques*, Vol. 41, No. 10, pp. 1730–1737, October 1993.

41. T. H. Lee, C. Y. Chi, J. R. East, G. M. Rebeiz, and G. I. Haddad, "A Quasi-Optical Subharmonically Pumped Receiver Using Separately Biased Schottky Diode Pairs," *IEEE MTT-S International Microwave Symposium Digest*, pp. 783–786 (1994).

42. T. H. Lee, C. Y. Chi, J. R. East, G. M. Rebeiz, and G. I. Haddad, "A Novel Biased Anti-Parallel Schottky Diode Structure for Subharmonic Mixing," *IEEE Microwave and Guided Wave Letters*, Vol. 4, No. 10, pp. 341–343, October 1994.

43. G. M. Rebeiz, Y. Guo, D. B. Rutledge, and D. P. Kasilingam, "Two Dimensional Horn Imaging Arrays," *12th International Conference on Infrared and Millimeter Waves*, pp. 224–225 (1987).

44. W. Y. Ali-Ahmad, G. V. Eleftheriades, L. P. Katehi, and G. M. Rebeiz, "94 GHz Integrated Horn Antenna: Impedance, Patterns and Double Polarized Applications," *15th International Conference on Infrared and Millimeter Waves*, pp. 614–616 (1990).

45. W. Y. Ali-Ahmad and G. M. Rebeiz, "92 GHz Dual-Polarized Integrated Horn Antennas," *IEEE Transactions on Antennas and Propagation*, Vol. 39, No. 6, pp. 820–825, June 1991.

46. G. V. Eleftheriades, W. Y. Ali-Ahmad, and G. M. Rebeiz, "A 20 dB Quasi-Integrated Horn Antenna," *IEEE Microwave and Guided Wave Letters*, Vol. 2, No. 2, pp. 73–75, February 1992.

47. W. Y. Ali-Ahmad and G. M. Rebeiz, "A 20 dB Quasi-Integrated Horn Antenna,"

IEEE MTT-S International Microwave Symposium Digest, pp. 1417–1420 (1992).

48. G. M. Rebeiz, D. P. Kasilingam, Y. Guo, P. A. Stimson, and D. B. Rutledge, "Monolithic Millimeter Wave Two-Dimensional Horn Imaging Arrays," *IEEE Transactions on Antennas and Propagation,* Vol. 38, No. 9, pp. 1473–1482, September 1990.

49. Y. Guo, K. Lee, P. Stimson, K. Potter, and D. Rutledge, "Aperture Efficiency of Integrated-Circuit Horn Antennas," *Microwave and Optical Technology Letters,* Vol. 4, No. 1, pp. 6–9, January 1991.

50. S. S. Gearhart, C. C. Ling, and G. M. Rebeiz, "Integrated 222-GHz Corner-Reflector Antennas," *Microwave and Optical Technology Letters,* Vol. 4, No. 1, pp. 12–15, January 1991.

51. S. S. Gearhart, C. C. Ling, and G. M. Rebeiz, "Integrated Millimeter-Wave Corner-Cube Antennas," *IEEE Transactions on Antennas and Propagation,* Vol. AP-39, No. 7, pp. 1000–1006, July 1991.

52. S. S. Gearhart, C. C. Ling, G. M. Rebeiz, H. Davee, and G. Chin, "Integrated 119-μm Linear Corner-Cube Array," *IEEE Microwave and Guided Wave Letters,* Vol. 1, No. 7, pp. 155–157, July 1991.

53. H. Ekstrom, G. Gearhart, P. R. Acharya, G. M. Rebeiz, E. L. Kollberg, and S. Jacobsson, "348 GHz Endfire Slotline Antennas on Thin Dielectric Membranes," *IEEE Microwave and Guided Wave Letters,* Vol. 2, No. 9, pp. 357–358, September 1992.

54. P. R. Acharya, H. Ekstrom, S. S. Gearhart, S. Jacobsson, J. F. Johansson, E. L. Kollberg, and G. M. Rebeiz, "Tapered Slotline Antennas at 802 GHz," *IEEE Transactions on Microwave Theory and Techniques,* Vol. 41, No. 10, pp. 1715–1719, October 1993.

55. P. A. Stimson, R. J. Dengler, P. H. Siegal and H. G. LeDuc, "A Planar Quasi-Optical SIS Receiver Suitable for Array Applications," *IEEE MTT-S International Microwave Symposium Digest,* pp. 1421–1424, 1992.

56. P. A. Stimson, R. J. Dengler, H. G. LeDuc, and P. H. Siegel, "A Prototype Quasi-Optical SIS Array Receiver Suitable for Array Applications," *IEEE MTT-S International Microwave Symposium Digest,* pp. 1337–1339 (1994).

57. W. C. Brown, "Experiments in the Transportation of Energy by Microwave Beam," *IEEE International Convention Record, Part 2,* pp. 8–17, 1964.

58. W. C. Brown, "Experimental Airborne Microwave Supported Platform," Final Report RADC-TR-65-188, December 1965.

59. W. C. Brown, "The History of Power Transmission by Radio Waves," *IEEE Transactions on Microwave Theory and Techniques,* Vol. 32, No. 9, pp. 1230–1242, September 1984.

60. E. M. Sabbagh and R. George, "Microwave Energy Conversion," WADD Technical Report, Part I, April 1961.

61. S. S. Bharj, R. Camisa, S. Grober, F. Wozniak, and E. Pendleton, "High Efficiency C-band 1000 Element Rectenna Array for Microwave Powered Applications," *IEEE MTT-S International Microwave Symposium Digest,* pp. 301–303 (1992).

62. S. M. Wentworth, J. M. Lewis, and D. P. Neikirk, "Antenna-Coupled Thermal Detectors of MM-Wave Radiation," *Microwave Journal,* Vol. 36, No. 1, pp. 94–103, January 1993.

63. T. Yoo, J. McSpadden, and K. Chang, "35 GHz Rectenna Implemented with a Patch and a Microstrip Dipole Antenna," *IEEE MTT-S International Microwave Symposium Digest*, pp. 345–348 (1992).

64. T. Yoo and K. Chang, "Theoretical and Experimental Development of 10 and 35 GHz Rectennas," *IEEE Transactions on Microwave Theory and Techniques*, Vol. 40, pp. 1259–1266, June 1992.

65. P. R. Herczfeld, "Special Issue on Applications of Lightwave Technology to Microwave Devices," *IEEE Transactions on Microwave Theory and Techniques*, Vol. 38, No. 5, pp. 465–466, May 1990.

66. W. Chew and H. R. Fetterman, "Printed Circuit Antennas with Integrated FET Detectors for Millimeter-Wave Quasi Optics," *IEEE Transactions on Microwave Theory and Techniques*, Vol. MTT-37, No. 3, pp. 593–597, March 1989.

CHAPTER TWELVE

Beam Steering for Active Antenna Arrays and Spatial Power Combiners

12.1 INTRODUCTION

Many systems require high output powers and large antennas with very directive radiation patterns. In applications such as radar, communiction links, and astronomy, the antennas may function as transmitters, receivers, or both simultaneously. To produce these pencil beams, large-reflector antennas have been used extensively. These beams are often required to be pointed in various directions. Figure 12.1 shows a couple of mechanical drive schemes used to point the antenna beam over several axes. For a radar the motor positions are correlated to return signals for tracking, while a communication link would obtain a peak in the received signal strength. Mechanical beam-pointing methods are relatively inexpensive and the antenna performance is not compromised throughout the scan, but they tend to be large, bulky, and slow. Instead of reflectors, large planar arrays which consist of several hundred to several thousand antennas can also provide narrow, high-gain pencil beams. They may be arranged in one, two, or three-dimensions and pointed mechanically. Since one has access to each radiator, electronic methods can also be employed for beam scanning.

Antenna arrays are not as bulky as comparable reflector antennas, but they often require complicated feeding networks. Feed networks which distribute energy to each radiator include corporate-, parallel-, series-, and space-fed arrays as shown in Figure 12.2. These networks increase complexity, cost, and losses in an array aperture, but they can be integrated with components to provide both beamwisth flexibility and electronic steering capability.

Although it is more complex and expensive, electronic beam steering is many orders of magnitudes faster than mechanical beam pointing. Electronic

Large Aperture Antenna
(Reflector, Horn, Slot Antenna Array...)

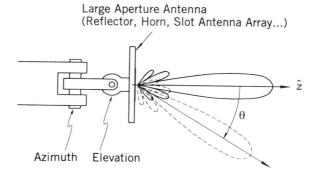

Azimuth   Elevation

Large Aperture Antenna

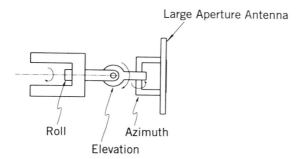

Roll   Azimuth
Elevation

FIGURE 12.1. Mechanical beam pointing.

beam steering can be achieved by using frequency scanning, beam switching, time delay, or phase delay. Electronic steering allows beam manipulations in tens of nanoseconds using compact, mechanically rigid structures. Due to the increase in complexity there is some loss in gain and pattern deterioration at large scan angles, but in many applications (e.g., military radar) beam steering agility and speed may be the most critical parameters.

12.2 INTERINJECTION-LOCKED PHASED ARRAYS

Phased arrays provide an interesting and challenging application for active integrated antennas. As early as 1968, an idea was introduced which used a single low-power solid-state source which could be amplified at each antena input for spatial power combining. Staiman et al. introduced the new power-combining scheme which used a corporate network to feed solid-state amplifiers and an array of dipole antennas [1]. Unlike other combiners, power combining occurs in free space instead of in a waveguide cavity. The rf losses

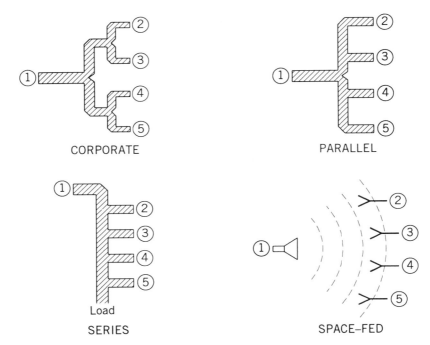

CORPORATE   PARALLEL

SERIES   SPACE–FED

FIGURE 12.2. Feed networks.

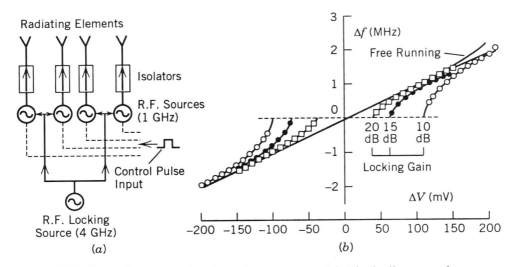

(a)   (b)

FIGURE 12.3. Beam steering in active antennas. (*a*) Block diagram of complete system. (*b*) Dependence of oscillator frequency output on base voltage change in free-running and harmonically locked modes. (*c*) (top) Antenna radiation pattern, broadside position. (bottom) Antenna radiation pattern, shifted position. (From Ref. 2 with permission from IEEE.)

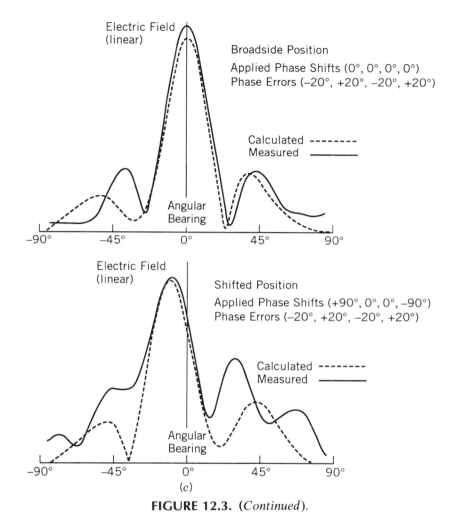

Electric Field (linear)

Broadside Position
Applied Phase Shifts (0°, 0°, 0°, 0°)
Phase Errors (–20°, +20°, –20°, +20°)

Calculated ---------
Measured ⎯⎯⎯

Angular Bearing

–90°   –45°   0°   45°   90°

Electric Field (linear)

Shifted Position
Applied Phase Shifts (+90°, 0°, 0°, –90°)
Phase Errors (–20°, +20°, –20°, +20°)

Calculated ---------
Measured ⎯⎯⎯

Angular Bearing

–90°   –45°   0°   45°   90°

(c)

FIGURE 12.3. (*Continued*).

incurred in the feed networks are at low power levels, which minimizes their effect on the overall losses. Synchronization depends on the parameters of the amplifiers and feed network. One hundred transistor amplifiers were used to provide a net gain of 4.75 dB at 410 MHz with 100 W of output power.

A similar approach which uses many low-power solid-state oscillators to feed antennas was shown in 1974 by Al-Ani et al. [2]. All of the sources are externally synchronized to a single master oscillator to ensure high power output. These investigators also unveiled a novel method for beam steering in a spatial power combiner. Figure 12.3a shows the configuration where several 1-GHz transistor oscillator are injection-locked to a master source operating

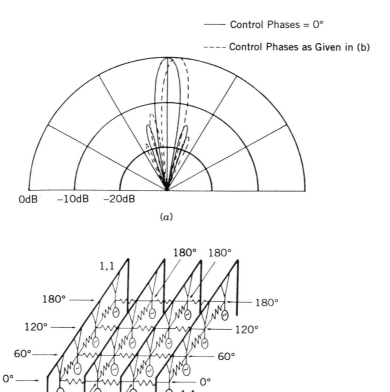

FIGURE 12.4. (*a*) *E*-plane pattern of hypothetical 4 × 4 array showing effect of control phase changes. (*b*) Control phases for 5.5° steered pattern of (*a*). (From Ref. 6 with permission from IEEE.)

at the 4 GHz. Steering of 18° was achieved using dc pulses applied to the bias of each transistor oscillator. In Figure 12.3*b* the frequency is plotted as a function of base voltage for both free and injection-locked oscillators. Figure 12.3*c* shows the radiation patterns for the array. Although novel and potentially useful, this approach stirred little interest. The state of solid-state device technology in the 1970s and the sparse use of millimeter wavelengths would not push for the use of spatial power combining for over a decade.

In 1986, Mink introduced the use of an array of radiating sources in a large open resonator for power combining [3]. The open resonator serves as a quasi-optical cavity for synchronization and power combining. The array spacing, number of oscillators, and taper distribution across the aperture determine the efficiency with which the energy couples to the modes of the open resonator. The open resonator has two basic forms: one form uses two metallic mirrors and the other uses one metallic and one dielectric (or partially reflecting) mirror. The first approach is, essentially, a high-power source, while the second approach also provides a fixed, directive antenna pattern.

The second approach is easily used with mechanical beam-pointing methods. However, if the mirrors of the open resonator are removed, the large array of active antennas can be synchronized via mutual coupling, circuit interconnections, a partially reflecting surface, or an external space feed. The power, frequency, and phase of each source can then be manipulated to alter the radiation characteristics of the entire array. In this way, spatial power combiners can serve to not only solve the need for power at millimeter wavelengths but to provide a method for beam steering, as shown by Stephan in 1986 [4]. He used distributed oscillators interinjection-locked together in a linear cascade for coherent power combining. Initially, three VHF transistors oscillators were studied and simulated [5].

State-variable analysis of the linearized equations led to closed-form solutions for one- and two-dimensional phased array [6]. The feasibility and behavior of a planar *spatial power combining phased arrays* was studied at 220 MHz. A 4 × 4 planar array as shown in Figure 12.4 was simulated at broadside with 5.5° of beam steering. The figure shows the schematic, simulated phase angles, and radiation patterns for the array. In this investigation, effects of device failures were simulated. Figure 12.5 shows the change in radiation

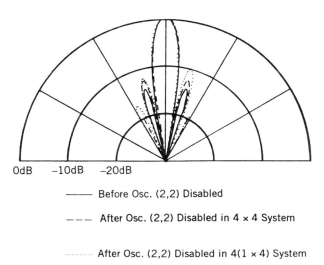

0dB   −10dB   −20dB

——— Before Osc. (2,2) Disabled

− − − After Osc. (2,2) Disabled in 4 × 4 System

············ After Osc. (2,2) Disabled in 4(1 × 4) System

FIGURE 12.5. *H*-plane cut (tilted 5°) of steered beam showing effects of oscillator (2, 2) failure upon 4 × 4 and 4(1 × 4) systems. (From Ref. 6 with permission from IEEE.)

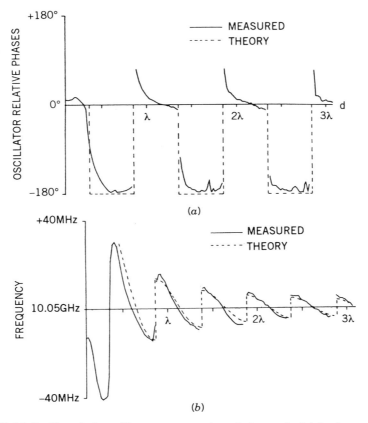

FIGURE 12.6. Coupled oscillator measured and theoretical (*a*) phase and (*b*) frequency versus separation *d*. (From Ref. 8 with permission from IEEE.)

patterns due to device failures in the planar array. The results are directly related to claims of graceful degradation in distributed systems which may not hold in beam-steerable spatial combiners.

A model was then developed to understand the interaction of two radiation-coupled oscillators [7]. Figure 12.6 shows the oscillator relative phase and frequency as a function of the separation. Measurements of the phase and frequency dependence of two coupled radiating oscillators versus separation allow a more accurate determination of the oscillator coupling characteristics [8]. During the investigation, X-band Gunn-integrated antennas were also developed.

X-band microstrip Gunn diode oscillators were used in conjuction with a 1 × 4 array of linear tapered slot antennas to demonstrate 7.4 W of effective isotropic radiated power (EIRP) at 10 GHz with ±5° of beam steering [9].

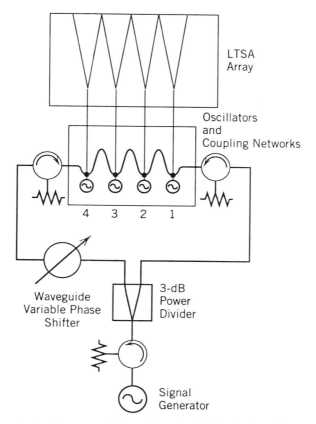

FIGURE 12.7. Block diagram of interinjection-locked phased array system. (From Ref. 10 with permission from IEEE.)

Figure 12.7 shows the interinjection-locked linear array configuration, and Figure 12.8 shows simulated and measured radiation patterns for the array. Several oscillators are interinjection-locked to an external source, and the single phase shifter was varied over 300° to induce the 10° of total beam steering [10].

12.3 SINGLE ACTIVE ANTENNA

Instead of multiple devices imbedded within a resonant structure, Stephan concentrated on coupling between several coupled oscillator circuits [4]. The coupling between these oscillators is carefully controlled to maintain a predictable phase progression required by a phased array. Beginning with the

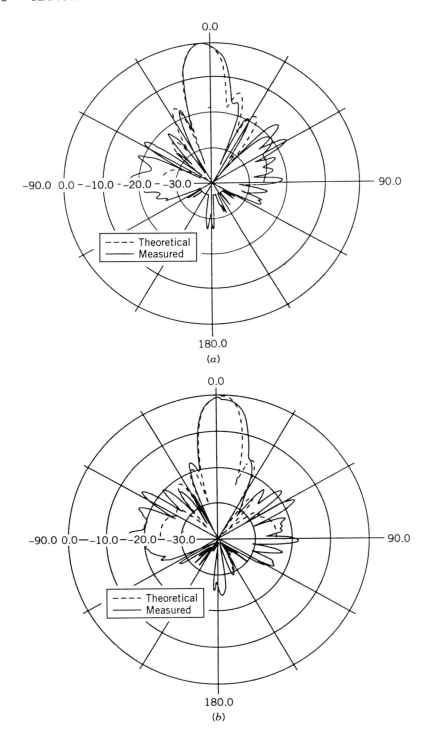

(a)

(b)

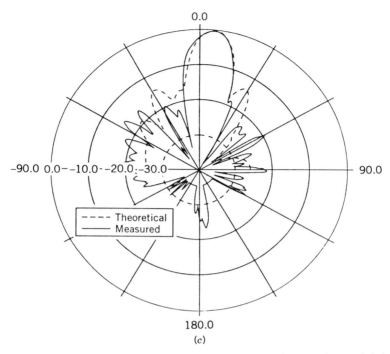

FIGURE 12.8. (*a*) Experimental and predicted patterns for maximum left beam shift of interinjection-locked phased array. (*b*) Experimental and predicted patterns for broadside beam of interinjection-locked phased array. (*c*) Experimental and predicted patterns for maximum right beam shift of interinjection-locked phased array. (From Ref. 10 with permission from IEEE.)

description of a single canonical oscillator, the theory was expanded to include the large number of these oscillators encountered in a spatial power combiner.

Stephan uses an active nonlinear element $(Y_D(A) = -G_D(A) + jB_D(A))$ whose characteristics depend on a voltage amplitude (A). The canonical oscillator circuit is shown in Figure 12.9. The injected current $(i(t))$ is the sum of the currents flowing through the individual components:

$$i(t) = C\frac{dv}{dt} + G_L v + \frac{1}{L}\int v\,dt + Y_D v \qquad (12.1)$$

where the tank circuit consists of an inductance (L), capacitance (C), and load conductance (G_L).

The voltage (v) in Equation (12.1) is assumed to be sinusoidal with angular frequency ω_i, magnitude $A(t)$, and instantaneous phase $\phi(t)$:

$$v = A(t)e^{j(\omega_i t + \phi(t))} \qquad (12.2)$$

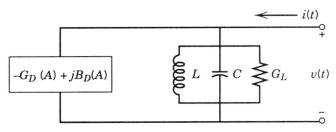

FIGURE 12.9. Oscillator circuit.

The magnitude and phase in Equation (12.2) are assumed to be slowly varying functions of time. This simplification allows us to omit higher order terms which arise from the insertion of Equation (12.2) into Equation (12.1). Integration by parts yields the following expression for the current [4]:

$$
\begin{aligned}
\mathrm{Re}(i(t)) = C\Bigg(&-A\left(\omega_i + \frac{d\phi}{dt}\right)\sin(\omega_i t + \phi) + \frac{dA}{dt}\cos(\omega_i t + \phi)\Bigg) \\
&+ (G_L - G_D)(A\cos(\omega_i t + \phi)) - B_D A\sin(\omega_i t + \phi) \\
&+ \frac{1}{L}\left(\left(\frac{A}{\omega_i} - \frac{A}{\omega_i^2}\frac{d\phi}{dt}\right)\sin(\omega_i t + \phi) + \frac{1}{\omega_i^2}\frac{dA}{dt}\cos(\omega_i t + \phi)\right)
\end{aligned}
\tag{12.3}
$$

The injected current is then assumed to have an in-phase cosinusoidal component of magnitude $I_c(t)$ and a quadrature sinusoidal component with magnitude $I_s(t)$. This is given by

$$
i(t) = I_c(t)\cos(\omega_i t + \phi(t)) + I_s(t)\sin(\omega_i t + \phi(t))
\tag{12.4}
$$

Equation (12.4) is inserted in the LHS of Equation (12.3). The principle of orthogonality allows us to reduce the expression into separate components. First, multiplying both sides by $\sin(\omega_i t + \phi(t))$ and integrating removes all cosine terms:

$$
\frac{d\phi}{dt}\left(C + \frac{1}{\omega_i^2 L}\right) + B_D + \omega_i C - \frac{1}{\omega_i L} = -\frac{I_s}{A}
\tag{12.5}
$$

Similarly, multiplying both sides of Equation (12.3) by $\cos(\omega_i t + \phi(t))$ and integrating removes all sine terms:

$$
\frac{dA}{dt}\left(C + \frac{1}{\omega_i^2 L}\right) + (G_L - G_D)A = I_c
\tag{12.6}
$$

Given the oscillator free-running frequency ($\omega_0 = 1/\sqrt{LC}$) and assuming the injected frequency ω_i is near ω_0, a frequency deviation is defined by $\Delta\omega = \omega_i - \omega_0$. The terms in Equations (12.5) and (12.6) are approximated by the following:

$$\omega_i C - \frac{1}{\omega_i L} \approx 2(\omega_i - \omega_0)C = 2\Delta\omega C \tag{12.7}$$

$$C + \frac{1}{\omega_i^2 L} \approx 2C \tag{12.8}$$

The simplifications shown in Equations (12.7) and (12.8) are substituted into Equations (12.5) and (12.6). The differential equations for the phase and amplitude variation of the oscillator voltage become

$$\frac{d\phi}{dt} = -\Delta\omega - \frac{B_D}{2C} - \frac{I_s}{2CA} \tag{12.9}$$

$$\frac{dA}{dt} = \frac{A}{2C}(G_D - G_L) + \frac{I_c}{2C} \tag{12.10}$$

In the absence of injection current ($I_c = I_s = 0$), the steady-state amplitude A_0 is reached when $G_D(A) - G_L = 0$, making $dA/dt = 0$. Stephan also noted that the in-phase component I_c of the injection current has a first-order effect on amplitude while the instantaneous frequency $d\phi/dt$ is primarily influenced by the quadrature component I_s.

12.4   LINEAR CASCADE OF INTERINJECTION-LOCKED OSCILLATORS

For a large number of oscillators, each oscillator will have its own frequency, magnitude, and phase components. If N oscillators are arranged in a ladder network as shown in Figure 12.10, then coupling within the inner elements can be assumed to be primarily due to its nearest neighbors. This coupling occurs through the admittances ($Y_c = G_c + jB_c$) and is assumed to be constant and identical throughout the array. For a more general case, this coupling could also be frequency dependent.

The first source is the reference (injection source) with a current I_1 and phase $\theta = 0$, while the last source has current I_N and phase θ_N. The resulting currents $i_1, i_2, \ldots, i_N$ and phase components can be calculated with respect to each oscillator. These constants and variables arranged in the linear cascade shown in Figure 12.10 can be used to calculate the instantaneous frequency and oscillator amplitudes over time [5].

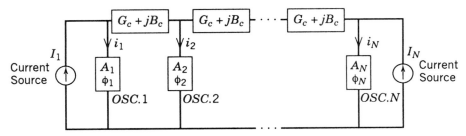

FIGURE 12.10. Linear cascade of interinjection-locked oscillators.

$$\frac{d\phi_i}{dt} = -\Delta\omega_i - \frac{B_{D_i}}{2C} - \frac{1}{2C}\left[2B_c + \frac{A_{i-1}}{A_i}\left(G_c\sin(\phi_i - \phi_{i-1}) - B_c\cos(\phi_i - \phi_{i-1})\right)\right.$$

$$\left. + \frac{I_i}{A_i}\sin(\phi_i - \theta_i) + \frac{A_{i+1}}{A_i}\left(G_c\sin(\phi_i - \phi_{i+1}) - B_c\cos(\phi_i - \phi_{i+1})\right)\right]$$

$$(12.11)$$

$$\frac{dA_i}{dt} = \frac{1}{2C}\left[A_i(G_{D_i} - G_{L_i} - 2G_c) + I_i\cos(\phi_i - \theta_i)\right]$$

$$+ \frac{1}{2C}\left[A_{i-1}(G_c\cos(\phi_i - \phi_{i-1}) + B_c\sin(\phi_i - \phi_{i-1}))\right]$$

$$+ \frac{1}{2C}\left[A_{i+1}(G_c\cos(\phi_i - \phi_{i+1}) + B_c\sin(\phi_i - \phi_{i+1}))\right] \qquad (12.12)$$

Equations (12.11) and (12.12) describe the change in magnitude and frequency for each oscillator in the array. They can be integrated to obtain a time-domain solution for the entire system. The accuracy of the analysis depends on the characterization of the individual oscillator admittance $(Y_D(A) = -G_D(A) + jB_D(A))$.

12.5   BEAM STEERING ACTIVE ANTENNA ARRAYS

Stephan developed a description and analysis for a set of distributed oscillators such that a single, stable power-combining frequency would result, and he showed how a progressive phase shift could be introduced in a linear cascade of oscillators to provide beam steering [5]. For a linear cascade of oscillators spaced φ electrical degrees apart, the progressive phase shift ($\Delta\phi$) which will steer the array beam ϕ_0 degrees away from boresight is given by

$$\phi_0 = \sin^{-1}\left(\frac{\Delta\phi}{\varphi}\right) = \sin^{-1}\left(\Delta\phi\,\frac{\lambda}{2\pi d}\right) \tag{12.13}$$

where d is the physical separation between sources at the wavelength operation (λ). For a given spacing, the maximum one-sided beam angle allowed ($\phi_{0,\mathrm{max}}$) depends on the maximum phase progression per element ($\Delta\phi_{\mathrm{max}}$). The progressive phase shift can be controlled by the free-running frequencies and, thus, the biases to the active antenna elements.

As the phase progression between elements approaches $90°$, coupling between adjacent oscillators approaches zero. This results in a loss of synchronization which may be due to oscillators with too high a Q-factor or low mutual coupling. For whatever reasons, these parameters reduce phase progression between sources and limit the maximum scan angle. As an example, if one assumes a maximum phase progression of $60°$, the maximum attainable scan angle for a typical half-wave array is $\pm 19.5°$.

Wider beam steering can be accomplished by using more densely packed sources. For the same phase progression, a quarter-wavelength spacing has a scan limit of $\pm 42°$. Rearranging Equation (12.13), we see that the required phase progression ($\Delta\phi$) between sources needed to steer a beam ϕ_0 degrees away from boresight is

$$\Delta\phi = \frac{2\pi d}{\lambda}\,\sin(\phi_0) \tag{12.14}$$

In Figure 12.8, Stephan demonstrated $\pm 5°$ of beam steering. Following Stephan's work, other investigators have used various methods to induce wider scan angles and improved performance. Table 12.1 shows their results in phased array demonstrations using active antennas.

In 1992, Fralich and Litva used a 1×11 active leaky wave antenna array to obtain beam steering as a function of oscillator frequency [11]. A total of $8°$ of steering was shown for an oscillator tuning range of $380\,\mathrm{MHz}$ at $11.5\,\mathrm{GHz}$. The configuration and measured results of the X-band linear imageline array are shown in Figure 12.11.

Kirk and Chang [12] simultaneously developed the same active imageline leaky wave antenna in 1992. The resulting theoretical beam angle for this periodic grating antenna depends on the operating wavelength (λ_0), guided wavelength of the structure (λ_g), and spacing of the grating (d):

$$\phi_0 = \sin^{-1}\left(\lambda_0\left(\frac{1}{\lambda_g}+\frac{n}{d}\right)\right) \tag{12.15}$$

where $n = 0, \pm 1, \pm 2, \ldots$ is the nth space harmonic of the Floquet modes for

TABLE 12.1. Beam Steering Demonstrations

Reference No.	Array Class	Antenna Type	Array Size	Device Type	Frequency (GHz)	Synchronization Method	Steering angle (deg)
2	Distributed sources	Conventional	1 × 4	JFET	1	External circuit: 4th harmonic	18
9, 10	Distributed sources	Conventional	1 × 4	JFET	10	External circuit	10
11	Single source	Parasitic leaky-wave	1 × 11	Gunn	11.5	N. A.	8
12	Single source	Parasitic leaky-wave	1 × 12	Gunn	8.1	N. A.	4
13	Distributed sources	Active	1 × 4	FET	2.3	External probe-fed	40
17	Distributed sources	Active	2 × 2	Gunn	9.5	Mutual coupling	15
18	Distributed sources	Active	1 × 4	Gunn	9.3	Mutual coupling	36
22, 23	Distributed sources	Active	1 × 4	MESFET	10	Mutual coupling	27.5
24	Distributed sources	Conventional	1 × 6	MESFET	4.2	External circuit	70
26	Distributed sources	Conventional	1 × 3	FET	6	External circuit (unilateral)	27
26	Distributed sources	Active	1 × 2	MESFET	10	External circuit (Phase-locked loop	33

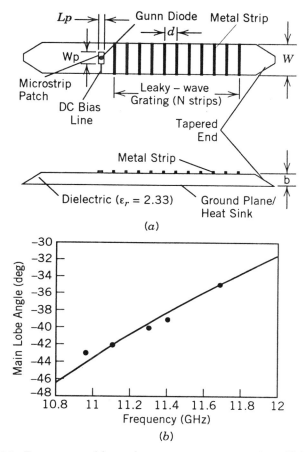

FIGURE 12.11. Beam-steerable active array antenna using dielectric wave-guide. (*a*) Active dielectric antenna with *N* metal strips. (*b*) Measured (•) and theoretical (—) frequency variation of main radiation pattern lobe direction with respect to array broadside. (From Ref. 11 with permission from *Electronics Letters.*)

the periodic structure. The dominant space harmonic is known to be $n = -1$. Figure 10.24 shows the 1×12 linear array configuration, and Figure 12.12 shows the simulated and measured results.

Hall and Haskins used two transistors and two external injection-locking inputs to demonstrate some of the most impressive beam steering demonstrations of active antennas [13]. An edge-fed dual-FET integrated microstrip patch antenna oscillator operating at a frequency of 2.28 GHz was integrated with two external injection-locking signals to provide phase-shifting capability.

Measured Versus Calculated ϕ_0

Frequency (GHz)	Measured ϕ_0(deg)	Calculated ϕ_0(deg)
9	-9.5	-10.0
10	0.0	-0.5
11	7.0	7.4

(a)

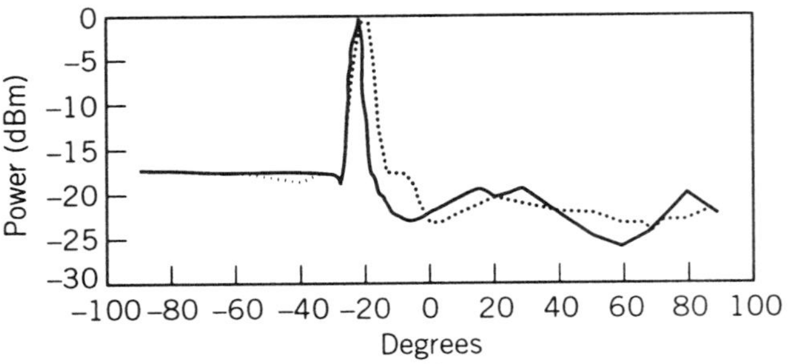

(b)

FIGURE 12.12. Beam-steerable imageline active antenna: (a) theoretical and measured beam angles; (b) radiation patterns for two different biases on Gunn diode; (solid line) 8 GHz, (dotted line) 8.1 GHz [12].

The probes at each patch location for the injection-locking signal provide accurate and stable injection-locking signals for dependable beam steering. Array patterns at 2.3 GHz are smooth with beam steering of over 40°, as shown in Figure 12.13.

Hall and Haskins also demonstrated a unique polarization agile active antenna, using a similar injection-locking scheme [14]. Figure 12.14 shows the configuration. The radiated power and phase versus locking frequency for a single active antenna were investigated. Also studied are the effects of mutual coupling on the relative phase difference between two oscillators and the radiated power and phase with locking frequency. Similar investigations have been carried out by Drew, Fusco, and McDowall [15, 16]. Figure 12.15 shows the results of measured and calculated radiated phase deviation versus normalized frequency [16].

Navarro and Chang introduced a 2 × 2 beam steering array in 1993, using active inverted stripline antennas [17, 18]. The array operated at 9.5 GHz with an EIRP of 3.8 W and an rf combining efficiency of 89%. The single element and radiation pattern are shown in Figure 10.15. Figuare 12.16 shows the configuration and the radiation patterns for the beam-steered array. A total of 15° of phase shift was demonstrated by controlling the bias voltages and thus the free-running frequencies.

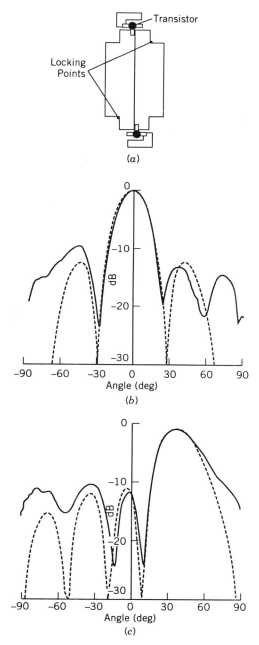

FIGURE 12.13. Microstrip active antennas with external injection locking. (*a*) Microstrip active patch for 180° to 360° phase control. (*b*) Radiation pattern of four-element array phased for 0° beam angle (element spacing = 74 mm). (*c*) Radiation pattern of four-element array phased for 40° beam angle (element spacing = 74 mm). (From Ref. 13 with permission from *Electronics Letters.*)

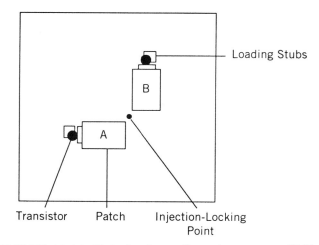

FIGURE 12.14. Polarization agile active antenna [14].

Four element E- and H-plane arrays of active Gunn integrated inverted stripline antennas were also used in beam steering demonstrations. The 1×4 E- and H-plane configurations are shown in Figure 12.17 along with their respective radiation pattern measurements. Maximum E-plane beam scanning demonstrated was 36°, while maximum H-plane beam scanning was 34°. Table 12.2 shows the frequency distribution of the two linear arrays along with the resulting scan angle.

The beam scanning shown used only mutual coupling to maintain injection locking between the four sources. There was no interconnections between active antennas or any external locking inputs. Although simple, the scan angle is limited by the low coupling levels. When using this method, lower-quality oscillators and increased mutual coupling can increase the scan angle. Mutual coupling due to space waves is increased by the use of a partially reflecting surface such as a low-dielectric substrate. The increase in mutual coupling between active antennas helps to maintain injection lock over a wider frequency range, which results in wider scanning angles.

The use of interconnecting transmission lines as shown by Stephan [9, 10] also increases the coupling level and provides a more reliable control mechanism. The array surface shown in Figure 12.17 can be easily modified to include such interconnections. However, the most reliable and effective method to increase scanning angle and active array performance was shown by Hall and Haskins [13]. The method uses an external signal which is fed to each active antenna and precisely locks each source to the desired frequencies. The method ensures good repeatable performance and provides the most control over the array. However, it is more intricate and complex than other approaches.

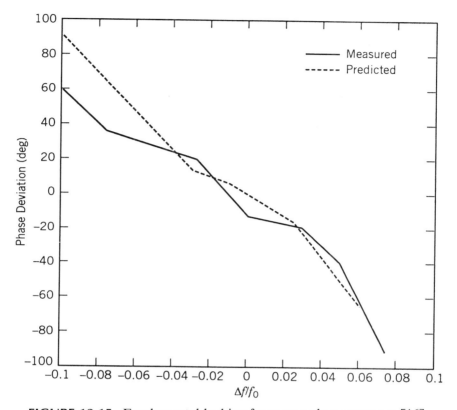

FIGURE 12.15. Fundamental locking frequency phase response [16].

Some of the most thorough analysis and experimental demonstrations in active antenna arrays was carried out by York and Compton [19]. In their investigations, they used both diode and transistor integrated microstrip patch antennas in linear and planar arrays. For their coupled-oscillator theory, the individual oscillators were modeled by simple, single-tuned resonant circuits with a negative lumped resistance for the active device. The coupling coefficient between oscilators i and j is assumed to be $\varepsilon_{ij}e^{-j\Phi_{ij}}$. For N oscillators, as shown in Figure 12.18, assuming that only nearest-neighbor interactions exist, a method-of-averages approach yields the following equations for the individual oscillator amplitudes and phase as a function of time [20]:

$$\frac{dA_i}{dt} = \frac{\omega_i}{2Q}\left[\mu(\alpha_i^2 - A_i^2)A_i + \sum_{j=1}^{N}\varepsilon_{ij}A_j\cos(\Phi_{ij} + \theta_i - \theta_j)\right] \qquad (12.16)$$

$$\frac{d\theta_i}{dt} = \left[\omega_i - \frac{\omega_i}{2Q}\sum_{j=1}^{N}\varepsilon_{ij}\frac{A_j}{A_i}\sin(\Phi_{ij} + \theta_i - \theta_j)\right], \qquad i = 1, 2, \ldots, N \quad (12.17)$$

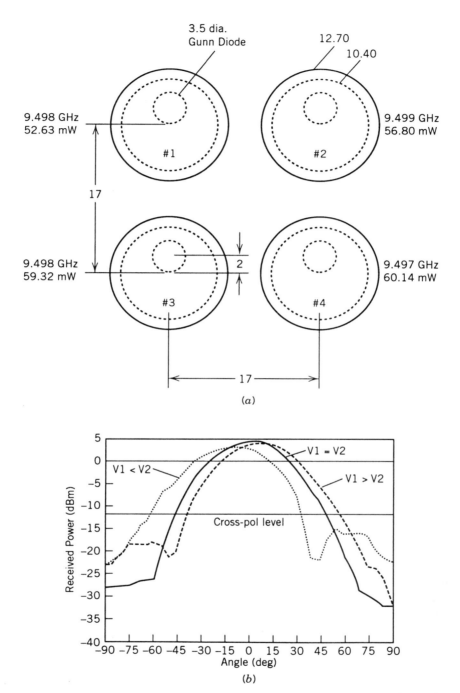

FIGURE 12.16. Beam steering for a 2 × 2 array using inverted stripline active antennas [17]: (a) Array configuration (antennas #1,3 are biased at V1; antennas #2,4 are biased at V2). All dimensions are in millimeters. (b) Beam steering of active antenna arrays without conventional phase shifters.

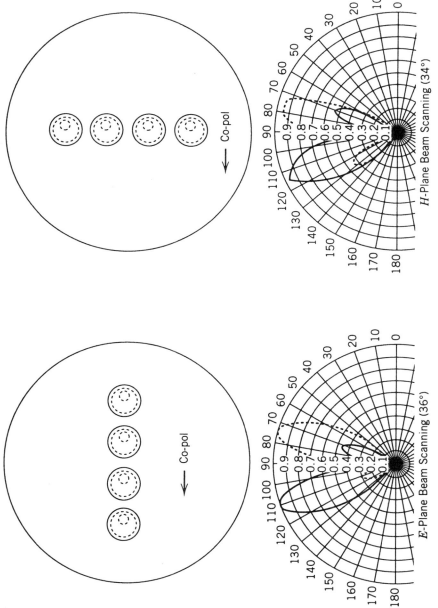

FIGURE 12.17. Active antenna beam scanning of linear 1×4 arrays.

349

TABLE 12.2. Linear Array Frequency Distributions for Beam Scanning

Linear 1 × 4 Arrays	Active Antenna 1 (GHz)	Active Antenna 2 (GHz)	Active Antenna 3 (GHz)	Active Antenna 4 (GHz)	Maximum Scan Angle (deg)
E-plane (1 × 4)	9.3307	9.3157	9.3157	9.3001	+20
	9.3097	9.3244	9.3247	9.3397	−16
H-plane (4 × 1)	9.3639	9.3422	9.3395	9.3179	+21
	9.3273	9.3410	9.3400	9.3553	−13

where A_i, ω_i, and α_i are the instantaneous amplitude, free-running angular frequency, and free-running oscillator amplitude of the ith element. The instantaneous phase of each oscillator is $\theta_i = \omega_i t + \phi_i$. Q is the quality factor of the oscillator embedding circuit. Equations (12.16) and (12.17) describe a set of coupled Van der Pol oscillators.

When the coupling between oscillators is weak, the individual amplitudes A_i remain very close to their free-running values, and the dynamics of the system are primarily governed by the phase description of Equation (12.17). If these free-running frequencies are within the locking bandwidths of the oscillators and the coupling levels are within the locking gain, the set of sources will injection-lock to a single frequency ($d\theta_i/dt \equiv \omega$) for all i. This is the basis for

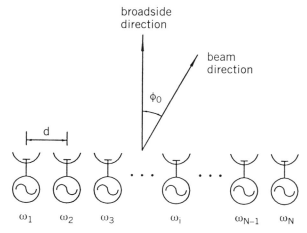

FIGURE 12.18. N-element active antenna array.

spatial power combining where the combiner frequency is given by

$$\omega = \omega_f = \omega_i \left[1 - \sum_{\substack{j=1 \\ j \neq i}}^{N} \frac{\varepsilon_{ij}}{2Q} \frac{A_j}{A_i} \sin(\theta_i - \theta_j + \Phi_{ij}) \right] \qquad (12.18)$$

where $i = 1, 2, \ldots, N$ and ω_f is the steady-state frequency. Given that one phase variable can be set to zero as a phase reference, Equation (12.18) is a set of N equations and N unknowns. It can be solved for the unknown phase distribution and the steady-state synchronized frequency.

Although 2^{N-1} phase distributions satisfy Equation (12.18), not all are stable solutions. A perturbation analysis can be used to investigate mode stability [21]. The equations are linearized around the various solutions, creating a stability matrix. The various solutions which make up the solution vector are perturbed by a small amount. The perturbed solution is stable only if it decays with time. This, for a linear array, occurs when all of the eigenvalues of the stability matrix have negative real parts. This requirement will eliminate most of the solutions found in the solution vector.

A linear array of loosely coupled oscillators was then used by Liao and York in 1993 to demonstrate beam steering of linear array of active antennas [22, 23]. From the analysis the oscillator frequencies were distributed to develop a phase progression across the array aperture. For N loosely coupled oscillators with a coupling phase (Φ), the following individual free-running frequencies (ω_i) achieve a constant phase progression required in phased arrays:

$$\omega_i = \omega_f \begin{cases} 1 + \dfrac{\varepsilon}{2Q} \sin(\Phi + \Delta\theta) & \text{if } i = 1 \\[2mm] 1 + \dfrac{\varepsilon}{Q} \sin(\Phi) \cos(\Delta\theta) & \text{if } 1 < i < N \\[2mm] 1 + \dfrac{\varepsilon}{2Q} \sin(\Phi - \Delta\theta) & \text{if } i = N \end{cases} \qquad (12.19)$$

Here, $\Delta\theta = \theta_i - \theta_{i-1} =$ progressive phase shift and we assume that $\varepsilon_{ij} = \varepsilon$, $\Phi_{ij} = \Phi$. The spacing which, as shown earlier, limits the total scan available with this method is used to achieve a mutual coupling angle (Φ) of $0°$.

Setting $\Phi = 0$ in Equation (12.19) simplifies the notation and reduces the distribution frequencies across the array aperture to

$$\omega_i = \omega_f \begin{cases} 1 + \dfrac{\varepsilon}{2Q} \sin(\Delta\theta) & \text{if } i = 1 \\[2mm] 1 & \text{if } 1 < i < N \\[2mm] 1 - \dfrac{\varepsilon}{2Q} \sin(\Delta\theta) & \text{if } i = N \end{cases} \qquad (12.20)$$

Therefore, the interelement phase shift is controlled by the free-running frequency of the end elements, and the synchronized frequency is equal to the free-running frequency of the centered oscillators. By adjusting the free-running frequency of the end elements in the opposite direction by an amount $(\varepsilon/2Q)\sin(\Delta\theta)$, one can steer the radiation direction electronically. The progressive phase shift along the array is

$$\Delta\theta = \sin^{-1}\left[\frac{2Q}{\varepsilon\omega_f}(\omega_i - \omega_f)\right] \tag{12.21}$$

The scan angle ϕ_0 is determined by the well-known equation

$$\phi_0 = \sin^{-1}\left(\frac{\lambda\Delta\theta}{2\pi d}\right) \tag{12.22}$$

where d is the spacing between two neighboring elements and λ is the wavelength.

From Equation (12.21) one can adjust the bias voltage to change ω_i and thus obtain different values of $\Delta\theta$. The scan angle ϕ_0 is thus controlled by varying bias voltages.

For the X-band demonstration, four FET integrated patch antennas were used. The configuration and radiation patterns are shown in Figure 12.19 [23]. Table 12.3 lists the frequencies and maximum scan angles realized.

The limitations of mutual coupling schemes led York to other synchronization methods which would allow more flexibility, tighter control, and improved performance. Following the interinjection-locked approach of Stephan, York developed a six-element beam-steerable array which scanned over 70° [24].

Several modifications to the original active antenna configuration are evident in Figure 12.20a. The FET is no longer integrated with the microstrip patch antenna. The source is realized in a well-defined microstrip oscillator which feeds a conventional patch at the nearest radiating edge. The radiating edge on the far side is used to interinjection-lock the sources together. The approach provides much stronger coupling between antennas, and the oscillators can be realized more consistently at the intended designed frequencies. Figure 12.20b shows the broadside and beam-steered radiation patterns for the 4-GHz steering array. The element separation at the operating frequency is about 0.3λ, which would limit the scan angle to ±59°. The array exhibited continuous H-plane beam scanning from −30° to 40° for this remarkable demonstration.

Other investigations have yielded some very good results. An eight-element interinjection-locked linear array of Gunn diode oscillators was demonstrated by Nogi et al. [25]. Lin et al. introduced a novel unilateral injection locking approach to bring about 33° of beam scanning in a two-element array and 27°

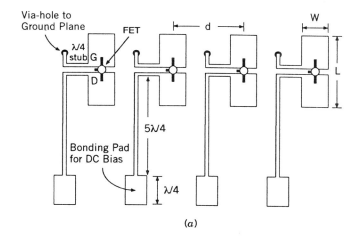

(a)

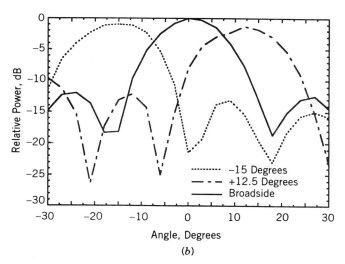

(b)

FIGURE 12.19. Four-element FET active array for beam steering. (a) Diagram illustrating the experimental four element FET array. A simple active patch design using zero gate bias was used, with dimensions chosen empirically for operation at 10 GHz. An array spacing of $d = 0.86\lambda_0$ was used to give the desired angle of coupling. (b) Comparison of measured radiation patterns at three different scan angles. Continuous beam scanning was possible from $-15°$ to $+12.5°$ by adjusting the end-element frequencies, which is close to the maximum $\pm 17°$ predicted by the theory. The scan range was limited by the large antenna spacing. (From Ref. 23 with permission from IEEE.)

TABLE 12.3. Frequency Distribution in Scanning Demonstration

Linear 1 × 4 Arrays	Active Antenna 1 (GHz)	Aactive Antenna 2 (GHz)	Active Antenna 3 (GHz)	Active Antenna 4 (GHz)	Maximum Scan Angle (deg)
H-plane (1 × 4)	10.0075	10.00	10.00	9.9925	+ 12.5
	9.9850	10.00	10.00	10.015	− 15

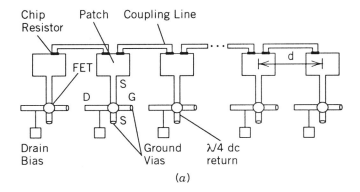

(a)

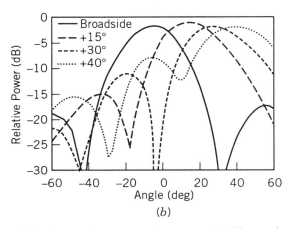

(b)

FIGURE 12.20. Six-element beam steering array. (*a*) Illustration of six-element beam-scanning array. Individual oscillators realized the lumped element circuit. (*b*) Sample measurements which illustrate continuous scanning from broadside to the scan limit of +40°. (From Ref. 24 with permission from IEEE.)

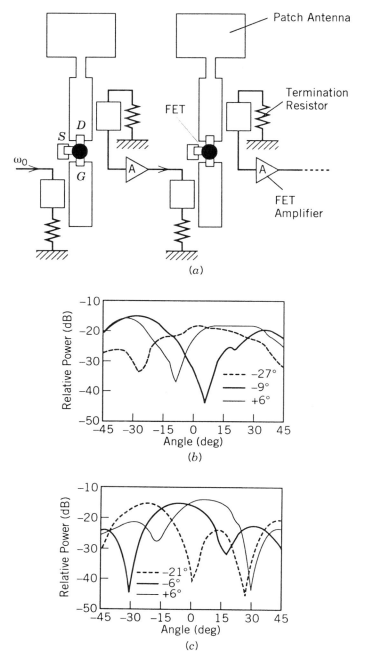

FIGURE 12.21. Unilateral beam-steering active arrays. (*a*) Circuit structure of the unilateral injection-locking type active phased array. (*b*) Beam scanning of the two-element active phased array with difference pattern. (*c*) Beam scanning of the three-element active phased array with sum pattern. (From Ref. 26 with permission from IEEE.)

Antennas

Coupler

Coupler

RF   LO

Voltage
Controlled
Oscillator

V_{tune}

V_{IF} IF

V_{off}

Reference
Oscillator

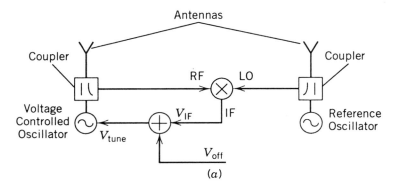

(a)

Microstrip
Substrate

IF
Voltage

Drain Bias

Gate Bias

GaAs
MESFET

RF
Signal

Two Mixer
Diodes
in Series

Balanced
LO Drive

LO
Bias

To
External
Source

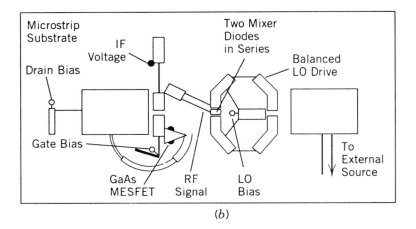

(b)

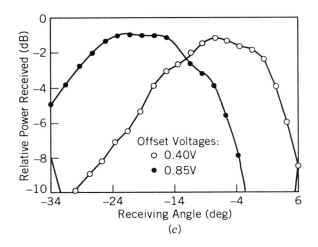

(c)

in a three-element linear array [26]. The coupling is enhanced with an amplifier which ensures strong coupling between sources. Kurokawa's injection-locking theory [27] was used to describe the induced phase shift and phase progression for beam steering:

$$\Delta\theta = \sin^{-1}\left(\frac{\omega_{\text{free}} - \omega_{\text{inj}}}{\Delta\omega_{\text{max}}}\right) \qquad (12.23)$$

where the maximum locking range is

$$\Delta\omega_{\text{max}} = \frac{\omega_{\text{inj}}}{Q_{\text{ext}}} \frac{G_s}{G_p} \sqrt{\frac{P_{\text{inj}}}{P_{\text{out}}}} \frac{1}{\sin(\alpha)} \qquad (12.24)$$

where ω_{free} is the free-running frequency, ω_{inj} is the injection-locking frequency, Q_{ext} is the external quality factor of the imbedding resonant circuit, P_{inj} is the injection-locking power, P_{out} is the oscillator output power, α is the angle between the impedance locus and device line, G_s is the maximum stable gain of the two-port oscillator, and G_p is the square root of the output power ratio of the two ports. The radiated beam angle induced by the phase progression is given in Equation (12.22). Figure 12.21 shows the unilateral interinjection-locked linear array configuration along with various radiation pattern measurements.

One of the latest demonstrations for active antenna beam scanning was shown by Martinez and Compton in 1994 [28]. The configuration uses a FET and an antenna for the free-running source which is coupled to a balanced mixer. A reference source is used to drive a second microstrip patch antenna which is electromagnetically coupled to pump the balanced mixer. An offset voltage is used to vary the phase difference between the two sources. Figure 12.22 shows the 1 × 2 configuration, equivalent schematic and received patterns at 10 GHz. Received beam steering of 15 degrees was demonstrated.

FIGURE 12.22. Beam-steering active arrays with phase-locked loop. (a) Phase-locked loop applied to a 1 × 2 oscillator array. Each oscillator has its own antenna and couplers deliver power to the mixer. Varying V_{off} changes the phase difference between the oscillators, which in turn changes the direction of the radiating beam. (b) Microstrip layout of X-band phase-locked loop. The MESFET and patch on the left form a voltage-controlled oscillator, and the patch on the right is driven by an external reference oscillator. Between the patches are circuits for the phase-locked loop. (c) Received power for main radiating beam versus the receiving angle. As the offset voltage changes, the radiating beam changes direction. By using basic array analysis, the adjustable phase difference between antennas is 122°. (From Ref. 28 with permission from IEEE.)

12.6 MODE LOCKING FOR PULSE SCANNED ACTIVE ANTENNA ARRAYS

Although primarily involved in spatial power combiners and phased arrays, York and Compton originally observed a mode-locking phenomena in an array of distributed sources [20]. Mode locking occurs when several oscillators operating at different frequencies can synchronize without pulling each other to a single oscillating frequency. The technique, which is used widely in short-pulse laser systems, produces a periodic train of pulses in time. The superposition of $N = 2n + 1$ different spectral modes is given by

$$E(t) = \sum_{i=-n}^{n} E_i \exp[j(\omega_i t + \phi_i)] \qquad (12.25)$$

For equally spaced frequencies at locked phases, $\omega_i = \omega_0 - i\Delta\omega$ and $\phi_i - \phi_{i-1} = \Delta\phi$ are given ($i = -n,\ldots,n$) and $\Delta\omega$ and $\Delta\phi$ are constants. For an equal-amplitude array ($E_i = E_0$), the electric field can be written as function of time from Equation (12.17) as

$$E(t) = E_0 \frac{\sin(\frac{N}{2}(\Delta\omega t + \Delta\phi))}{\sin(\frac{1}{2}(\Delta\omega t + \Delta\phi))} \exp[j\omega_0 t] \qquad (12.26)$$

Here $E(t)$ has the form of a carrier signal at ω_0 modulated by a train of pulses at a repetition frequency of $\Delta\omega$.

Two oscillators operating at different frequencies just outside of their individual locking bandwidths will produce a spectrum of equally spaced frequencies due to beating effects. Additional frequency components will occur at integer multiples of the differences between the free-running frequencies. The additional spectral components will have a constant phase progression which can be used to injection-lock other oscillators.

For active antennas spaced nearly one wavelength apart at a carrier frequency ω_0, the mode-locking phenomena provides a means to achieve the pulsed scanning concept. Given the directive gain function $G(\theta)$ of similar antennas, the radiated electric field can be expressed as a function of time by

$$E(r, \theta, t) = \sum_{i=-n}^{n} E_i G(\theta) \exp[j(\omega_i t + \phi_i + ik_0\Delta d \sin(\theta))]$$

$$= G(\theta) \frac{\sin(\frac{N}{2}(\Delta\omega t + k_0\Delta d \sin(\theta)))}{\sin(\frac{1}{2}(\Delta\omega t + k_0\Delta d \sin(\theta)))} \exp[j\omega_0 t] \qquad (12.27)$$

Figure 12.23 shows simulations for a linear five-element array. The measured time dependence of the outgoing power pulse is shown along with a simulated pulse scanning pattern.

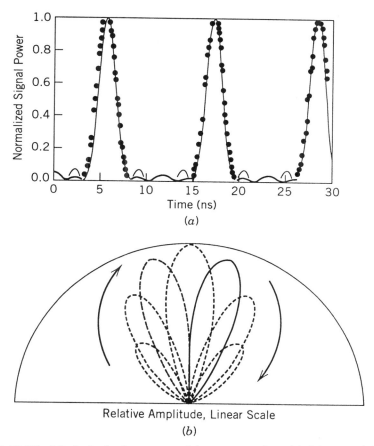

FIGURE 12.23. Mode-locked arrays and beam scanning. (*a*) Measured (dots) and theoretical (solid line) time dependence of the output signal power envelope of five mode-locked oscillators. (*b*) Polar antenna plot simulating pulse scanning for a five-element mode-locked array, using patch antennas. The elements are spaced one-half wavelength apart. Only the main lobes have been drawn for clarity, at equally spaced time increments over one cycle. (From Ref. 20 with permission from IEEE.)

12.7   CONCLUSION

The electronic beam steering demonstrations in this chapter represent the latest, leading-edge approaches for achieving beam scanning. Beam steering of active antennas can be achieved by varying the bias to individual element. When difference between the self-oscillating frequencies is within the locking bandwidth for a giving locking gain, the antennas will injection-lock at a single oscillating frequency. The difference between self-oscillating frequencies of each active antenna to the injection-locked signal introduces a phase shift. This phase shift can be used to electronically steer the beam of an active antenna array. This technique avoids the use of conventional phase shifters and has low cost and reduced complexity.

REFERENCES

1. D. Staiman, M. E. Breese, and W. T. Patton, "New Technique for Combining Solid-State Sources," *IEEE Journal of Solid-State Circuits*, Vol. SC-3, No. 3, pp. 238–243, September 1968.

2. A. H. Al-Ani, A. L. Cullen, and J. R. Forrest, "A Phase-Locking Method for Beam Steering in Active Array Antennas," *IEEE Transactions on Microwave Theory and Techniques*, Vol. MTT-22, No. 6, pp. 698–703, June 1974.

3. J. W. Mink, "Quasi-Optical Power Combining of Solid-State Millimeter-Wave Sources," *IEEE Transactions on Microwave Theory and Techniques*, Vol. MTT-34, No. 2, pp. 273–279, February 1986.

4. K. D. Stephan, "Inter-Injection-Locked Oscillators with Applications to Spatial Power Combining and Phased Arrays," *IEEE MTT-S International Microwave Symposium Digest*, pp. 159–162 (1986).

5. K. D. Stephan, "Inter-Injection-Locked Oscillators for Power Combining and Phased Arrays," *IEEE Transactions Microwave Theory and Techniques*, Vol. 34, No. 10, pp. 1017–1025, October 1986.

6. K. D. Stephan and W. A. Morgan, "Analysis of Inter-Injection-Locked Oscillators for Integrated Phased Arrays," *IEEE Transactions on Antennas and Propagation*, Vol. 35, No. 7, pp. 771–781, July 1987.

7. S. L. Young and K. D. Stephan, "Radiation Coupling of Inter-Injection Locked Oscillators," *SPIE, Vol. 791, Millimeter Wave Technology IV and Radio Frequency Power Sources*, pp. 69–76, 1987.

8. K. D. Stephan and S. L. Young, "Mode Stability of Radiation-Coupled Inter-Injection-Locked Oscillators for Integrated Phased Arrays," *IEEE Transactions on Microwave Theory and Techniques*, Vol. 36, No. 5, pp. 921–924, May 1988.

9. W. A. Morgan Jr. and K. D. Stephan, "Inter-Injection Locking — A Novel Phase Control Technique for Monolithic Phased Arrays," *12th International Conference on Infrared and Millimeter Waves*, pp. 81–82, December 1987.

10. W. A. Morgan and K. D. Stephan, "An X-Band Experimental Model of a Millimeter-Wave Inter-Injection-Locked Phased Array System," *IEEE Transactions on Antennas and Propagation*, Vol. 36, No. 11, pp. 1641–1645, November 1988.

11. R. Fralich and J. Litva, "Beam-Steerable Active Array Antenna," *Electronics Letters*, Vol. 28, No. 2, pp. 184–185, January 1992.

12. A. M. Kirk and K. Chang, "Integrated Image Line Steerable Active Antennas," *International Journal of Infrared and Millimeter Waves*, Vol. 13, No. 6, pp. 841–851, June 1992.

13. P. S. Hall and P. M. Haskins, "Microstrip Active Patch Array with Beam Scanning," *Electronics Letters*, Vol. 28, No. 22, pp. 2056–2057, October 1992.

14. P. S. Hall, I. L. Morrow, P. M. Haskins, and J. S. Dahele, "Phase Control in Injection Locked Microstrip Antennas," *IEEE MTT-S International Microwave Symposium Digest*, pp. 1227–1230 (1994).

15. S. Drew and V. F. Fusco, "Phase Modulated Active Antenna," *Electronics Letters*, Vol. 29, No. 10, pp. 835–836, 1993.

16. V. F. Fusco, S. Drew and D. S. McDowall, "Injection Locking Phenomena in an Active Microstrip Antenna," IEE Eighth International Conference on Antennas and Propagation, pp. 295–298, Edinburgh, UK, April 1993.

17. J. A. Navarro and K. Chang, "Electronic Beam Steering of Active Antenna Arrays," *Electronics Letters*, Vol. 29, No. 3, pp. 302–304, February 1993.

18. J. A. Navarro, L. Fan, and K. Chang, "Active Inverted Stripline Circular Patch Antennas for Spatial Power Combining," *IEEE Transactions on Microwave Theory and Techniques*, Vol. 41, No. 10, pp. 1856–1863, October 1993.

19. R. A. York and R. C. Compton, "Quasi-Optical Power Combining Using Mutually Synchronized Oscillator Arrays," *IEEE Transactions on Microwave Theory and Techniques*, Vol. 39, No. 6, pp. 1000–1009, June 1991.

20. R. A. York and R. C. Compton, "Coupled-Oscillator Arrays for Millimeter-Wave Power-Combining and Mode-Locking," *IEEE MTT-S International Microwave Symposium Digest*, pp. 429–432, Albuquerque, New Mexico, June 1992.

21. R. A. York, "Nonlinear Analysis of Phase Relationships in Quasi-Optical Oscillator Arrays," *IEEE Transactions on Microwave Theory and Techniques*, Vol. 41, No. 10, pp. 1799–1809, October 1993.

22. P. Liao and R. A. York, "Phase-Shifterless Beam-Scanning Using Coupled-Oscillators: Theory and Experiment," *IEEE Antennas and Propagation International Symposium*, pp. 668–671 (1993).

23. P. Liao and R. A. York, "A New Phase-Shifterless Beam-Scanning Technique Using Arrays of Coupled Oscillators," *IEEE Transactions on Microwave Theory and Techniques*, Vol. 41, No. 10, pp. 1810–1815, October 1993.

24. P. Liao and R. A. York, "A Six-Element Beam-Scanning Array," *IEEE Microwave and Guided Wave Letters*, Vol. 4, No. 1, pp. 20–22, January 1994.

25. S. Nogi, J. Lin, and T. Itoh, "Mode Analysis and Stabilization of a Spatial Power Combining Array with Strongly Coupled Oscillators," *IEEE Transactions on Microwave Theory and Techniques*, Vol. 41, No. 10, pp. 1827–1837, October 1993.

26. J. Lin, S. T. Chew, and T. Itoh, "A Unilateral Injection-Locking Type Active Phased Array for Beam Scanning," *IEEE MTT-S International Microwave Symposium Digest*, pp. 1231–1234, San Diego, California, June 1994.

27. K. Kurokawa, "Injection Locking of Microwave Solid-State Oscillators," *Proceedings of the IEEE*, Vol. 61, No. 10, pp. 1386–1410, October 1973.

28. R. D. Martinez and R. C. Compton, "Electronic Beamsteering of Active Arrays with Phase-Locked Loops," *IEEE Microwave and Guided Wave Letters*, Vol. 4, No. 6, pp. 166–168, June 1994.

Index